Jie Xiao

Q_α Analysis on Euclidean Spaces

Advances in Analysis and Geometry

Editor in Chief
Jie Xiao, Memorial University, Canada

Volume 1

Jie Xiao

Q_α Analysis on Euclidean Spaces

DE GRUYTER

Mathematics Subject Classification 2010
Primary: 42B37; Secondary: 35Q30

Author
Prof. Jie Xiao
Memorial University
Department of Mathematics & Statistics
St. John's, NL A1C 5S7
Canada
jxiao@mun.ca

ISBN 978-3-11-060112-1
e-ISBN (PDF) 978-3-11-060028-5
e-ISBN (EPUB) 978-3-11-060010-0
ISSN 2511-0438

Library of Congress Control Number: 2018966447

Bibliographic information published by the Deutsche Nationalbibliothek
The Deutsche Nationalbibliothek lists this publication in the Deutsche Nationalbibliografie; detailed bibliographic data are available on the Internet at http://dnb.dnb.de.

Typesetting: VTeX UAB, Lithuania
Printing and binding: CPI books GmbH, Leck

www.degruyter.com

Preface

In their 2000 paper [22], M. Essén, S. Janson, L. Peng and J. Xiao introduced such a new function space in several real variables that if $\alpha \in \mathbb{R} = (-\infty, \infty)$ then $Q_\alpha(\mathbb{R}^n)$ comprises all Lebesgue measurable functions

$$f : \mathbb{R}^n = \underbrace{\mathbb{R} \times \cdots \times \mathbb{R}}_{2 \le n \text{ copies}} \to \mathbb{R}$$

with the bounded double-integral-mean of a fractional difference quotient

$$\sup_{(r,x)\in\mathbb{R}_+\times\mathbb{R}^n} \left(\fint_{B(x,r)} \fint_{B(x,r)} \left(\frac{|f(y) - f(z)|}{|r^{-1}y - r^{-1}z|^{\frac{n}{2}+\alpha}} \right)^2 dydz \right)^{\frac{1}{2}} < \infty,$$

where:

$$\begin{cases} \mathbb{R}_+ = (0, \infty); \\ B(x,r) = \{y \in \mathbb{R}^n : |y - x| < r\} \quad \& \quad B(o,r) = \{y \in \mathbb{R}^n : |y| < r\} = B(0,r); \\ dz, dy, dx = \text{the Lebesgue measures associated with } z, y, x \in \mathbb{R}^n; \\ \fint = \text{the integral mean under Lebesgue measure on } \mathbb{R}^n. \end{cases}$$

Needless to say, $Q_\alpha(\mathbb{R}^n)$ module of real constants, denoted by $K(\mathbb{R}^n)$, is a Banach space, in particular,

▷ if $\alpha \in (-\infty, 0) = \mathbb{R}_-$ then $Q_\alpha(\mathbb{R}^n)$ coincides with John–Nirenberg's $BMO(\mathbb{R}^n)$ (cf. [38]) of all Lebesgue measurable functions $f : \mathbb{R}^n \to \mathbb{R}$ enjoying

$$\sup_{(r,x)\in\mathbb{R}_+\times\mathbb{R}^n} |f - f_{B(x,r)}|_{B(x,r)} = \sup_{(r,x)\in\mathbb{R}_+\times\mathbb{R}^n} \fint_{B(x,r)} |f(y) - f_{B(x,r)}| \, dy < \infty,$$

where $f_{B(x,r)}$ stands for the integral mean of f on $B(x,r)$ with respect to the Lebesgue measure on $\mathbb{R}^n$.

▷ If $\alpha \in [1, \infty)$, then $Q_\alpha(\mathbb{R}^n)$ is identified with $K(\mathbb{R}^n)$.

▷ $[0, 1) \ni \alpha \mapsto Q_\alpha(\mathbb{R}^n)$ is a decreasing map, that is,

$$0 \le \alpha_1 < \alpha_2 < 1 \Rightarrow BMO(\mathbb{R}^n) \supsetneq Q_{\alpha_1}(\mathbb{R}^n) \supsetneq Q_{\alpha_2}(\mathbb{R}^n) \supsetneq K(\mathbb{R}^n).$$

The above facts actually induced a list of five open problems in the final section of [22], the initially-BMO-oriented study of $Q_\alpha(\mathbb{R}^n)$. Since 2000, those problems have received a lot of attention. Not only have the problems been essentially resolved, but also they have been applied to an investigation of the evolutionary incompressible Navier–Stokes system and its stationary formulation. This has driven me to write a relatively self-contained book documenting almost all major results obtained in the past years. Below is an outline of the book.

https://doi.org/10.1515/9783110600285-201

- ▷ Chapter 1 introduces the reader to the basic building blocks of $Q_\alpha(\mathbb{R}^n)$—definition and invariance, monotonicity and relationship with Besov spaces, mean oscillation via successive dyadic partitions and wavelet decomposition.
- ▷ Chapter 2 uses the dyadic divisions of a cube to show a John–Nirenberg inequality with precise constants for $BMO(\mathbb{R}^n)$ and a John–Nirenberg inequality for $Q_\alpha(\mathbb{R}^n)$ with coarse constants.
- ▷ Chapter 3 gives a short account of the functional Hausdorff content and its Choquet integration, Carleson measure and tent space.
- ▷ Chapter 4 focuses on lifting $Q_\alpha(\mathbb{R}^n)$ to $\mathbb{R}^{n+1}_+$ via the Gauss integral or the Poisson integral, characterizing the divergence of $(Q_\alpha(\mathbb{R}^n))^n$ and linking $Q_\alpha(\mathbb{R}^n)$ up with the square Campanato space $C^{2,2\alpha}(\mathbb{R}^n)$.
- ▷ Chapter 5 looks at the predual of either $Q_\alpha(\mathbb{R}^n)$ or $Q^{-1}_\alpha(\mathbb{R}^n)$ via the Hardy–Hausdorff space or the Hardy–Hausdorff–Sobolev space, and decomposes $Q_\alpha(\mathbb{R}^n)$ through the Fefferman–Stein type based on $L^\infty(\mathbb{R}^n) \cap Q_\alpha(\mathbb{R}^n)$.
- ▷ Chapter 6 applies $Q_\alpha(\mathbb{R}^n)$ and $Q^{-1}_\alpha(\mathbb{R}^n)$, as well as their relatives: conformal Morrey space $M^{2,2}(\mathbb{R}^n)$, $BMO(\mathbb{R}^n)$ and conformal Besov space $\dot{B}^{1}_{\infty,\infty}(\mathbb{R}^n)$, to studying the mild solution of the evolutionary incompressible Navier–Stokes system in $\mathbb{R}^{n+1}_+$ and the zero solution of the stationary incompressible Navier–Stokes system in $\mathbb{R}^n$.
- ▷ Chapter 7 deals with the left or right composition problem for $Q_\alpha(\mathbb{R}^n)$, thereby finding a relation between $Q_\alpha(\mathbb{R}^n)_*$ and $\mathscr{A}_p(\mathbb{R}^n)$ and resolving [22, problem 8.4] via the local/global self-similar Minkowski dimension and the $\mathscr{A}_1$-class living on a degeneracy set, and leading to an investigation of the well-posedness of the transport equation in $Q_\alpha(\mathbb{R}^n)$.
- ▷ Chapter 8 settles the extension or restriction problem for $Q_\alpha(\mathbb{R}^n)$ posed in [22, problem 8.5] through the Hardy–Littlewood maximal operator which is actually bounded on $Q_\alpha(\mathbb{R}^n)$.

Below are two fundamental conventions which will be frequently used in this book:

- ▷ $X \approx Y$ represents $X \lesssim Y \lesssim X$; namely, there is a constant $\kappa > 0$ such that $\kappa^{-1}X \le Y \le \kappa X$;
- ▷ $1\!\!1_S$ stands for the characteristic function of a subset $S \subseteq \mathbb{R}^n$.

It is my hope that this book will be useful for researchers, graduate students and even senior undergraduate students working in functional-harmonic-potential analysis and partial differential equations as well as the closely related fields.

Last but not least, it is a great pleasure to thank my coauthors who communicated directly or indirectly with me in this endeavor. This work was supported in part by the Natural Sciences and Engineering Research Council of Canada (#20171864).

Fall 2017–Summer 2018, St. John's, Canada Jie Xiao

Contents

1 Groundwork

This chapter contains four basic topics: definition and invariance, monotonicity and relationship with Besov spaces, mean oscillation via successive dyadic partitions and wavelet decomposition, about $Q_\alpha(\mathbb{R}^n)$ under $(\alpha, n-1) \in \mathbb{R} \times \mathbb{N} = (-\infty, \infty) \times \{1, 2, 3, \ldots\}$.

1.1 Definition and invariance

The history of Q_α analysis on $\mathbb{R}^n$ started in year 2000 with the following concept.

Definition 1.1. For $(\alpha, n-1) \in \mathbb{R}\times\mathbb{N}$, let $Q_\alpha(\mathbb{R}^n)$ be the space of all Lebesgue measurable functions $f : \mathbb{R}^n \to \mathbb{R}$ that satisfy

$$\|f\|_{Q_\alpha(\mathbb{R}^n)} = \sup_{I\in\mathscr{C}(\mathbb{R}^n)} \left(\ell(I)^{2\alpha-n} \int_I \int_I \frac{|f(y)-f(z)|^2}{|y-z|^{n+2\alpha}} \, dydz \right)^{\frac{1}{2}} < \infty,$$

where $\mathscr{C}(\mathbb{R}^n)$ is used as the family of all coordinate cubes I in $\mathbb{R}^n$ with edge-length $\ell(I)$. Evidently, a Lebesgue measurable function f on $\mathbb{R}^n$ belongs to $Q_\alpha(\mathbb{R}^n)$ if and only if

$$\|f\|_{Q_\alpha(\mathbb{R}^n),b} = \sup_{(r,x)\in\mathbb{R}_+\times\mathbb{R}^n} \left(r^{2\alpha-n} \int_{B(x,r)} \int_{B(x,r)} \frac{|f(y)-f(z)|^2}{|y-z|^{n+2\alpha}} \, dydz \right)^{\frac{1}{2}} < \infty.$$

This definition induces a fundamental invariance principle as stated below.

Theorem 1.2. *Let $(\alpha, n-1) \in \mathbb{R} \times \mathbb{N}$. Then $Q_\alpha(\mathbb{R}^n)$ is invariant under translations, rotations, reflections and dilations of $\mathbb{R}^n$. Moreover, the following three statements are equivalent for any Lebesgue measurable function f on $\mathbb{R}^n$:*

(i) $f \in Q_\alpha(\mathbb{R}^n)$;

(ii)
$$\sup_{I\in\mathscr{C}(\mathbb{R}^n)} \ell(I)^{2\alpha-n} \int_{|y|<\ell(I)} \left(\int_I |f(x+y)-f(x)|^2 \, dx \right) \frac{dy}{|y|^{n+2\alpha}} < \infty;$$

(iii) *$(x,t) \mapsto F(x,t) = f(x)$ defined on $\mathbb{R}^n \times \mathbb{R} = \mathbb{R}^{n+1}$ belongs to $Q_\alpha(\mathbb{R}^{n+1})$ under $\alpha \in (-\frac{n}{2}, \infty)$.*

Proof. Upon replacing $f(\cdot)$ by $f(\cdot - x_0)$ or $f(r_0\cdot)$ or $f \circ \mathcal{R}(\cdot)$ with $x_0 \in \mathbb{R}^n$ or $r_0 \in \mathbb{R}_+$ or rotation $\mathcal{R}$ of $\mathbb{R}^n$ in $\|f\|_{Q_\alpha(\mathbb{R}^n),b}$, we can easily see that $\|f\|_{Q_\alpha(\mathbb{R}^n),b}$ is not affected by any translation, dilation and rotation of $\mathbb{R}^n$.

Given a point $a \in \mathbb{R}^n$, the formula for the reflection in the hyperplane through the origin o, orthogonal to a, is given by

$$x \mapsto \mathcal{R}_a(x) = x - 2\frac{x \cdot a}{|a|^2} a,$$

https://doi.org/10.1515/9783110600285-001

where $x \cdot a$ denotes the dot product of x and a. It is plain to check that $\mathcal{R}$ is an isometry:

$$|\mathcal{R}_a(x) - \mathcal{R}_a(x)| = |x - y| \quad \forall x, y \in \mathbb{R}^n,$$

and so

$$\|f \circ \mathcal{R}_a\|_{\mathcal{Q}_\alpha(\mathbb{R}^n)} = \|f\|_{\mathcal{Q}_\alpha(\mathbb{R}^n)},$$

thereby deriving the reflection-invariance of $\mathcal{Q}_\alpha(\mathbb{R}^n)$.

Next, let us verify the equivalence among (i)–(ii)–(iii).

(i)⇔(ii) By the change of variable from y to $x + y$ and an elementary geometric calculation with any $I \in \mathscr{C}(\mathbb{R}^n)$ it follows that

$$\begin{aligned} &\ell(I)^{2\alpha-n} \int_{|y|<\ell(I)} \Big(\int_I |f(x+y) - f(x)|^2 \, dx \Big) \frac{dy}{|y|^{n+2\alpha}} \\ &\le \ell(3I)^{2\alpha-n} \int_{3I}\int_{3I} \frac{|f(y)-f(z)|^2}{|y-z|^{n+2\alpha}} \, dydz \end{aligned}$$

and

$$\begin{aligned} &\ell(\sqrt{n}I)^{2\alpha-n} \int_{|y|<\ell(\sqrt{n}I)} \Big(\int_{\sqrt{n}I} |f(x+y) - f(x)|^2 \, dx \Big) \frac{dy}{|y|^{n+2\alpha}} \\ &\ge \ell(I)^{2\alpha-n} \int_{I}\int_{I} \frac{|f(y)-f(z)|^2}{|y-z|^{n+2\alpha}} \, dydz \end{aligned}$$

where given $c \in \mathbb{R}_+$ the symbol cI represents the cube with the same center as I and the edge-length $c\ell(I)$.

(i)⇔(iii) Note that a cube in $\mathscr{C}(\mathbb{R}^{n+1})$ can be formulated as

$$I \times [a, a + \ell(I)] \quad \text{where } (I, a) \in \mathscr{C}(\mathbb{R}^n) \times \mathbb{R}.$$

So

$$\begin{aligned} \|F\|^2_{\mathcal{Q}_\alpha(\mathbb{R}^{n+1})} &= \sup_{(I,a)\in\mathscr{C}(\mathbb{R}^n)\times\mathbb{R}} \int_I\int_I\int_a^{a+\ell(I)}\int_a^{a+\ell(I)} \Big(\frac{|f(y)-f(z)|^2}{|(y,t)-(z,s)|^{n+1+2\alpha}} \Big) \frac{dtdsdydz}{\ell(I)^{2(n+1)-2\alpha}} \\ &= \sup_{I\in\mathscr{C}(\mathbb{R}^n)} \int_I\int_I\int_0^{\ell(I)}\int_0^{\ell(I)} \Big(\frac{|f(y)-f(z)|^2}{|(y,t)-(z,s)|^{n+1+2\alpha}} \Big) \frac{dtdsdydz}{\ell(I)^{2(n+1)-2\alpha}} \\ &\approx \sup_{I\in\mathscr{C}(\mathbb{R}^n)} \int_I\int_I\int_0^{\ell(I)}\int_0^{\ell(I)} \Big(\frac{|f(y)-f(z)|^2}{|y-z|^{n+1+2\alpha} + |s-t|^{n+1+2\alpha}} \Big) \frac{dtdsdydz}{\ell(I)^{2(n+1)-2\alpha}}. \end{aligned}$$

Now, an application of the equivalence

$$\begin{aligned}\frac{\ell(I)}{|y-z|^{n+2\alpha}} &\lesssim \iint\limits_{(t,s)\in[0,\ell(I)]\times[0,\ell(I)]\ \&\ |t-s|\le|x-y|} \frac{dtds}{|y-z|^{n+1+2\alpha}+|s-t|^{n+1+2\alpha}}\\ &\lesssim \int_0^{\ell(I)}\int_0^{\ell(I)} \frac{dtds}{|y-z|^{n+1+2\alpha}+|s-t|^{n+1+2\alpha}}\\ &\lesssim \frac{\ell(I)}{|y-z|^{n+2\alpha}} \quad \forall\,\alpha > -2^{-1}n\end{aligned}$$

derives

$$\|F\|_{Q_\alpha(\mathbb{R}^{n+1})} \approx \|f\|_{Q_\alpha(\mathbb{R}^n)}.$$

□

1.2 Monotonicity and relationship with Besov space

On the one hand, we show that $Q_\alpha(\mathbb{R}^n)$ is always a subspace of $\mathrm{BMO}(\mathbb{R}^n)$. To do so, recall that a Lebesgue measurable function f is said to be of bounded mean oscillation on $\mathbb{R}^n$ (in short $f \in \mathrm{BMO}(\mathbb{R}^n)$) provided (cf. [38])

$$\|f\|_{\mathrm{BMO}(\mathbb{R}^n)} = \sup_{I\in\mathscr{C}(\mathbb{R}^n)} \Big(\fint_I |f(y)-f_I|^2\,dy\Big)^{\frac{1}{2}} < \infty$$

where

$$f_E = \fint_E f$$

is the integral mean of f over $E \subseteq \mathbb{R}^n$ with respect to the Lebesgue measure, in particular, –

$$f_I = \ell(I)^{-n}\int_I f(x)\,dx.$$

Equivalently,

$$\|f\|_{\mathrm{BMO}(\mathbb{R}^n)} \approx \sup_{(r,x)\in\mathbb{R}_+\times\mathbb{R}^n} \Big(\fint_{B(x,r)} |f(y)-f_{B(x,r)}|^2\,dy\Big)^{\frac{1}{2}} = |||f|||_{\mathrm{BMO}(\mathbb{R}^n)}.$$

Theorem 1.3. *Let $(\alpha, n-1) \in \mathbb{R}\times\mathbb{N}$ and $\mathrm{K}(\mathbb{R}^n)$ be the class of all real constants on $\mathbb{R}^n$. Then:*

(i) $\alpha \mapsto Q_\alpha(\mathbb{R}^n)$ *is a decreasing map on* $\mathbb{R}$;

(ii)

$$Q_\alpha(\mathbb{R}^n) = \begin{cases} \mathrm{BMO}(\mathbb{R}^n) & \text{as } \alpha \in (-\infty, 0) \\ \text{a proper space between } \mathrm{K}(\mathbb{R}^n) \,\&\, \mathrm{BMO}(\mathbb{R}^n) & \text{as } \alpha \in [0, 1) \\ \mathrm{K}(\mathbb{R}^n) & \text{as } \alpha \in [1, \infty); \end{cases}$$

(iii) $Q_\alpha(\mathbb{R}^n)$ *modulo* $\mathrm{K}(\mathbb{R}^n)$ *is a Banach space.*

Proof. (i) This follows from Definition 1.1 and the basic inequality

$$|y - z| \le \sqrt{n}\ell(I) \quad \forall y, z \in I \in \mathscr{C}(\mathbb{R}^n).$$

(ii) Four cases are considered below.

▷ If $-2^{-1}n \le \alpha < 0$, then (i) implies

$$Q_\alpha(\mathbb{R}^n) \subseteq Q_{-2^{-1}n}(\mathbb{R}^n) = \mathrm{BMO}(\mathbb{R}^n).$$

Meanwhile, if

$$f \in \mathrm{BMO}(\mathbb{R}^n),$$

then for

$$I \in \mathscr{C}(\mathbb{R}^n) \quad \text{and} \quad y \in \mathbb{R}^n \quad \text{with} \quad |y| < \ell(I)$$

one has

$$\int_I |f(x+y) - f(x)|^2 \, dx \lesssim \int_{2I} |f(x) - f_{2I}|^2 \, dx \lesssim \ell(I)^n \|f\|^2_{\mathrm{BMO}(\mathbb{R}^n)}$$

and thus, due to $\alpha < 0$,

$$\begin{aligned} &\int_{|y|<\ell(I)} \Big(\int_I |f(x+y) - f(x)|^2 \, dx \Big) \frac{dy}{|y|^{n+2\alpha}} \\ &\quad \lesssim \ell(I)^n \|f\|^2_{\mathrm{BMO}(\mathbb{R}^n)} \int_{|y|<\ell(I)} \frac{dy}{|y|^{n+2\alpha}} \\ &\quad \lesssim \ell(I)^{n-2\alpha} \|f\|^2_{\mathrm{BMO}(\mathbb{R}^n)}. \end{aligned}$$

This, along with Theorem 1.2, yields $f \in Q_\alpha(\mathbb{R}^n)$, whence

$$Q_\alpha(\mathbb{R}^n) = \mathrm{BMO}(\mathbb{R}^n) \quad \forall \alpha \in [-2^{-1}n, 0).$$

▷ If $-\infty < \alpha < -2^{-1}n$, then (i) is utilized to derive

$$\mathsf{Q}_\alpha(\mathbb{R}^n) \supseteq \mathsf{BMO}(\mathbb{R}^n).$$

To get the reverse inclusion of the last one, suppose

$$f \in \mathsf{Q}_\alpha(\mathbb{R}^n);\ I \in \mathscr{C}(\mathbb{R}^n);\ x, y \in I.$$

Then the Lebesgue measure of the following set:

$$\{z \in I : \min\{|x-z|, |y-z|\} > 2^{-3}\ell(I)\}$$

is at least $2^{-1}\ell(I)^n$. Accordingly,

$$-2\alpha > n \Rightarrow \int_I \min\{|x-z|^{-2\alpha-n}, |y-z|^{-2\alpha-n}\}\,dz \gtrsim \ell(I)^{-2\alpha}.$$

This in turn derives

$$\begin{aligned}
&\ell(I)^{-2n}\int_I\int_I |f(x)-f(y)|^2\,dxdy\\
&\quad\lesssim \ell(I)^{2(\alpha-n)}\int_I\int_I \frac{\min\{|x-z|^{-2\alpha-n}, |y-z|^{-2\alpha-n}\}}{(|f(x)-f(z)|^2+|f(y)-f(z)|^2)^{-1}}\,dxdydz\\
&\quad\lesssim \ell(I)^{2\alpha-n}\int_I\int_I \frac{|f(y)-f(z)|^2}{|y-z|^{n+2\alpha}}\,dydz,
\end{aligned}$$

whence

$$f \in \mathsf{BMO}(\mathbb{R}^n).$$

Consequently,

$$\mathsf{BMO}(\mathbb{R}^n) = \mathsf{Q}_\alpha(\mathbb{R}^n) \quad \forall\, \alpha \in (-\infty, -2^{-1}n).$$

▷ If $2^{-1}n < \alpha < \infty$, then

$$f \in \mathsf{Q}_\alpha(\mathbb{R}^n) \Rightarrow \int_I\int_I \frac{|f(y)-f(z)|^2}{|y-z|^{n+2\alpha}}\,dydz \le \ell(I)^{n-2\alpha}\|f\|^2_{\mathsf{Q}_\alpha(\mathbb{R}^n)} \to 0$$
$$\text{as } \mathscr{C}(\mathbb{R}^n) \ni I \to \mathbb{R}^n,$$

and hence

$$\int_{\mathbb{R}^n}\int_{\mathbb{R}^n} \frac{|f(y)-f(z)|^2}{|y-z|^{n+2\alpha}}\,dydz = 0,$$

which ensures that f is constant a. e. on $\mathbb{R}^n$.

▷ If $1 \le \alpha \le 2^{-1}n$, then $f \in \mathcal{Q}_\alpha(\mathbb{R}^n)$ implies that

$$\int_I \int_I \left(\frac{|f(y)-f(z)|^2}{|y-z|^{n+2}}\right) \frac{dydz}{\ell(I)^{2(\alpha-1)}} \lesssim \int_I \int_I \frac{|f(y)-f(z)|^2}{|y-z|^{n+2\alpha}}\, dydz \lesssim \frac{\|f\|^2_{\mathcal{Q}_\alpha(\mathbb{R}^n)}}{\ell(I)^{2\alpha-n}}$$

holds for any $I \in \mathscr{C}(\mathbb{R}^n)$, and hence

$$\inf_{(y,z)\in I\times I} \left(\frac{|f(y)-f(z)|}{|y-z|}\right)^2 \int_I \int_I \frac{dydz}{|y-z|^n} \lesssim \int_I \int_I \left(\frac{|f(y)-f(z)|}{|y-z|}\right)^2 \frac{dydz}{|y-z|^n}$$
$$\lesssim \|f\|^2_{\mathcal{Q}_\alpha(\mathbb{R}^n)} \ell(I)^{n-2}$$

holds for any $I \in \mathscr{C}(\mathbb{R}^n)$. Note that

$$\int_I \int_I \frac{dydz}{|y-z|^n} = \infty \quad \& \quad n-2 \ge 0.$$

So the last estimation forces

$$\inf_{(y,z)\in I\times I} \frac{|f(y)-f(z)|}{|y-z|} = 0 \quad \forall I \in \mathscr{C}(\mathbb{R}^n),$$

thereby deriving that f is constant a. e. on $\mathbb{R}^n$.

To check the proper inclusion, we construct an appropriate function. Suppose that $\{a_k\}$ is a sequence of real numbers with

$$\begin{cases} \sum_{k=1}^{\infty} a_k^2 < \infty \\ f(x) = \phi(x)g(x) \\ g(x) = \sum_{k=1}^{\infty} a_k \cos(2^k x_1) \\ x = (x_1,\dots,x_n) \in \mathbb{R}^n, \end{cases}$$

where $\phi \in \mathcal{S}(\mathbb{R}^n)$ (the class of all Schwarz functions in $\mathbb{R}^n$) is a fixed function such that its Fourier transform

$$\hat{\phi}(x) = \int_{\mathbb{R}^n} \phi(y) \exp(2\pi i x \cdot y)\, dy$$

is supported in the unit ball $B(o,1)$ and $\phi \ne 0$ on the cube $[-3\pi, 3\pi]^n$. In accordance with [22, pp. 9–10], we have

$$\infty > \|f\|_{\mathcal{Q}_\alpha(\mathbb{R}^n)} \Leftrightarrow \infty > \begin{cases} \sum_{k=1}^{\infty} |a_k|^2 & \text{as } \alpha \in (-\infty, 0) \\ \sum_{k=1}^{\infty} k|a_k|^2 & \text{as } \alpha = 0 \\ \sum_{k=1}^{\infty} 2^{2k\alpha}|a_k|^2 & \text{as } \alpha \in (0,1), \end{cases}$$

thereby reaching the desired strict inclusion

$$\mathrm{K}(\mathbb{R}^n) \subsetneq \mathrm{Q}_\alpha(\mathbb{R}^n) \subsetneq \mathrm{BMO}(\mathbb{R}^n) \quad \forall\, \alpha \in [0,1)$$

via choosing a suitable sequence $\{a_k\}$.

(iii) Suppose that $\{f_j\}$ is a Cauchy sequence in $\mathrm{Q}_\alpha(\mathbb{R}^n)$. From (i)–(ii) above, it follows that $\{f_j\}$ is a Cauchy sequence in $\mathrm{BMO}(\mathbb{R}^n)$ and so that f_j converges to some $f \in \mathrm{BMO}(\mathbb{R}^n)$. An application of Fatou's lemma gives

$$\|f - f_k\|_{\mathrm{Q}_\alpha(\mathbb{R}^n)} \le \limsup_{j\to\infty} \|f_j - f_k\|_{\mathrm{Q}_\alpha(\mathbb{R}^n)} \quad \forall\, k \in \mathbb{N},$$

where

$$\lim_{j\to\infty} \|f_j - f\|_{\mathrm{Q}_\alpha(\mathbb{R}^n)} = 0. \qquad \square$$

On the other hand, we find that $\mathrm{Q}_\alpha(\mathbb{R}^n)$ exists a space bigger than some scale-invariant Besov spaces and their limiting cases—the conformal Sobolev energy spaces. For

$$(p,q,\alpha) \in [1,\infty) \times [1,\infty] \times (0,1)$$

let $\dot{\mathrm{B}}^{\alpha}_{p,q}(\mathbb{R}^n)$ be the homogeneous Besov space of all Lebesgue measurable functions $f : \mathbb{R}^n \to \mathbb{R}$ with

$$\infty > \|f\|_{\dot{\mathrm{B}}^{\alpha}_{p,q}(\mathbb{R}^n)} = \begin{cases} \left(\int_{\mathbb{R}^n}\left(\int_{\mathbb{R}^n} |f(x+y) - f(x)|^p \,dx\right)^{\frac{q}{p}} \frac{dy}{|y|^{n+\alpha q}}\right)^{\frac{1}{p}} & \text{as } q \in [1,\infty) \\ \sup_{y\in\mathbb{R}^n} |y|^{-\alpha}\left(\int_{\mathbb{R}^n} |f(x+y) - f(x)|^p \,dx\right)^{\frac{1}{p}} & \text{as } q = \infty. \end{cases}$$

Clearly, if

$$\mathcal{T}_{r_0,x_0}(x) = r_0 x + x_0,$$

then

$$\|f \circ \mathcal{T}_{r_0,x_0}\|_{\dot{\mathrm{B}}^{\alpha}_{p,q}(\mathbb{R}^n)} = r_0^{\alpha - \frac{n}{p}} \|f\|_{\dot{\mathrm{B}}^{\alpha}_{p,q}(\mathbb{R}^n)}.$$

Moreover, if $p = q$, then there are two constants $c_{p,n,0}$ and $c_{p,n,1}$ such that (cf. [62, 8])

$$\lim_{\alpha\to 0} \alpha^{\frac{1}{p}} \|f\|_{\dot{\mathrm{B}}^{\alpha}_{p,p}(\mathbb{R}^n)} = c_{p,n,0}\|f\|_{L^p(\mathbb{R}^n)} \quad \& \quad \lim_{\alpha\to 1}(1-\alpha)^{\frac{1}{p}} \|f\|_{\dot{\mathrm{B}}^{\alpha}_{p,p}(\mathbb{R}^n)} = c_{p,n,1}\|\nabla f\|_{L^p(\mathbb{R}^n)}$$

hold for any f in the class $\mathrm{C}_0^\infty(\mathbb{R}^n)$ of all infinitely differentiable functions

$$g : \mathbb{R}^n \to \mathbb{R}$$

with compact support

$$\mathrm{supp}(g) \subseteq \mathbb{R}^n.$$

Of course, $\|\cdot\|_{L^p(\mathbb{R}^n)}$ stands for the Lebesgue p-norm on $\mathbb{R}^n$; ∇ is the standard gradient in spacial variable $x \in \mathbb{R}^n$.

Theorem 1.4. *Let* $(\alpha, n-1) \in [0,1) \times \mathbb{N}$. *Then:*

(i)

$$\dot{\mathrm{B}}^{\beta}_{\frac{n}{\beta},q}(\mathbb{R}^n) \subsetneq \begin{cases} \dot{\mathrm{B}}^{\beta}_{\frac{n}{\beta},2}(\mathbb{R}^n) & \text{as } 0<\alpha=\beta<1\le q\le 2 \\ \dot{\mathrm{B}}^{\beta}_{\frac{n}{\beta},\infty}(\mathbb{R}^n) & \text{as } 0<\alpha<\beta<1\le q<\infty \end{cases} \subsetneq \mathrm{Q}_\alpha(\mathbb{R}^n);$$

(ii)

$$\dot{\mathrm{W}}^{1,n}(\mathbb{R}^n) \subseteq \mathrm{CIS}(\mathbb{R}^n) \subsetneq \mathrm{Q}_\alpha(\mathbb{R}^n)$$

exists as the limit $\beta = \frac{n}{q} \to 1$ *in (i), where a weakly differentiable function* $f : \mathbb{R}^n \to \mathbb{R}$ *lies in the conformal Sobolev energy space* $\dot{\mathrm{W}}^{1,n}(\mathbb{R}^n)$ *or the conformal invariant Sobolev space* $\mathrm{CIS}(\mathbb{R}^n)$ *if and only if*

$$\begin{cases} \|f\|_{\dot{\mathrm{W}}^{1,n}(\mathbb{R}^n)} = \left(\int_{\mathbb{R}^n} |\nabla f(x)|^n \,dx\right)^{\frac{1}{n}} < \infty \quad \text{or} \\ \|f\|_{\mathrm{CIS}(\mathbb{R}^n)} = \sup_{I\in\mathscr{C}(\mathbb{R}^n)} \left(\ell(I)^{2-n} \int_I |\nabla f(x)|^2 \,dx\right)^{\frac{1}{2}} < \infty. \end{cases}$$

Proof. (i) Note that (cf. [69, Chapter 3. Theorems 3–4])

$$\dot{\mathrm{B}}^{\beta}_{\frac{n}{\beta},q}(\mathbb{R}^n) \subsetneq \begin{cases} \dot{\mathrm{B}}^{\beta}_{\frac{n}{\beta},2}(\mathbb{R}^n) & \text{as} \quad 0<\beta=\alpha<1\le q\le 2 \\ \dot{\mathrm{B}}^{\beta}_{\frac{n}{\beta},\infty}(\mathbb{R}^n) & \text{as} \quad 0<\alpha<\beta<1\le q<\infty. \end{cases}$$

So, it is enough to verify

$$\mathrm{Q}_\alpha(\mathbb{R}^n) \supsetneq \begin{cases} \dot{\mathrm{B}}^{\beta}_{\frac{n}{\beta},2}(\mathbb{R}^n) & \text{as} \quad 0<\beta=\alpha<1 \\ \dot{\mathrm{B}}^{\beta}_{\frac{n}{\beta},\infty}(\mathbb{R}^n) & \text{as} \quad 0<\alpha<\beta<1. \end{cases}$$

To get the inclusion, we consider the following two cases.

▷ If $0<\beta=\alpha<1$, then Theorem 1.2 and Hölder's inequality with exponents $\frac{n}{2\alpha}$ and $\frac{n}{n-2\alpha}$ give

$$\begin{aligned} \|f\|^2_{\mathrm{Q}_\alpha(\mathbb{R}^n)} &\approx \sup_{I\in\mathscr{C}(\mathbb{R}^n)} \ell(I)^{2\alpha-n} \int_{|y|<\ell(I)} \left(\int_I |f(x+y)-f(x)|^2 \,dx\right) \frac{dy}{|y|^{n+2\alpha}} \\ &\lesssim \sup_{I\in\mathscr{C}(\mathbb{R}^n)} \ell(I)^{n-2\alpha} \int_{\mathbb{R}^n} \left(\int_I |f(x+y)-f(x)|^{\frac{n}{\alpha}} \,dx\right)^{\frac{2\alpha}{n}} \frac{dy}{|y|^{n+2\alpha}} \\ &\lesssim \|f\|^2_{\dot{\mathrm{B}}^{\alpha}_{\frac{n}{\alpha},2}(\mathbb{R}^n)}. \end{aligned}$$

▷ If $0 < \alpha < \beta < 1$, then Theorem 1.2 and Hölder's inequality with exponents $\frac{n}{2\beta}$ and $\frac{n}{n-2\beta}$ derive

$$\begin{aligned}
\|f\|^2_{Q_\alpha(\mathbb{R}^n)} &\approx \sup_{I\in\mathscr{C}(\mathbb{R}^n)} \ell(I)^{2\alpha-n} \int\limits_{|y|<\ell(I)} \Big(\int\limits_I |f(x+y)-f(x)|^2\,dx\Big)\frac{dy}{|y|^{n+2\alpha}} \\
&\lesssim \sup_{I\in\mathscr{C}(\mathbb{R}^n)} \ell(I)^{n-2\beta} \int\limits_{\mathbb{R}^n}\Big(\int\limits_I |f(x+y)-f(x)|^{\frac{n}{\beta}}\,dx\Big)^{\frac{2\beta}{n}} \frac{dy}{|y|^{n+2\alpha}} \\
&\lesssim \sup_{I\in\mathscr{C}(\mathbb{R}^n)} \ell(I)^{n-2\beta} \int\limits_{|y|<\ell(I)} |y|^{2(\beta-\alpha)-n}\,dy \|f\|^2_{\dot{B}^\beta_{\frac{n}{\beta},\infty}(\mathbb{R}^n)} \\
&\lesssim \|f\|^2_{\dot{B}^\beta_{\frac{n}{\beta},\infty}(\mathbb{R}^n)}.
\end{aligned}$$

To check the proper inclusions, we use the function $f = \phi g$ constructed in Theorem 1.3(ii) and [69, p. 51] to obtain

$$\infty > \|f\|_{\dot{B}^\beta_{p,q}(\mathbb{R}^n)} \Leftrightarrow \infty > \begin{cases} \sum_{k=1}^{\infty} 2^{qk\beta}|a_k|^q & \text{as } q \in [1,\infty) \\ \sup_{k\in\mathbb{N}} 2^{k\beta}|a_k| & \text{as } q = \infty \end{cases}$$

and

$$\|f\|_{Q_\alpha(\mathbb{R}^n)} < \infty \Leftrightarrow \sum_{k=1}^{\infty} 2^{2k\alpha}|a_k|^2 < \infty \quad \text{under } \alpha \in (0,1),$$

whence arriving at the desired properness via selecting an appropriate sequence $\{a_k\}$.

(ii) An application of Hölder's inequality gives

$$\dot{W}^{1,n}(\mathbb{R}^n) \subseteq \mathrm{CIS}(\mathbb{R}^n)$$

with equality if $n = 2$. Although

$$x \mapsto \ln|x|$$

is an element of $\mathrm{CIS}(\mathbb{R}^n)$, it does not belong to $\dot{W}^{1,n\geq 3}(\mathbb{R}^n)$.

Now, for

$$(f, I) \in \mathrm{CIS}(\mathbb{R}^n) \times \mathscr{C}(\mathbb{R}^n)$$

we utilize

$$|f(z+y) - f(y)| \leq \int\limits_0^1 |\nabla f(y+tz)||z|\,dt$$

and the Minkowski inequality to estimate

$$
\begin{aligned}
&\Big(\int_I\int_I \frac{|f(x)-f(y)|^2}{|x-y|^{n+2\alpha}}\,\mathrm{d}x\mathrm{d}y\Big)^{\frac12}\\
&\le \Big(\int_I\int_{|x-y|\le\sqrt{n}\ell(I)} \Big(\frac{|f(x)-f(y)|}{|x-y|}\Big)^2|x-y|^{2(1-\alpha)-n}\,\mathrm{d}x\mathrm{d}y\Big)^{\frac12}\\
&\le \Big(\int_I\int_{|z|\le\sqrt{n}\ell(I)} \Big(\frac{|f(z+y)-f(y)|}{|z|}\Big)^2|z|^{2(1-\alpha)-n}\,\mathrm{d}z\mathrm{d}y\Big)^{\frac12}\\
&\le \Big(\int_I\int_{|z|\le\sqrt{n}\ell(I)} \Big(\int_0^1|\nabla f(y+tz)|\,\mathrm{d}t\Big)^2|z|^{2(1-\alpha)-n}\,\mathrm{d}z\mathrm{d}y\Big)^{\frac12}\\
&\le \int_0^1\Big(\int_I\int_{|z|\le\sqrt{n}\ell(I)} |\nabla f(y+tz)|^2|z|^{2(1-\alpha)-n}\,\mathrm{d}z\mathrm{d}y\Big)^{\frac12}\mathrm{d}t\\
&\lesssim \ell(I)^{1-\alpha}\Big(\int_{(1+\sqrt{n})I}|\nabla f(w)|^2\,\mathrm{d}w\Big)^{\frac12}\\
&\lesssim \ell(I)^{\frac{n-2\alpha}{2}}\|f\|_{\mathrm{CIS}(\mathbb{R}^n)},
\end{aligned}
$$

namely, $f\in Q_\alpha(\mathbb{R}^n)$.

Moreover, in accordance with the construction of a function in $Q_\alpha(\mathbb{R}^n)$ presented in Theorem 1.3(ii), we see that $\mathrm{CIS}(\mathbb{R}^n)$ is a proper subspace of $Q_{0\le\alpha<1}(\mathbb{R}^n)$. □

1.3 Mean oscillation via successive dyadic partition

For

$$I\in\mathscr{C}(\mathbb{R}^n)\quad\&\quad k+1\in\mathbb{N},$$

let $\mathscr{D}_k(I)$ be the class of all cubes with side-length $2^{-k}\ell(I)$ and

$$\mathscr{D}(I)=\bigcup_{k=0}^{\infty}\mathscr{D}_k(I).$$

The following assertion shows the bounded mean oscillation of a function in $Q_\alpha(\mathbb{R}^n)$ via

$$\{\mathscr{D}(I)\}_{I\in\mathscr{C}(\mathbb{R}^n)},$$

which comprises all successive dyadic partitions of $I\in\mathscr{C}(\mathbb{R}^n)$.

Theorem 1.5. *Let* $(\alpha, n-1) \in \mathbb{R} \times \mathbb{N}$ *and* $f : \mathbb{R}^n \to \mathbb{R}$ *be Lebesgue measurable. Then the following three statements are equivalent:*

(i)
$$f \in \mathcal{Q}_\alpha(\mathbb{R}^n);$$

(ii)
$$\sup_{I \in \mathscr{C}(\mathbb{R}^n)} \sum_{k=0}^{\infty} 2^{(2\alpha-n)k} \sum_{J \in \mathscr{D}_k(I)} \ell(J)^{-n} \int_J |f(x) - c_J|^2 \, dx < \infty$$

holds for a sequence of constants $\{c_J\}$ *depending on* f *and* $J \in \mathscr{D}_k(I)$*;*

(iii)
$$\sup_{I \in \mathscr{C}(\mathbb{R}^n)} \sum_{k=0}^{\infty} 2^{(2\alpha-n)k} \sum_{J \in \mathscr{D}_k(I)} \ell(J)^{-n} \int_0^{\infty} t \mathrm{m}_{f,J}(t) dt < \infty$$

holds with

$$\mathrm{m}_{f,J}(t) = \int_{\{x \in J : |f(x) - f_J| > t\}} dy.$$

Proof. This consists of three implications.

(i)⇒(iii) For simplicity, given

$$(I, k+1, J) \in \mathscr{C}(\mathbb{R}^n) \times \mathbb{N} \times \mathscr{D}_k(I)$$

let

$$\begin{cases} \Theta_{f,\alpha}(I) = \ell(I)^{2\alpha-n} \int_I \int_I \frac{|f(y) - f(z)|^2}{|y-z|^{n+2\alpha}} \, dy dz \\ \Phi_f(J) = \ell(J)^{-n} \int_J |f(x) - f_J|^2 \, dx \\ \Psi_{f,\alpha}(I) = \sum_{k=0}^{\infty} 2^{(2\alpha-n)k} \sum_{J \in \mathscr{D}_k(I)} \Phi_f(J). \end{cases}$$

Upon writing

$$\Psi_{f,\alpha}(I) = \int_{\mathbb{R}^n} \int_{\mathbb{R}^n} \mathbb{K}_{I,\alpha}(x,y) |f(x) - f(y)|^2 \, dx dy$$

where

$$\mathbb{K}_{I,\alpha}(x,y) = 2^{-1} \sum_{k=0}^{\infty} \sum_{J \in \mathscr{D}_k(I)} 2^{(2\alpha+n)k} \ell(I)^{-2n} \mathbb{1}_J(x) \mathbb{1}_J(y),$$

we consider two situations as below:

▷ If $\alpha > -2^{-1}n$, then

$$x, y \in J \in \mathscr{D}_k(I) \Rightarrow 2^k \leq \ell(I)|x-y|_\infty^{-1} \lesssim \ell(I)|x-y|^{-1}$$

where

$$|x-y|_\infty = |(x_1-y_1,\dots,x_n-y_n)|_\infty = \max_{1\leq j\leq n}|x_j-y_j|,$$

and hence

$$\mathbb{K}_{I,\alpha}(x,y) \leq \sum_{2^k\leq \ell(I)|x-y|_\infty^{-1}} 2^{(2\alpha+n)k}\ell(I)^{-2n} \lesssim \ell(I)^{2\alpha-n}|x-y|^{-2\alpha-n}.$$

Consequently,

$$\Psi_{f,\alpha}(I) \lesssim \Theta_{f,\alpha}(I).$$

▷ If $\alpha \leq -2^{-1}n$, then

$$\begin{cases}\Psi_{f,\alpha}(I) \approx \Phi_f(I)\\ \ell(I)^{-2n}\int_I\int_I |f(x)-f(y)|^2\,dxdy \lesssim \Theta_{f,\alpha}(I),\end{cases}$$

and hence

$$\Psi_{f,\alpha}(I) \lesssim \Theta_{f,\alpha}(I).$$

Accordingly,

$$\sup_{I\in\mathscr{C}(\mathbb{R}^n)}\Psi_{f,\alpha}(I) \lesssim \sup_{I\in\mathscr{C}(\mathbb{R}^n)}\Theta_{f,\alpha}(I) = \|f\|^2_{Q_\alpha(\mathbb{R}^n)}.$$

(iii)⇒(ii) This is trivial.
(ii)⇒(i) This follows from the following two-fold treatment.
On the one hand, it is evident to get the inequality

$$\Psi_{f,\alpha}(I) \lesssim \sum_{k=0}^{\infty} 2^{(2\alpha-n)k}\sum_{J\in\mathscr{D}_k(I)}\ell(J)^{-n}\int_J |f(x)-c_J|^2\,dx \quad \forall I\in\mathscr{C}(\mathbb{R}^n).$$

On the other hand, we are about to prove the following inequality

$$\Theta_{f,\alpha}(I) \lesssim \Psi_{f,\alpha}(I) + \ell(I)^{-n}\int_{|z|_\infty<\ell(I)} \Psi_{f,\alpha}(I+z)\,dy \quad \forall I\in\mathscr{C}(\mathbb{R}^n),$$

where

$$I+z = \{x+z : x\in I\}$$

is the z-translation.

To do so, note that

$$\int_{|z|_\infty<\ell(I)} \Psi_{f,\alpha}(I+z)\,\mathrm{d}z = \int_{\mathbb{R}^n}\int_{\mathbb{R}^n}\Big(\int_{|z|_\infty<\ell(I)} \mathbb{K}_{I+z,\alpha}(x,y)\,\mathrm{d}z\Big)|f(x)-f(y)|^2\,\mathrm{d}x\mathrm{d}y.$$

So, it is enough to establish the following inequality

$$\frac{\ell(I)^{2\alpha}}{|x-y|^{2\alpha+n}} \lesssim \int_{|z|_\infty<\ell(I)} \mathbb{K}_{I+z,\alpha}(x,y)\,\mathrm{d}z + \mathbb{K}_{I,\alpha}(x,y) \quad \forall\, x,y\in I\in\mathscr{C}(\mathbb{R}^n).$$

In fact, for

$$x,y\in I \quad \& \quad |x-y|_\infty \le 2^{-1}\ell(I)$$

let

$$l-1\in\mathbb{N} \quad \& \quad 2^{-l-2}\ell(I) < |x-y|_\infty \le 2^{-l-1}\ell(I).$$

Then an application of the implication

$$x\notin I+z \Rightarrow \mathbb{K}_{I+z,\alpha}(x,y)=0 \quad \forall\,|z|>\ell(I)$$

yields

$$\begin{aligned}\ell(I)^{-n}\int_{|z|_\infty<\ell(I)} \mathbb{K}_{I+z,\alpha}(x,y)\,\mathrm{d}z &\gtrsim 2^{(2\alpha+n)l}\ell(I)^{-3n}\sum_{J\in\mathscr{D}_l(I)}\int_{\mathbb{R}^n} 1\!\!1_{J+z}(x)1\!\!1_{J+z}(y)\,\mathrm{d}z\\ &\gtrsim |x-y|^{-2\alpha-n}\ell(I)^{2(\alpha-n)}\sum_{J\in\mathscr{D}_l(I)}\int_{\mathbb{R}^n} 1\!\!1_{J+z}(x)1\!\!1_{J+z}(y)\,\mathrm{d}z\\ &\gtrsim |x-y|^{-2\alpha-n}\ell(I)^{2\alpha-n}.\end{aligned}$$

Meanwhile, if

$$x,y\in I \quad \text{with } |x-y|>2^{-1}\ell(I),$$

then

$$\mathbb{K}_{I,\alpha}(x,y)\ge 2^{-1}\ell(I)^{-2n}\gtrsim \ell(I)^{2\alpha-n}|x-y|^{-2\alpha-n},$$

and hence the desired inequality follows. □

Especially, let $\mathscr{D}_0(\mathbb{R}^n)$ be the class of unit cubes whose vertices have integer coordinates, and for

$$\pm k\in\mathbb{N}\cup\{0\}$$

let

$$\mathscr{D}_k(\mathbb{R}^n) = \{2^{-k}I : I \in \mathscr{D}_0(\mathbb{R}^n)\} \quad \& \quad \mathscr{D}(\mathbb{R}^n) = \cup_{k\in\mathbb{Z}}\mathscr{D}_k(\mathbb{R}^n).$$

Also, define the dyadic distance

$$\delta(x,y) = \inf\{\ell(I) : x,y \in I \in \mathscr{D}(\mathbb{R}^n)\}$$

between x and y in $\mathbb{R}^n$.

Theorem 1.6. *For $(\alpha, n-1) \in \mathbb{R}\times\mathbb{N}$, let $\mathrm{Q}^d_\alpha(\mathbb{R}^n)$ be the space of all Lebesgue measurable functions $f : \mathbb{R}^n \to \mathbb{R}$ such that*

$$\|f\|_{\mathrm{Q}^d_\alpha(\mathbb{R}^n)} = \sup_{I\in\mathscr{D}(\mathbb{R}^n)} \left(\ell(I)^{2\alpha-n}\int_I\int_I \frac{|f(y)-f(z)|^2}{\delta(y,z)^{n+2\alpha}}\,\mathrm{d}y\mathrm{d}z\right)^{\frac{1}{2}} < \infty.$$

Then:

(i)
$$f \in \mathrm{Q}^d_\alpha(\mathbb{R}^n) \Leftrightarrow \sup_{I\in\mathscr{D}(\mathbb{R}^n)} \Psi_{f,\alpha}(I) < \infty;$$

(ii)
$$f \in \mathrm{Q}_\alpha(\mathbb{R}^n) \Leftrightarrow \sup_{z\in\mathbb{R}^n} \|f(\cdot - z)\|_{\mathrm{Q}^d_\alpha(\mathbb{R}^n)} < \infty;$$

(iii)
$$\mathrm{Q}_\alpha(\mathbb{R}^n) = \mathrm{Q}^d_\alpha(\mathbb{R}^n) \cap \mathrm{BMO}(\mathbb{R}^n) \quad as \quad \alpha \in (-\infty, 2^{-1}).$$

Proof. (i) Two situations are handled below.

▷ If $\alpha > -2^{-1}n$, then using the equivalence

$$x,y \in J \text{ for some } J \in \mathscr{D}_k(I) \Leftrightarrow \delta(x,y) \le 2^{-k}\ell(I) \quad \forall\, (I,k+1) \in \mathscr{D}(\mathbb{R}^n)\times\mathbb{N},$$

we get

$$\mathbb{K}_{I,\alpha}(x,y) = 2^{-1}\sum_{2^k\le\ell(I)/\delta(x,y)} 2^{(2\alpha+n)k}\ell(I)^{-2n} \approx \ell(I)^{2\alpha-n}\delta(x,y)^{-2\alpha-n},$$

whence

$$\begin{aligned}\Psi_{f,\alpha}(I) &= \int_{\mathbb{R}^n}\int_{\mathbb{R}^n} \mathbb{K}_{I,\alpha}(x,y)|f(x)-f(y)|^2\,\mathrm{d}x\mathrm{d}y\\ &\approx \ell(I)^{2\alpha-n}\int_I\int_I \frac{|f(y)-f(z)|^2}{\delta(y,z)^{n+2\alpha}}\,\mathrm{d}y\mathrm{d}z.\end{aligned}$$

▷ If $\alpha \le -2^{-1}n$, then the above analysis, along with

$$|x-y| \lesssim \delta(x,y)$$

and the argument for Theorem 1.5, gives

$$\Psi_{f,\alpha}(I) \approx \ell(I)^{2\alpha-n}\int_I\int_I \frac{|f(y)-f(z)|^2}{\delta(y,z)^{n+2\alpha}}\,\mathrm{d}y\mathrm{d}z \quad \forall I \in \mathscr{D}(\mathbb{R}^n).$$

(ii) Noticing two facts:

- ▷ $\mathrm{Q}_\alpha(\mathbb{R}^n)$ actually equals the space of all Lebesgue measurable functions

$$f : \mathbb{R}^n \to \mathbb{R}$$

such that

$$\sup_I \Psi_{f,\alpha}(I) < \infty$$

where the supremum ranges over $I \in \mathscr{C}(\mathbb{R}^n)$ with dyadic edge-length;

- ▷ $I \in \mathscr{C}(\mathbb{R}^n)$ with $\ell(I)$ being dyadic is the translate of a cube in $\mathscr{D}(\mathbb{R}^n)$,

we use the just-verified (i) to derive

$$\|f\|^2_{\mathrm{Q}_\alpha(\mathbb{R}^n)} \approx \sup_{z\in\mathbb{R}^n} \sup_{I\in\mathscr{D}(\mathbb{R}^n)} \Psi_{f,\alpha}(I - z) \approx \sup_{z\in\mathbb{R}^n} \|f(\cdot - z)\|^2_{\mathrm{Q}^d_\alpha(\mathbb{R}^n)}.$$

(iii) The inclusion

$$\mathrm{Q}^d_\alpha(\mathbb{R}^n) \cap \mathrm{BMO}(\mathbb{R}^n) \supseteq \mathrm{Q}_\alpha(\mathbb{R}^n)$$

follows readily from definition and

$$|x - y| \lesssim \delta(x, y).$$

For the reverse inclusion, suppose that

$$f \in \mathrm{Q}^d_\alpha(\mathbb{R}^n) \cap \mathrm{BMO}(\mathbb{R}^n)$$

and $\ell(I)$ of $I \in \mathscr{C}(\mathbb{R}^n)$ is dyadic. Then there is a family of 2^n dyadic cubes I_i of the same size as I such that $\cup_i I_i$ is a cube J with

$$\ell(J) = 2\ell(I) \quad \& \quad I \subseteq J.$$

Accordingly,

$$\Psi_{f,\alpha}(I) \lesssim \Psi_{f,\alpha}(J) \lesssim \Phi_f(J) + \sum_{i=1}^{2^n} \Psi_{f,\alpha}(I_i) \lesssim \|f\|^2_{\mathrm{Q}^d_\alpha(\mathbb{R}^n)} + \|f\|^2_{\mathrm{BMO}(\mathbb{R}^n)}.$$

This in turn implies $f \in \mathrm{Q}_\alpha(\mathbb{R}^n)$. □

1.4 Wavelet decomposition

This section is to show that the well-known characterization of $\mathrm{BMO}(\mathbb{R}^n)$ by means of an orthonormal wavelet basis (cf. [63, p. 154]) can be extended to $\mathrm{Q}_\alpha(\mathbb{R}^n)$. From now on, set

$$\psi_{j,\Bbbk,l}(x) = 2^{\frac{jn}{2}}\psi_l(2^j x - \Bbbk) \quad \forall\,(j,\Bbbk,l) \in \mathbb{Z}\times\mathbb{Z}^n\times\{1,\dots,2^n-1\}$$

be a 1-regular orthonormal wavelet basis as in [63, Chapter 3]. For simplicity, we utilize only the wavelets of compact support and then assume that the wavelets obey the conditions in [63, p. 108], in particular, there is a constant $\kappa > 0$ such that

$$\operatorname{supp}(\psi_{j,\Bbbk,l}) \subseteq \kappa I(j,\Bbbk,l)$$

holds for the dyadic cube

$$I(j,\Bbbk,l) = \{x \in \mathbb{R}^n : 2^j x - \Bbbk \in [0,1)^n\}.$$

Theorem 1.7. *For*

$$(\alpha, n-1, I) \in (0,1)\times\mathbb{N}\times\mathcal{D}(\mathbb{R}^n)$$

and a sequence $\{a(j,\Bbbk,l)\}$ *of real numbers, let*

$$T_{\{a(j,\Bbbk,l)\},\alpha}(I) = \ell(I)^{-n}\sum_{I(j,\Bbbk,l)\subseteq I}\left(\frac{\ell(I)}{\ell(I(j,\Bbbk,l))}\right)^{2\alpha}|a(j,\Bbbk,l)|^2.$$

(i) *If* $f \in \mathrm{Q}_\alpha(\mathbb{R}^n)$, *then the sequence of its wavelet coefficients*

$$a(j,\Bbbk,l) = \langle f, \psi_{j,\Bbbk,l}\rangle = \int_{\mathbb{R}^n} f(x)\psi_{j,\Bbbk,l}(x)\,dx$$

satisfies

$$\sup_{I\in\mathcal{D}(\mathbb{R}^n)} T_{\{a(j,\Bbbk,l)\},\alpha}(I) < \infty.$$

(ii) *Conversely, every sequence* $\{a(j,\Bbbk,l)\}$ *obeying the last inequality is the sequence of wavelet coefficients of a unique* $f \in \mathrm{Q}_\alpha(\mathbb{R}^n)$; *moreover,*

$$\|f\|_{\mathrm{Q}_\alpha(\mathbb{R}^n)} \approx \sup_{I\in\mathcal{D}(\mathbb{R}^n)}\left(T_{\{a(j,\Bbbk,l)\},\alpha}(I)\right)^{\frac{1}{2}}.$$

Proof. (i) Let

$$(f,I) \in \mathrm{Q}_\alpha(\mathbb{R}^n)\times\mathcal{D}(\mathbb{R}^n).$$

For $J \in \mathscr{D}_k(I)$, write

$$f = f_{\kappa J} + (f - f_{\kappa J})\mathbb{1}_{\kappa J} + (f - f_{\kappa J})\mathbb{1}_{\mathbb{R}^n \setminus \kappa J} = f_1 + f_2 + f_3.$$

Thanks to

$$\operatorname{supp}(\psi_{(j,\Bbbk,l)}) \subseteq \kappa I(j,\Bbbk,l),$$

we have

$$I(j,\Bbbk,l) \subseteq J \Rightarrow \langle f_3, \psi_{(j,\Bbbk,l)}\rangle = 0.$$

Also, the integral of each wavelet $\psi_{j,\Bbbk,l}$ vanishes. So

$$\langle f, \psi_{j,\Bbbk,l}\rangle = \langle f_2, \psi_{j,\Bbbk,l}\rangle.$$

Consequently,

$$\begin{aligned}\sum_{I(j,\Bbbk,l)\subseteq J} |\langle f, \psi_{j,\Bbbk,l}\rangle|^2 &\le \sum_{j,\Bbbk,l} |\langle f_2, \psi_{j,\Bbbk,l}\rangle|^2 \\ &\lesssim \ell(\kappa J)^{-n} \int_{\kappa J}\int_{\kappa J} |f(x) - f(y)|^2 \,\mathrm{d}x\mathrm{d}y.\end{aligned}$$

This in turn implies

$$T_{\{a(j,\Bbbk,l)\},0}(J) \lesssim \big(\ell(J)\ell(\kappa J)\big)^{-n} \int_{\kappa J}\int_{\kappa J} |f(x) - f(y)|^2 \,\mathrm{d}x\mathrm{d}y \quad \forall\,(k,J) \in \mathbb{N}\times\mathscr{D}_k(I),$$

and so

$$\begin{aligned}T_{\{a(j,\Bbbk,l)\},\alpha}(I) &\lesssim \sum_{k=0}^{\infty} 2^{k(2\alpha-n)} \sum_{J\in\mathscr{D}_k(I)} T_{\{a(j,\Bbbk,l)\},0}(J) \\ &\lesssim \int_{\kappa I}\int_{\kappa I} |f(x) - f(y)|^2 \sum_{k=0}^{\infty}\sum_{J\in\mathscr{D}_k(I)} \frac{\mathbb{1}_{\kappa J}(x)\mathbb{1}_{\kappa J}(y)}{2^{pk}\ell(J)^{2n}} \,\mathrm{d}x\mathrm{d}y \\ &\lesssim \ell(I)^{2\alpha-n} \int_{\kappa I}\int_{\kappa I} \frac{|f(x) - f(y)|^2}{|x-y|^{n+2\alpha}} \,\mathrm{d}x\mathrm{d}y \\ &\lesssim \|f\|^2_{Q_\alpha(\mathbb{R}^n)}.\end{aligned}$$

(ii) Let

$$\sup_{I\in\mathscr{D}(\mathbb{R}^n)} \big(T_{\{a(j,\Bbbk,l)\},\alpha}(I)\big)^{\frac{1}{2}} < \infty.$$

Via multiplying f by a constant, we may assume

$$T_{\{a(j,\mathbb{k},l)\},\alpha}(I) \le 1 \quad \forall I \in \mathscr{D}(\mathbb{R}^n),$$

where

$$T_{\{a(j,\mathbb{k},l)\},0}(I) \le T_{\{a(j,\mathbb{k},l)\},\alpha}(I) \le 1 \quad \forall I \in \mathscr{D}(\mathbb{R}^n)$$

and so

$$f = \sum_{j,\mathbb{k},l} a(j,\mathbb{k},l)\psi_{j,\mathbb{k},l} \in \mathrm{BMO}(\mathbb{R}^n),$$

with the sum converging, for example, in the weak-$*$ topology on $\mathrm{BMO}(\mathbb{R}^n)$. We are about to show $f \in Q_\alpha(\mathbb{R}^n)$.

Fix $I \in \mathscr{C}(\mathbb{R}^n)$ of dyadic length and consider $J \in \mathscr{D}(I)$. Let

$$\Lambda_0(J) = \{(j,\mathbb{k},l) : \kappa I(j,\mathbb{k},l) \cap J \neq \emptyset\}$$

and split this set into

$$\begin{cases} \Lambda_1 = \{(j,\mathbb{k},l) \in \Lambda_0(J) : \ell(I(j,\mathbb{k},l)) \le \ell(J)\} \\ \Lambda_2 = \{(j,\mathbb{k},l) \in \Lambda_0(J) : \ell(J) < \ell(I(j,\mathbb{k},l)) \le \ell(I)\} \\ \Lambda_3 = \{(j,\mathbb{k},l) \in \Lambda_0(J) : \ell(I) < \ell(I(j,\mathbb{k},l))\}. \end{cases}$$

Since

$$\psi_{j,\mathbb{k},l} = 0 \quad \text{on} \quad J \quad \text{unless} \ (j,\mathbb{k},l) \in \Lambda_0,$$

we have

$$f = f_1 + f_2 + f_3 \quad \text{on} \quad J,$$

where

$$f_i = \sum_{(j,\mathbb{k},l)\in\Lambda_i} a(j,\mathbb{k},l)\psi_{j,\mathbb{k},l} \quad \forall i \in \{1,2,3\}.$$

Now, an application of the Cauchy–Schwarz inequality derives

$$\Phi_f(J) \lesssim \Phi_{f_1}(J) + \Phi_{f_2}(J) + \Phi_{f_3}(J).$$

▷ For f_1, one has

$$\Phi_{f_1}(J) \le \ell(J)^{-n}\|f_1\|^2_{L^2(\mathbb{R}^n)} = \ell(J)^{-n} \sum_{(j,\mathbb{k},l)\in\Lambda_1} |a(j,\mathbb{k},l)|^2.$$

▷ For f_2, one has

$$|\nabla\psi_{j,\Bbbk,l}| \lesssim \ell(I(j,\Bbbk,l))^{-\frac{3n}{2}},$$

and thus

$$|f_2(x) - f_2(y)| \lesssim \sum_{(j,\Bbbk,l)\in\Lambda_2} |a(j,\Bbbk,l)|\ell(I(j,\Bbbk,l))^{-\frac{2+n}{2}}|x-y|.$$

Accordingly, by letting $\epsilon = 1-\alpha$ and using the Cauchy–Schwarz inequality it follows that

$$\begin{aligned}\Phi_{f_2}(J) &\lesssim \ell(J)^2\Big(\sum_{(j,\Bbbk,l)\in\Lambda_3} \frac{|a(j,\Bbbk,l)|}{\ell(I(j,\Bbbk,l))^{\frac{3n}{2}}}\Big)^2\\ &\lesssim \ell(J)^2 \sum_{(j,\Bbbk,l)\in\Lambda_2} \frac{|a(j,\Bbbk,l)|^2}{\ell(I(j,\Bbbk,l))^{2+n}}\Big(\frac{\ell(I(j,\Bbbk,l))}{\ell(J)}\Big)^{\epsilon} \sum_{(j,\Bbbk,l)\in\Lambda_2}\Big(\frac{\ell(J)}{\ell(I(j,\Bbbk,l))}\Big)^{\epsilon}.\end{aligned}$$

If

$$(j,\Bbbk,l)\in\Lambda_2,$$

then $I(j,\Bbbk,l)$ is a dyadic cube contained in a cube with the same center as J and edge-length

$$(\kappa+1)\ell(I(j,\Bbbk,l)) + \ell(J) \le \ell((\kappa+2)J).$$

Hence, for each $k\in\mathbb{N}$, there are at most $(\kappa+2)^n$ such cubes $I(j,\Bbbk,l)$ with

$$\ell(I(j,\Bbbk,l)) = 2^k\ell(J).$$

Furthermore, there exists a constant number of different $(j,\Bbbk,l)$ for each such cube, and so

$$\#\{(j,\Bbbk,l)\in\Lambda_2 : \ \ell(I(j,\Bbbk,l)) = 2^k\ell(J)\} \lesssim 1 \quad \forall k\in\mathbb{N}.$$

Accordingly,

$$\Phi_{f_2}(J) \lesssim \sum_{(j,\Bbbk,l)\in\Lambda_2} |a(j,\Bbbk,l)|^2\Big(\frac{\ell(J)}{\ell(I(j,\Bbbk,l))}\Big)^{2-\epsilon}\ell(I(j,\Bbbk,l))^{-n}.$$

▷ For f_3, one similarly has

$$|f_3(x) - f_3(y)| \lesssim |x-y| \sum_{(j,\Bbbk,l)\in\Lambda_3} \ell(I(j,\Bbbk,l))^{-1}$$

by using

$$|a(j,\Bbbk,l)|\ell(I(j,\Bbbk,l))^{-\frac{n}{2}} \le (T_{\{a(j,\Bbbk,l)\},\alpha}(I(j,\Bbbk,l)))^{\frac{1}{2}} \le 1.$$

Again, there exists a bounded number of terms for each $\ell(I(j,\Bbbk,l))$, and now

$$\ell(I(j,\Bbbk,l)) = 2^k\ell(I) \quad \forall k \in \mathbb{N};$$

hence

$$|f_3(x)-f_3(y)| \lesssim |x-y|\ell(I)^{-1} \quad \& \quad \Phi_{f_3}(J) \lesssim \ell(J)^2\ell(I)^{-2}.$$

Consequently,

$$\Phi_f(J) \lesssim \sum_{(j,\Bbbk,l)\in\Lambda_1} \frac{|a(j,\Bbbk,l)|^2}{\ell(J)^n} + \sum_{(j,\Bbbk,l)\in\Lambda_2} \frac{|a(j,\Bbbk,l)|^2}{\ell(I(j,\Bbbk,l))^n}\left(\frac{\ell(J)}{\ell(I(j,\Bbbk,l))}\right)^{2-\epsilon} + \left(\frac{\ell(J)}{\ell(I)}\right)^2.$$

Upon summing over $J \in \mathscr{D}(I)$, we achieve

$$\begin{aligned}\Psi_{f,\alpha}(I) &= \sum_{J\in\mathscr{D}(I)} \left(\frac{\ell(J)}{\ell(I)}\right)^{n-2\alpha} \Phi_f(J)\\ &\lesssim \sum_{J\in\mathscr{D}(I)} \sum_{(j,\Bbbk,l)\in\Lambda_1(J)} |a(j,\Bbbk,l)|^2\ell(J)^{-2\alpha}\ell(I)^{2\alpha-n}\\ &\quad + \sum_{J\in\mathscr{D}(I)} \sum_{(j,\Bbbk,l)\in\Lambda_2(J)} |a(j,\Bbbk,l)|^2\ell(J)^{n-2\alpha+2-\epsilon}\ell(I(j,\Bbbk,l))^{\epsilon-2-n}\ell(I)^{2\alpha-n}\\ &\quad + \sum_{J\in\mathscr{D}(I)} \left(\frac{\ell(J)}{\ell(I)}\right)^{n-2\alpha+2}.\end{aligned}$$

The final sum equals a constant. In the two double sums, we interchange the order of summation. If $(j,\Bbbk,l)$ occurs there, then

$$\ell(I(j,\Bbbk,l)) \le \ell(I) \quad \& \quad \kappa I(j,\Bbbk,l) \cap I \ne \emptyset,$$

and hence, if we set $\mathcal{E}(I)$ be the class of all dyadic cubes J of the same size as I with

$$\kappa J \cap I \ne \emptyset$$

then

$$I(j,\Bbbk,l) \in \mathscr{D}_1(J) \quad \text{for some} \quad J \in \mathcal{E}(I).$$

Fix such a

$$(j,\Bbbk,l) \quad \text{with} \quad \ell(I(j,\Bbbk,l)) = 2^{-k}\ell(I).$$

For each $j \le k$, there exist at most finitely many cubes

$$J \in \mathcal{D}_j(I) \quad \text{with} \quad (j, \Bbbk, l) \in \Lambda_1(J),$$

each contributing

$$2^{2\alpha j}\ell(I)^{-n}|a(j, \Bbbk, l)|^2$$

to the first double sum in the last chain of inequalities. Similarly, for each integer $j > k$, there exist at most $c2^{(j-k)}$ (for an absolute constant $c > 0$) cubes

$$J \in \mathcal{D}_j(I) \quad \text{with} \quad (j, \Bbbk, l) \in \Lambda_2(J),$$

each contributing

$$2^{-j(n-2\alpha+2-\epsilon)-k(\epsilon-n-2)}\ell(I)^{-n}|a(j, \Bbbk, l)|^2$$

to the second double sum mentioned above. Putting them together yields at most

$$2^{2\alpha k}|a(j, \Bbbk, l)|^2\ell(I)^{-n}.$$

Consequently,

$$\Psi_{f,\alpha}(I) \lesssim \sum_{J\in\mathcal{E}(I)}\sum_{k=0}^{\infty}\sum_{I(j,\Bbbk,l)\in\mathcal{D}_k(J)} 2^{2\alpha k}\ell(I)^{-n}|a(j, \Bbbk, l)|^2 + 1.$$

We have proved that

$$\Psi_{f,\alpha}(I) \lesssim 1$$

holds for any cube I of dyadic edge-length. Since the same estimate applies to any translate $I + z$, it follows that

$$\ell(I)^{2\alpha-n}\int_I\int_I \frac{|f(x) - f(y)|^2}{|x - y|^{n+2\alpha}}\,dx\,dy \lesssim \sup_{|z|_\infty<\ell(I)} \Psi_{f,\alpha}(I + z)$$

is valid for every cube I of dyadic edge-length. Therefore, $f \in \mathrm{Q}_\alpha(\mathbb{R}^n)$.

The uniqueness of f follows from the uniqueness in $\mathrm{BMO}(\mathbb{R}^n)$: if two elements

$$f, g \in \mathrm{Q}_\alpha(\mathbb{R}^n) \subseteq \mathrm{BMO}(\mathbb{R}^n)$$

have the same wavelet coefficients, then they define the same linear functional on the predual space of $\mathrm{BMO}(\mathbb{R}^n)$, and thus $f = g$ as elements of $\mathrm{BMO}(\mathbb{R}^n)$; see also [63, Section 5.6] once again. □

1.5 Notes

Below are some historical remarks on Chapter 1.

Section 1.1. Definition 1.1 is more or less motivated by [80]. Theorem 1.2 comes from the second section of [22].

Section 1.2. Theorems 1.3–1.4 come from the second section of [22] and the fourth section of [91] which improves the well-known Poincaré embedding (cf. [76, p. 34])

$$\dot{W}^{1,n}(\mathbb{R}^n) \subsetneq \mathrm{BMO}(\mathbb{R}^n).$$

Section 1.3. Theorems 1.5–1.6 come from the fifth and seventh sections of [22] and the second section of [18]. Moreover, [22, Lemma 5.8] reveals that

$$\Theta_{f,\alpha}(I) \approx \Psi_{f,\alpha}(I) \quad \forall I \in \mathscr{C}(\mathbb{R}^n)$$

holds for $\alpha \in (-\infty, 2^{-1})$, but leaves the case $\alpha \in [2^{-1}, 1)$ unsolved; of course, the main issue is to prove or disprove

$$\Theta_{f,\alpha}(I) \lesssim \Psi_{f,\alpha}(I) \quad \forall I \in \mathscr{C}(\mathbb{R}^n)$$

under $\alpha \in [2^{-1}, 1)$. Also, it is interesting to decide the truth of

$$\mathrm{Q}_\alpha(\mathbb{R}^n) = \mathrm{Q}_\alpha^d(\mathbb{R}^n) \cap \mathrm{BMO}(\mathbb{R}^n) \quad \text{as} \quad \alpha \in [2^{-1}, 1).$$

Nevertheless, according to [105, Remark 1.6(ii)] we have

$$\|f\|^2_{\mathrm{Q}_\alpha(\mathbb{R}^n)} \approx \sup_{I \in \mathscr{D}(\mathbb{R}^n)} \ell(I)^{2\alpha-n} \int_I \sum_{k\in\mathbb{Z}} 2^{2k\alpha} |f(x) - f_{B(x,2^{-k})}|^2 \, dx \quad \text{as} \quad \alpha \in (0,1).$$

Section 1.4. Theorem 1.7 comes from [22, Section 6]. Moreover, an atomic decomposition of $\mathrm{Q}_\alpha(\mathbb{R}^n)$ based on $\mathrm{Q}_\alpha^d(\mathbb{R}^n)$ is given in [18, Theorem 3.7]; more precisely, a function a_I on $\mathbb{R}^n$ is called a $(0,\infty)$-atom provided that there are a cube $I \subseteq \mathbb{R}^n$ and a pair

$$(N_1, N_2 + 1) \in \mathbb{N} \times \mathbb{N}$$

such that

$$\begin{cases} \operatorname{supp}(a_I) \subseteq 3I \\ |D^\gamma a_I(x)| \le \ell(I)^{-|\gamma|} \quad \text{for } D^\gamma = \partial_{x_1}^{\gamma_1} \cdots \partial_{x_n}^{\gamma_n} \;\&\; |\gamma| = \sum_{j=1}^n \gamma_j \le N_1 \\ \int_{\mathbb{R}^n} x^\gamma a_I(x) \, dx = 0 \quad \text{or} \quad |\gamma| \le N_2. \end{cases}$$

Under $(\alpha, n-1) \in (-\infty, 1) \times \mathbb{N}$, we have the following assertions.

▷ If $\{a_I\}_{I\in\mathcal{D}(\mathbb{R}^n)}$ is a sequence of $(0,\infty)$-atoms, and $\{s_I\}_{I\in\mathcal{D}(\mathbb{R}^n)}$ is a sequence obeying

$$\|\{s_I\}_{I\in\mathcal{D}(\mathbb{R}^n)}\|_{\mathscr{C}_\alpha} = \sup_{I\in\mathcal{D}(\mathbb{R}^n)}\left(\sum_{J\subseteq I}\left(\frac{\ell(J)}{\ell(I)}\right)^{n-2\alpha}|s_J|^2\right)^{\frac{1}{2}} < \infty,$$

then there is a function $f \in Q_\alpha(\mathbb{R}^n)$ such that

$$f = \sum_{I\in\mathcal{D}(\mathbb{R}^n)} s_I a_I = \lim_{k\to-\infty, l\to\infty} \sum_{I\in\mathcal{D}(\mathbb{R}^n),\, 2^k\le\ell(I)\le 2^l} s_I a_I,$$

where the convergence is in the dual of $S(\mathbb{R}^n)$ modulo $K(\mathbb{R}^n)$ and in the weak-$*$ topology in $Q_\alpha(\mathbb{R}^n)$.

▷ Conversely, if $f \in Q_\alpha(\mathbb{R}^n)$ then there is a sequence $\{a_I\}_{I\in\mathcal{D}(\mathbb{R}^n)}$ of $(0,\infty)$-atoms, and a $\mathscr{C}_\alpha$-sequence $\{s_I\}_{I\in\mathcal{D}(\mathbb{R}^n)}$ such that the above limit holds. Moreover,

$$\|f\|_{Q_\alpha(\mathbb{R}^n)} \approx \|\{s_I\}_{I\in\mathcal{D}(\mathbb{R}^n)}\|_{\mathscr{C}_\alpha}.$$

In the case $\alpha \in (-\infty, 0)$, the above atomic decomposition amounts to the $BMO(\mathbb{R}^n)$-case shown in [86, Lemmas 3.1–3.4], in particular,

$$\|\{s_I\}_{I\in\mathcal{D}(\mathbb{R}^n)}\|_{\mathscr{C}_{-\infty<\alpha<0}} \approx \sup_{I\in\mathcal{D}(\mathbb{R}^n)}\left(\sum_{J\subseteq I}\left(\frac{\ell(J)}{\ell(I)}\right)^{n}|s_J|^2\right)^{\frac{1}{2}} \approx \left\|f = \sum_{I\in\mathcal{D}(\mathbb{R}^n)} s_I a_I\right\|_{BMO(\mathbb{R}^n)}.$$

2 Estimation of John–Nirenberg type

Having set down some basic facts about $Q_\alpha(\mathbb{R}^n)$ in Chapter 1, we now turn to the first deeper property of $Q_\alpha(\mathbb{R}^n)$-functions—their exponential integrability—this is indeed devoted to settling [22, Problem 8.1] on the distribution estimation for $Q_{0\le\alpha<1}(\mathbb{R}^n)$ extending the celebrated John–Nirenberg (J-N) inequality for $\mathrm{BMO}(\mathbb{R}^n)$ (cf. [38])

$$\mathrm{m}_{f,I}(t) \lesssim \ell(I)^n \exp(-ct\|f\|^{-1}_{\mathrm{BMO}(\mathbb{R}^n)}) \quad \forall\, 0 < \|f\|_{\mathrm{BMO}(\mathbb{R}^n)} < \infty$$

for a constant $c > 0$ and any $(I, t) \in \mathscr{C}(\mathbb{R}^n) \times \mathbb{R}_+$.

2.1 J-N-BMO inequality with special constant

One of the fundamental theorems in $\mathrm{BMO}(\mathbb{R}^n)$ is the John–Nirenberg inequality with precise constants as seen below.

Theorem 2.1. *Let*

$$\mathrm{m}(E) = \int_E \mathrm{d}x$$

stand for the n-dimensional Lebesgue measure of $E \subseteq \mathbb{R}^n$.

(i) *If*

$$\|f\|_{\mathrm{BMO}(\mathbb{R}^n)} < \infty,$$

then

$$\mathrm{m}(\{x \in I : |f(x) - f_I| > t\}) \le \mathrm{m}(I)\exp\Big(1 - \frac{t}{e2^n\|f\|_{\mathrm{BMO}(\mathbb{R}^n)}}\Big) \quad \forall I \in \mathscr{C}(\mathbb{R}^n),$$

and hence

$$\fint_I \exp\Big(\frac{c|f(x) - f_I|}{\|f\|_{\mathrm{BMO}(\mathbb{R}^n)}}\Big)\,\mathrm{d}x \le \frac{ec}{(e2^n)^{-1} - c} \quad \forall I \in \mathscr{C}(\mathbb{R}^n)$$

holds with $0 < c < (e2^n)^{-1}$.

(ii) *Conversely, if there are two constants $c_1, c_2 > 0$ such that*

$$\fint_I \exp\Big(\frac{|f(x) - f_I|}{c_1}\Big)\,\mathrm{d}x \le c_2 \quad \forall I \in \mathscr{C}(\mathbb{R}^n),$$

then

$$\|f\|_{\mathrm{BMO}(\mathbb{R}^n)} \le c_1 c_2.$$

https://doi.org/10.1515/9783110600285-002

Proof. (i) By scaling, without loss of generality, we may assume

$$\|f\|_{\mathrm{BMO}(\mathbb{R}^n)} = 1,$$

recall

$$\mathrm{m}_{f,I}(t) = \mathrm{m}(\{x \in I : |f(x) - f_I| > t\}) \quad \forall I \in \mathscr{C}(\mathbb{R}^n),$$

and define

$$\tau(t) = \sup_{I \in \mathscr{C}(\mathbb{R}^n)} \frac{\mathrm{m}_{f,I}(t)}{\mathrm{m}(I)}.$$

For

$$t \geq e \quad \& \quad I \in \mathscr{C}(\mathbb{R}^n),$$

let $\mathscr{F}(I,t)$ be the family of all elements $J \in \mathscr{D}(I)$ with

$$\fint_J |f(x) - f_I|\,\mathrm{d}x > t.$$

Then $I \notin \mathscr{F}(I,t)$ due to

$$\fint_I |f(x) - f_I|\,\mathrm{d}x \leq \|f\|_{\mathrm{BMO}(\mathbb{R}^n)} = 1 < e \leq t.$$

Next, we introduce the dyadic maximal function

$$\mathcal{M}_{I,d}(|f - f_I|)(x) = \sup\left\{\fint_J |f(y) - f_I|\,\mathrm{d}y : x \in J \in \mathscr{D}(I)\right\},$$

thereby finding

$$\{x \in I : \mathcal{M}_{I,d}(|f - f_I|)(x) > t\} = \bigcup_{J \in \mathscr{F}(I,t)} J.$$

Since

$$t < |f(x) - f_I| \leq \mathcal{M}_{I,d}(|f - f_I|)(x),$$

it follows that

$$\{x \in I : |f(x) - f_I| > t\} \subseteq \bigcup_{J \in \mathscr{F}(I,t)} J.$$

Note that if $J \in \mathscr{F}(I,t)$, then it is contained in a cube J^* of twice the side-length, which is not in $\mathscr{F}(I,t)$. So

$$\fint_{J^*} |f(x) - f_I| \,dx \le t.$$

This in turn implies

$$t < \fint_J |f(y) - f_I| \,dy| \le \frac{m(J^*)}{m(J)} \fint_{J^*} |f(y) - f_I| \,dy| \le 2^n t \quad \forall J \in \mathscr{F}(I,t),$$

and

$$|f(x) - f_I| \le |f(x) - f_J| + \fint_J |f(y) - f_I| \,dy \le |f(x) - f_J| + 2^n t \quad \forall J \in \mathscr{F}(I,t).$$

Consequently, if

$$|f(x) - f_I| > s > 2^n t,$$

then

$$|f(x) - f_J| > s - 2^n t \quad \forall J \in \mathscr{F}(I,t),$$

and hence

$$\begin{aligned}
m_{f,I}(s) &= m(\{x \in I : |f(x) - f_I| > s\} \cap \{x \in I : |f(x) - f_I| > t\}) \\
&\le \sum_{J \in \mathscr{F}(I,t)} m(\{x \in I : |f(x) - f_I| > s\} \cap J) \\
&\le \sum_{J \in \mathscr{F}(I,t)} \left(\frac{m(\{x \in J : |f(x) - f_J| > s - 2^n t\})}{m(J)} \right) m(J) \\
&\le \tau(s - 2^n t) \sum_{J \in \mathscr{F}(I,t)} m(J) \\
&\le \tau(s - 2^n t) t^{-1} \sum_{J \in \mathscr{F}(I,t)} \int_J |f(y) - f_I| \,dy \\
&\le \tau(s - 2^n t) t^{-1} m(I).
\end{aligned}$$

In short, we have

$$\tau(s) \le t^{-1} \tau(s - 2^n t) \quad \text{as} \quad s > 2^n t.$$

In what follows, let us choose $t = e$. Note that

$$\tau(s) \le 1 \quad \forall s > 0.$$

So the last inequality on τ gives

$$\tau(s) \le \exp\left(1 - \frac{s}{e2^n}\right) \quad \forall s \in (0, e2^n].$$

Consequently, if

$$s \in (e2^n, 2e2^n],$$

then

$$\tau(s - 2^n e) \le \exp\left(1 - \frac{s - e2^n}{e2^n}\right) \quad \text{as} \quad s > e2^n,$$

and hence

$$\tau(s) \le e^{-1}\tau(s - 2^n e) \le \exp\left(1 - \frac{s}{e2^n}\right) \quad \text{as} \quad s \in (e2^n, 2e2^n].$$

Since

$$(0, \infty) = \bigcup_{k=0}^{\infty} (ke2^n, (k+1)e2^n],$$

if

$$\tau(s) \le \exp\left(1 - \frac{s}{e2^n}\right) \quad \text{as} \quad s \in (ke2^n, (k+1)e2^n],$$

then

$$\begin{aligned}
\tau(s) &\le e^{-1}\tau(s - e2^n) \\
&\le e^{-1}\exp\left(1 - \frac{s - e2^n}{e2^n}\right) \\
&= \exp\left(1 - \frac{s}{e2^n}\right) \quad \text{as} \quad s \in ((k+1)e2^n, (k+2)e2^n],
\end{aligned}$$

and hence

$$\frac{\mathrm{m}_{f,I}(s)}{\mathrm{m}(I)} \le \tau(s) \le \exp\left(1 - \frac{s}{e2^n}\right) \quad \text{as} \quad s \in (0, \infty).$$

For the second part, we utilize $0 < c < (e2^n)^{-1}$ and

$$\int_I \exp\left(\frac{c|f(x) - f_I|}{\|f\|_{\mathrm{BMO}(\mathbb{R}^n)}}\right) \mathrm{d}x = \left(\frac{c}{\|f\|_{\mathrm{BMO}(\mathbb{R}^n)}}\right) \int_0^{\infty} \exp\left(\frac{ct}{\|f\|_{\mathrm{BMO}(\mathbb{R}^n)}}\right) \mathrm{m}_{f,I}(t)\, dt$$

$$\le \left(\frac{ec}{\|f\|_{\mathrm{BMO}(\mathbb{R}^n)}}\right)\mathrm{m}(I)\int_0^\infty \exp\left(-\frac{t((e2^n)^{-1}-c)}{\|f\|_{\mathrm{BMO}(\mathbb{R}^n)}}\right)\mathrm{d}t$$
$$= \left(\frac{ec}{(e2^n)^{-1}-c}\right)\mathrm{m}(I)$$

to get the desired inequality.

(ii) It follows from

$$e^t \ge t \quad \forall t \ge 0. \qquad \square$$

Proposition 2.2. *For a Lebesgue measurable function f on $\mathbb{R}^n$, let*

$$\|f\|_{\mathrm{L}^p(\mathbb{R}^n)} = \begin{cases} (\int_{\mathbb{R}^n} |f(x)|^p\,\mathrm{d}x)^{\frac{1}{p}} & \text{as} \quad 1 \le p < \infty \\ \operatorname{ess\,sup}_{x\in\mathbb{R}^n} |f(x)| & \text{as} \quad p = \infty, \end{cases}$$

and

$$\mathcal{I}_\alpha f(x) = \begin{cases} \int_{\mathbb{R}^n} \frac{f(y)}{|y-x|^{n-\alpha}}\,\mathrm{d}y & \text{as} \quad 0 < \alpha < n \\ f(x) & \text{as} \quad \alpha = 0. \end{cases}$$

Then:

(i)
$$\|\mathcal{I}_{\frac{n}{p}} f\|_{\mathrm{BMO}(\mathbb{R}^n)} \lesssim \|f\|_{\mathrm{L}^p(\mathbb{R}^n)} \quad \text{for} \quad 1 < p \le \infty;$$

(ii)
$$\|f\|_{\mathrm{L}^q(\mathbb{R}^n)} \lesssim \|f\|_{\mathrm{BMO}(\mathbb{R}^n)}^{1-\frac{p}{q}} \|f\|_{\mathrm{L}^p(\mathbb{R}^n)}^{\frac{p}{q}} \quad \text{for} \quad 1 \le p < q < \infty.$$

Proof. (i) Since $p = \infty$ gives

$$\mathcal{I}_{\frac{n}{p}} \mathrm{L}^p(\mathbb{R}^n) = \mathrm{L}^\infty(\mathbb{R}^n) \subseteq \mathrm{BMO}(\mathbb{R}^n),$$

it remains to verify the case $1 < p < \infty$. For any ball $B(x,r)$ and any function $f \in \mathrm{L}^p(\mathbb{R}^n)$ write

$$f = f\mathbb{1}_{B(x,2r)} + f\mathbb{1}_{\mathbb{R}^n\setminus B(x,2r)} \quad \& \quad c_{B(x,2r)} = \int_{\mathbb{R}^n\setminus B(x,2r)} f(y)|y-x|^{\frac{n}{p}-n}\,\mathrm{d}y.$$

Then the L^p-boundedness of $\mathcal{I}_{\frac{n}{p}}$ and the Hölder inequality are used to derive

$$\int_{B(x,r)} |\mathcal{I}_{\frac{n}{p}} f(y) - (\mathcal{I}_{\frac{n}{p}} f)_{B(x,r)}|\,\mathrm{d}y \lesssim \int_{B(x,r)} |\mathcal{I}_{\frac{n}{p}} f(y) - c_{B(x,r)}|\,\mathrm{d}y$$
$$\lesssim \int_{B(x,r)} |\mathcal{I}_{\frac{n}{p}}(f(y)\mathbb{1}_{B(x,2r)})|\,\mathrm{d}y$$

$$
\begin{aligned}
&+ \mathrm{m}(B(x,r))|\mathcal{I}_{\frac{n}{p}}(f(y)\mathbb{1}_{\mathbb{R}^n\setminus B(x,2r)}) - c_{B(x,r)}| \\
&\lesssim \mathrm{m}(B(x,r))^{\frac{p-1}{p}}\Big(\int_{|y|<3r} |y|^{\frac{n}{p}-n}\,\mathrm{d}y\Big)\|f\|_{L^p(\mathbb{R}^n)} \\
&+ r\mathrm{m}(B(x,r)) \int_{|y-x|\ge 2r} |y-x|^{\frac{n}{p}-n-1}|f(y)|\,\mathrm{d}y \\
&\lesssim \mathrm{m}(B(x,r))\|f\|_{L^p(\mathbb{R}^n)},
\end{aligned}
$$

as requested.

(ii) Suppose

$$
0 < \|f\|_{\mathrm{BMO}(\mathbb{R}^n)} < \infty \quad \& \quad \|f\|_{L^p(\mathbb{R}^n)} < \infty.
$$

According to the classical Calderón–Zygmund decomposition theorem, there are a sequence of disjoint cubes

$$
\{I_j\} \subseteq \mathcal{C}(\mathbb{R}^n)
$$

and a set $E \subseteq \mathbb{R}^n$ such that

$$
\begin{cases}
\mathbb{R}^n = E \cup \Big(\bigcup_{j=1}^{\infty} I_j\Big) \\
|f| \le \|f\|_{\mathrm{BMO}(\mathbb{R}^n)} \quad \text{a.e. on} \quad E \\
\|f\|^p_{\mathrm{BMO}(\mathbb{R}^n)} \le \fint_{I_j} |f(x)|^p\,\mathrm{d}x \le 2^n\|f\|^p_{\mathrm{BMO}(\mathbb{R}^n)}.
\end{cases}
$$

The last estimate and the Hölder inequality give

$$
\begin{cases}
\sum_{j=1}^{\infty} \mathrm{m}(I_j) \le \Big(\frac{\|f\|_{L^p(\mathbb{R}^n)}}{\|f\|_{\mathrm{BMO}(\mathbb{R}^n)}}\Big)^p \\
|f|_{I_j} \le 2^{\frac{n}{p}}\|f\|_{\mathrm{BMO}(\mathbb{R}^n)}.
\end{cases}
$$

Accordingly, we utilize Theorem 2.1(i) to get two constants $c_1, c_2 > 0$ such that

$$
\begin{aligned}
\mathrm{m}(\{x \in \mathbb{R}^n : |f(x)| > t\}) &= \mathrm{m}\Big(\bigcup_{j=1}^{\infty}\{x \in I_j : |f(x)| > t\}\Big) \\
&\le \sum_{j=1}^{\infty} \mathrm{m}(\{x \in I_j : |f(x) - f_{I_j}| > t - |f_{I_j}|\}) \\
&\le \sum_{j=1}^{\infty} c_1 \exp\Big(-\frac{c_2(t - 2^{\frac{n}{p}}\|f\|_{\mathrm{BMO}(\mathbb{R}^n)})}{\|f\|_{\mathrm{BMO}(\mathbb{R}^n)}}\Big)\mathrm{m}(I_j)
\end{aligned}
$$

$$\le c_1 \exp\left(-\frac{c_2(t-2^{\frac{n}{p}}\|f\|_{\mathrm{BMO}(\mathbb{R}^n)})}{\|f\|_{\mathrm{BMO}(\mathbb{R}^n)}}\right)\left(\frac{\|f\|_{\mathrm{L}^p(\mathbb{R}^n)}}{\|f\|_{\mathrm{BMO}(\mathbb{R}^n)}}\right)^p.$$

This, along with the standard layer-cake formula, yields

$$\begin{aligned}\|f\|^q_{\mathrm{L}^q(\mathbb{R}^n)} &= \int_0^{2^{\frac{n}{p}}\|f\|_{\mathrm{BMO}(\mathbb{R}^n)}} \mathrm{m}(\{x\in\mathbb{R}^n:\ |f(x)|>t\})\,dt^q \\ &\quad + \int_{2^{\frac{n}{p}}\|f\|_{\mathrm{BMO}(\mathbb{R}^n)}}^{\infty} \mathrm{m}(\{x\in\mathbb{R}^n:\ |f(x)|>t\})\,dt^q \\ &\lesssim \int_0^{2^{\frac{n}{p}}\|f\|_{\mathrm{BMO}(\mathbb{R}^n)}} t^{-p}\|f\|^p_{\mathrm{L}^p(\mathbb{R}^n)}\,dt^q \\ &\quad + \int_{2^{\frac{n}{p}}\|f\|_{\mathrm{BMO}(\mathbb{R}^n)}}^{\infty} c_1\exp\left(-\frac{c_2(t-2^{\frac{n}{p}}\|f\|_{\mathrm{BMO}(\mathbb{R}^n)})}{\|f\|_{\mathrm{BMO}(\mathbb{R}^n)}}\right)\left(\frac{\|f\|_{\mathrm{L}^p(\mathbb{R}^n)}}{\|f\|_{\mathrm{BMO}(\mathbb{R}^n)}}\right)^p dt^q \\ &\lesssim \|f\|^{q-p}_{\mathrm{BMO}(\mathbb{R}^n)}\|f\|^p_{\mathrm{L}^p(\mathbb{R}^n)}.\end{aligned}$$

□

2.2 J-N-Q_α inequality with general constant

Although Theorem 2.1(i) is globally applicable for

$$Q_\alpha(\mathbb{R}^n)\subseteq \mathrm{BMO}(\mathbb{R}^n),$$

it is not the exact form for $Q_\alpha(\mathbb{R}^n)$. In [22, Problem 8.1] it is conjectured that if

$$\begin{cases} f\in Q_\alpha(\mathbb{R}^n)\\ \alpha\in(0,1)\\ t\in\mathbb{R}_+,\end{cases}$$

then there is a constant $c>0$ such that

$$\sup_{I\in\mathscr{C}(\mathbb{R}^n)}\sum_{k=0}^{\infty}2^{(2\alpha-n)k}\sum_{J\in\mathscr{D}_k(I)}\frac{\mathrm{m}_{f,J}(t)}{\mathrm{m}(J)}\lesssim t^{-1}\exp\left(-\frac{ct}{\|f\|_{Q_\alpha(\mathbb{R}^n)}}\right).$$

But, according to Theorem 1.5, if there is a constant $c>0$ such that

$$\sup_{I\in\mathscr{C}(\mathbb{R}^n)}\sum_{k=0}^{\infty}2^{(2\alpha-n)k}\sum_{J\in\mathscr{D}_k(I)}\frac{\mathrm{m}_{f,J}(t)}{\mathrm{m}(J)}\lesssim t^{-1}\exp(-ct),$$

then $f\in Q_\alpha(\mathbb{R}^n)$.

Theorem 2.3. *Let* $(\alpha, n-1) \in [0,1) \times \mathbb{N}$.

(i) *If* $f \in Q_\alpha(\mathbb{R}^n)$, *then there is a constant* $c > 0$ *such that*

$$\sup_{I\in\mathscr{C}(\mathbb{R}^n)} \sum_{k=0}^{\infty} 2^{(2\alpha-n)k} \sum_{J\in\mathscr{D}_k(I)} \frac{\mathrm{m}_{f,J}(t)}{\mathrm{m}(J)} \lesssim \frac{1+(t^{-1}\|f\|_{Q_\alpha(\mathbb{R}^n)})^2}{\exp(ct\|f\|^{-1}_{Q_\alpha(\mathbb{R}^n)})} \quad \forall t > 0;$$

(ii) *Conversely, if for* $p \in [0,2)$ *there is a constants* $c > 0$ *such that*

$$\sup_{I\in\mathscr{C}(\mathbb{R}^n)} \sum_{k=0}^{\infty} 2^{(2\alpha-n)k} \sum_{J\in\mathscr{D}_k(I)} \frac{\mathrm{m}_{f,J}(t)}{\mathrm{m}(J)} \lesssim \frac{1+t^{-p}}{\exp(ct)} \quad \forall t > 0,$$

then $f \in Q_\alpha(\mathbb{R}^n)$.

Proof. (i) If

$$0 = \|f\|_{Q_\alpha(\mathbb{R}^n)},$$

then the result is trivial, and hence in the sequel it is always assumed that

$$0 < \|f\|_{Q_\alpha(\mathbb{R}^n)} < \infty.$$

Two cases are considered for

$$|||f|||_{Q_\alpha(\mathbb{R}^n)} = \sup_{I\in\mathscr{C}(\mathbb{R}^n)} \left(\Psi_{f,\alpha}(I)\right)^{\frac{1}{2}} \approx \|f\|_{Q_\alpha(\mathbb{R}^n)}.$$

Case $t \le |||f|||_{Q_\alpha(\mathbb{R}^n)}$. This ensures

$$1 \lesssim \exp(-ct\|f\|^{-1}_{Q_\alpha(\mathbb{R}^n)}) \quad \forall c > 0.$$

Upon applying

$$\mathrm{m}_{f,J}(t) \le t^{-2} \fint_J |f(x) - f_J|^2 \, dx \quad \forall t > 0,$$

we get

$$\begin{aligned}\sup_{I\in\mathscr{C}(\mathbb{R}^n)} \sum_{k=0}^{\infty} 2^{(2\alpha-n)k} \sum_{J\in\mathscr{D}_k(I)} \frac{\mathrm{m}_{f,J}(t)}{\mathrm{m}(J)} &\le \sup_{I\in\mathscr{C}(\mathbb{R}^n)} \Psi_{f,\alpha}(I) \\ &\lesssim t^{-2}\|f\|^2_{Q_\alpha(\mathbb{R}^n)} \\ &\lesssim \frac{1+(t^{-1}\|f\|_{Q_\alpha(\mathbb{R}^n)})^2}{\exp(ct\|f\|^{-1}_{Q_\alpha(\mathbb{R}^n)})}.\end{aligned}$$

Case $t > \|\|f\|\|_{Q_\alpha(\mathbb{R}^n)}$. Given

$$\begin{cases} I \in \mathscr{C}(\mathbb{R}^n) \\ k-1 \in \mathbb{N} \\ r \in (0,\infty), \end{cases}$$

we apply the classical Calderón–Zygmund decomposition theorem to

$$|f(x) - f_J| \quad \text{on each} \quad J \in \mathscr{D}_k(I),$$

thereby getting

$$J = \Omega_{f,J}(r) \cup (J \setminus \Omega_{f,J}(r))$$

such that:

▷ $$\Omega_{f,J}(r) = \bigcup_{l=1}^{\infty} J_l,$$

namely, the union of cubes $J_l \in \mathscr{D}(I)$ whose interiors are disjoint;

▷ $$|f(x) - f_J| \le r \quad \text{for a. e. } x \in J \setminus \Omega_{f,J}(r);$$

▷ $$r < \fint_{J_l} |f(x) - f_J|\, dx \le 2^n r \quad \forall J_l;$$

▷ $$r\mathrm{m}(U) \le \int_U |f(x) - f_J|\, dx \le 2^n r\mathrm{m}(U) \quad \text{for any union } U \text{ of some cubes in } \{J_l\}.$$

Note that

$$\|f\|_{\mathrm{BMO}(\mathbb{R}^n)} = \sup_{I \in \mathscr{C}(\mathbb{R}^n)} \fint_I |f(x) - f_I|\, dy \le \|\|f\|\|_{Q_\alpha(\mathbb{R}^n)}.$$

So for each cube K in the decomposition of $\Omega_{f,J}(r)$ there is $K^* \subseteq J$ such that

$$\ell(K^*) = 2\ell(K) \quad \& \quad |f_{K^*} - f_J| \le \fint_{K^*} |f(x) - f_J|\, dx \le r,$$

and thus

$$r \le \fint_K |f(x) - f_J|\, dx$$

$$\leq \fint_K |f(x) - f_{K^*}|\,dx + |f_{K^*} - f_J|$$
$$\leq 2^n \fint_{K^*} |f(x) - f_{K^*}|\,dx + r$$
$$\leq 2^n |||f|||_{Q_\alpha(\mathbb{R}^n)} + r.$$

Meanwhile, an application of the Cauchy–Schwarz inequality gives that if U is the union of some cubes in $\{J_l\}$ then

$$r^2 \mathrm{m}(U) \leq \int_U |f(x) - f_J|^2\,dx$$

and hence

$$r^2 \mathrm{m}(\Omega_{f,J}(r)) \leq \int_{\Omega_{f,J}(r)} |f(x) - f_J|^2\,dx.$$

Although

$$r_1 < r_2 \Rightarrow \Omega_{f,J}(r_2) \subseteq \Omega_{f,J}(r_1),$$

we want to establish the following inequality

$$\mathrm{m}(\Omega_{f,J}(r + 2^{n+1}|||f|||_{Q_\alpha(\mathbb{R}^n)})) \leq 2^{-n}\mathrm{m}(\Omega_{f,J}(r)).$$

To do so, taking $K \in \{J_l\}$ we have

$$\fint_K |f(x) - f_J|\,dx \leq 2^n |||f|||_{Q_\alpha(\mathbb{R}^n)} + r < 2^{n+1}|||f|||_{Q_\alpha(\mathbb{R}^n)} + r,$$

thereby finding that K does not belong to the family of cubes making

$$\Omega_{f,J}(r + 2^{n+1}|||f|||_{Q_\alpha(\mathbb{R}^n)}).$$

If

$$K \cap \Omega_{f,J}(r + 2^{n+1}|||f|||_{Q_\alpha(\mathbb{R}^n)}) \neq \emptyset,$$

then this set must be the union of some cubes from the decomposition of

$$\Omega_{f,J}(r + 2^{n+1}|||f|||_{Q_\alpha(\mathbb{R}^n)}),$$

and hence

$$r + 2^{n+1}|||f|||_{Q_\alpha(\mathbb{R}^n)} \leq \fint_{K\cap\Omega_{f,J}(r+2^{n+1}|||f|||_{Q_\alpha(\mathbb{R}^n)})} |f(x) - f_J|\,dx$$

$$\leq \fint_{K\cap\Omega_{f,J}(r+2^{n+1}|||f|||_{Q_\alpha(\mathbb{R}^n)})} |f(x) - f_K|\,dx + |f_K - f_J|$$
$$\leq \left(\frac{\mathrm{m}(K)}{\mathrm{m}(K\cap\Omega_{f,J}(r+2^{n+1}|||f|||_{Q_\alpha(\mathbb{R}^n)}))}\right)|||f|||_{Q_\alpha(\mathbb{R}^n)} + 2^n|||f|||_{Q_\alpha(\mathbb{R}^n)} + r.$$

Consequently, we read off

$$\mathrm{m}(K\cap\Omega_{f,J}(r+2^{n+1}|||f|||_{Q_\alpha(\mathbb{R}^n)})) \leq 2^{-n}\mathrm{m}(K).$$

This, along with a summation of those K and the fact that

$$\Omega_{f,J}(r+2^{n+1}|||f|||_{Q_\alpha(\mathbb{R}^n)}) = \Omega_{f,J}(r)\cap\Omega_{f,J}(r+2^{n+1}|||f|||_{Q_\alpha(\mathbb{R}^n)}),$$

derives the desired inequality. Consequently, the following inequality

$$\sum_{k=0}^{\infty} 2^{(2\alpha-n)k} \sum_{J\in\mathscr{D}_k(I)} \frac{\mathrm{m}(\Omega_{f,J}(r+2^{n+1}|||f|||_{Q_\alpha(\mathbb{R}^n)}))}{\mathrm{m}(J)}$$
$$\leq 2^{-n}\sum_{k=0}^{\infty} 2^{(2\alpha-n)k} \sum_{J\in\mathscr{D}_k(I)} \frac{\mathrm{m}(\Omega_{f,J}(r))}{\mathrm{m}(J)}$$

is valid.

Note that

$$\mathrm{m}_{f,J}(t) \leq \mathrm{m}(\Omega_{f,J}(t)).$$

So upon letting j be the integer part of

$$\frac{t - |||f|||_{Q_\alpha(\mathbb{R}^n)}}{2^{n+1}|||f|||_{Q_\alpha(\mathbb{R}^n)}}$$

and setting

$$s = (1 + j2^{n+1})|||f|||_{Q_\alpha(\mathbb{R}^n)},$$

we find

$$|||f|||_{Q_\alpha(\mathbb{R}^n)} \leq s \leq t,$$

thereby estimating

$$\sum_{k=0}^{\infty} 2^{(2\alpha-n)k} \sum_{J\in\mathscr{D}_k(I)} \frac{\mathrm{m}_{f,J}(t)}{\mathrm{m}(J)} \leq \sum_{k=0}^{\infty} 2^{(2\alpha-n)k} \sum_{J\in\mathscr{D}_k(I)} \frac{\mathrm{m}_{f,J}(s)}{\mathrm{m}(J)}$$
$$\leq \sum_{k=0}^{\infty} 2^{(2\alpha-n)k} \sum_{J\in\mathscr{D}_k(I)} \frac{\mathrm{m}(\Omega_{f,J}(s))}{\mathrm{m}(J)}$$

$$\leq 2^{-n} \sum_{k=0}^{\infty} 2^{(2\alpha-n)k} \sum_{J \in \mathcal{D}_k(I)} \frac{\mathrm{m}(\Omega_{f,J}(1+(j-1)2^{n+1}|||f|||_{\mathrm{Q}_\alpha(\mathbb{R}^n)}))}{\mathrm{m}(J)}.$$

An iteration with j times derives

$$\begin{aligned}\sum_{k=0}^{\infty} 2^{(2\alpha-n)k} \sum_{J \in \mathcal{D}_k(I)} \frac{\mathrm{m}_{f,J}(t)}{\mathrm{m}(J)} &\leq 2^{-nj} \sum_{k=0}^{\infty} 2^{(2\alpha-n)k} \sum_{J \in \mathcal{D}_k(I)} \frac{\mathrm{m}(\Omega_{f,J}(|||f|||_{\mathrm{Q}_\alpha(\mathbb{R}^n)}))}{\mathrm{m}(J)} \\ &\leq 2^{-jn} \\ &\leq 2^{-n(\frac{t-|||f|||_{\mathrm{Q}_\alpha(\mathbb{R}^n)}}{2^{n+1}|||f|||_{\mathrm{Q}_\alpha(\mathbb{R}^n)}}-1)} \\ &\leq (2^{-\frac{nt}{2^{n+1}|||f|||_{\mathrm{Q}_\alpha(\mathbb{R}^n)}}})2^{\frac{n}{2^n+1}+n}.\end{aligned}$$

Putting

$$c = 2^{-n-1} n \ln 2$$

produces

$$\sum_{k=0}^{\infty} 2^{(2\alpha-n)k} \sum_{J \in \mathcal{D}_k(I)} \frac{\mathrm{m}_{f,J}(t)}{\mathrm{m}(J)} \lesssim \exp(-ct|||f|||^{-1}_{\mathrm{Q}_\alpha(\mathbb{R}^n)}),$$

and a positive constant c_* depending on c such that

$$\sup_{I \in \mathcal{C}(\mathbb{R}^n)} \sum_{k=0}^{\infty} 2^{(2\alpha-n)k} \sum_{J \in \mathcal{D}_k(I)} \frac{\mathrm{m}_{f,J}(t)}{\mathrm{m}(J)} \lesssim \frac{1+(t^{-1}\|f\|_{\mathrm{Q}_\alpha(\mathbb{R}^n)})^2}{\exp(c_* t\|f\|^{-1}_{\mathrm{Q}_\alpha(\mathbb{R}^n)})}.$$

(ii) Suppose that there is a constant $c > 0$ obeying

$$\sup_{I \in \mathcal{C}(\mathbb{R}^n)} \sum_{k=0}^{\infty} 2^{(2\alpha-n)k} \sum_{J \in \mathcal{D}_k(I)} \frac{\mathrm{m}_{f,J}(t)}{\mathrm{m}(J)} \lesssim \frac{1+t^{-p}}{\exp(ct)} \quad \text{for} \quad p \in [0,2).$$

Then

$$\sup_{I \in \mathcal{C}(\mathbb{R}^n)} \sum_{k=0}^{\infty} 2^{(2\alpha-n)k} \sum_{J \in \mathcal{D}_k(I)} \mathrm{m}(J)^{-1} \int_0^{\infty} t\mathrm{m}_{f,J}(t)\mathrm{d}t \lesssim \int_0^{\infty} \frac{t(1+t^{-p})}{\exp(ct)}\,\mathrm{d}t < \infty,$$

and hence $f \in \mathrm{Q}_\alpha(\mathbb{R}^n)$ follows from Theorem 1.5. □

Not only Proposition 2.2(i) under $p = n$ can be slightly improved, but also Proposition 2.2(ii) under $q \to p = 2$ can be understood under $\mathrm{Q}_\alpha(\mathbb{R}^n)$-structure.

Proposition 2.4. *Let* $(\alpha, \beta, \gamma, n-1) \in (-\infty,1) \times (-\infty,1) \times (-\infty,1) \times \mathbb{N}$. *Then:*

(i)

$$\|\mathcal{I}_1 f\|_{\mathrm{Q}_\alpha(\mathbb{R}^n)} \lesssim \|f\|_{L^n(\mathbb{R}^n)};$$

(ii)
$$\|f\|_{Q_\alpha(\mathbb{R}^n)} \lesssim \|f\|_{Q_\beta(\mathbb{R}^n)}\|f\|_{Q_\gamma(\mathbb{R}^n)} \quad \text{for} \quad \alpha = \frac{\beta+\gamma}{2}.$$

Proof. (i) Note that if

$$\mathcal{R}_j(f)(x) = \frac{\Gamma(\frac{n+1}{2})}{\pi^{\frac{n+1}{2}}} \text{p. v.} \int_{\mathbb{R}^n} \frac{(x_j - y_j)f(y)}{|x-y|^{n+1}}\,dy$$

is the j-th Riesz transform of f with $\Gamma(\cdot)$ being the gamma function then $(R_j(f))_{j=1}^n$ exists as the gradient of the 1-order Riesz singular integral

$$\mathcal{I}_1 f(x) = \int_{\mathbb{R}^n} f(y)|y-x|^{1-n}\,\mathrm{d}y,$$

up to a dimensional constant, namely

$$(\mathcal{R}_1(f), \ldots, \mathcal{R}_n(f)) = \frac{\Gamma(\frac{n+1}{2})}{(1-n)\pi^{\frac{n+1}{2}}}\nabla \mathcal{I}_1 f.$$

According to Theorem 1.4(ii), we have

$$\|f\|_{Q_\alpha(\mathbb{R}^n)} \lesssim \|\nabla f\|_{L^n(\mathbb{R}^n)}$$

thereby utilizing the boundedness of the Riesz transformation

$$(\mathcal{R}_j, \ldots, \mathcal{R}_n) \quad \text{on} \quad L^n(\mathbb{R}^n)$$

to obtain

$$\|\mathcal{I}_1 f\|_{Q_\alpha(\mathbb{R}^n)} \lesssim \|\nabla \mathcal{I}_1 f\|_{L^n(\mathbb{R}^n)} \lesssim \sum_{j=1}^{n} \|\mathcal{R}_j(f)\|_{L^n(\mathbb{R}^n)} \lesssim \|f\|_{L^n(\mathbb{R}^n)},$$

namely,

$$\mathcal{I}_1 L^n(\mathbb{R}^n) \subseteq Q_\alpha(\mathbb{R}^n).$$

(ii) Since

$$\alpha = 2^{-1}(\beta+\gamma),$$

an application of the Cauchy–Schwarz inequality gives that

$$\int_I\int_I \frac{|f(y)-f(z)|^2}{|y-z|^{n+2\alpha}}\,\mathrm{d}y\mathrm{d}z \le \Big(\int_I\int_I \frac{|f(y)-f(z)|^2}{|y-z|^{n+2\beta}}\,\mathrm{d}y\mathrm{d}z\Big)^{\frac{1}{2}}\Big(\int_I\int_I \frac{|f(y)-f(z)|^2}{|y-z|^{n+2\gamma}}\,\mathrm{d}y\mathrm{d}z\Big)^{\frac{1}{2}}$$

holds for any $I \in \mathscr{C}(\mathbb{R}^n)$ so the desired inequality follows. □

2.3 Notes

Below are some historical remarks on Chapter 2.

Section 2.1. Theorem 2.1 comes from [53] and Proposition 2.2 comes from a combination of two notes [77, 20]. Theorem 2.1 indeed induces the so-called $\mathrm{BMO}(\gamma,\mathbb{R}^n)$ (cf. [82]) consisting of all locally integrable functions f on $\mathbb{R}^n$ with

$$\sup_{p\in[1,\infty)} p^{-\frac{1}{\gamma}} \sup_{I\in\mathscr{C}(\mathbb{R}^n)} \left(\fint_I |f(x)-f_I|^p \,\mathrm{d}x \right)^{\frac{1}{p}} < \infty,$$

equivalently, $\exists$ a constant $c>0$ such that

$$\sup_{I\in\mathscr{C}(\mathbb{R}^n)} \fint_I (\exp|c|f(x)-f_I|^\gamma)\,\mathrm{d}x < \infty.$$

Since

$$\gamma\in(1,\infty) \Rightarrow \mathrm{BMO}(\gamma,\mathbb{R}^n) \subseteq \mathrm{BMO}(\mathbb{R}^n),$$

an interesting problem is to investigate the relationship between $\mathrm{Q}_\alpha(\mathbb{R}^n)$ and $\mathrm{BMO}(\gamma,\mathbb{R}^n)$ under $(\alpha,\gamma)\in[0,1)\times(1,\infty)$.

Section 2.2. Theorem 2.3 comes from a modification of the first and third sections of [103] and Proposition 2.4 comes from another look at [91, Theorem 4.1]. Moreover, [103, Theorem 3] shows that under $\alpha\in(0,2^{-1})$:

- ▷ if $p\in[0,2)$, then there exists no constant $c>0$ such that

$$\sup_{I\in\mathscr{C}(\mathbb{R}^n)} \sum_{k=0}^{\infty} 2^{(2\alpha-n)k} \sum_{J\in\mathscr{D}_k(I)} \frac{\mathrm{m}_{f,J}(t)}{\mathrm{m}(J)} \lesssim \frac{1+(t^{-1}\|f\|_{\mathrm{Q}_\alpha(\mathbb{R}^n)})^p}{\exp(ct\|f\|^{-1}_{\mathrm{Q}_\alpha(\mathbb{R}^n)})} \quad \forall t>0$$

  holds for any function $f\in\mathrm{Q}_\alpha(\mathbb{R}^n)$;
- ▷ there is a function $f\notin\mathrm{Q}_\alpha(\mathbb{R}^n)$ such that

$$\sup_{I\in\mathscr{C}(\mathbb{R}^n)} \sum_{k=0}^{\infty} 2^{(2\alpha-n)k} \sum_{J\in\mathscr{D}_k(I)} \frac{\mathrm{m}_{f,J}(t)}{\mathrm{m}(J)} \lesssim \frac{1+t^{-2}}{\exp(ct)} \quad \forall t>0$$

  holds for some positive constant c.

3 Functional Hausdorff content

In order to solve [22, problems 8.2–8.3] via an appropriate lift of $f \in \mathcal{Q}_\alpha(\mathbb{R}^n)$ to $\mathbb{R}^{n+1}_+$, we are led to consider a brief theory of functional Hausdorff content.

3.1 Hausdorff capacity and Choquet integration

Due to the fractional phenomenon driven by Theorem 1.4, it is very natural to bring the Hausdorff capacity and the associated Choquet integration into play.

Definition 3.1. Let $d \in (0, n]$ and $E \subseteq \mathbb{R}^n$.

(i) The d-dimensional Hausdorff capacity (or content) of E is defined by

$$\Lambda_d^{(\infty)}(E) = \inf\left\{\sum_j r_j^d : E \subseteq \bigcup_{j=1}^{\infty} B(x_j, r_j)\right\},$$

where the infimum is taken over all coverings of E by countable families of balls with radii r_j.

(ii) The d-dimensional Hausdorff measure of E is given by

$$\Lambda_d(E) = \lim_{\epsilon \to 0} \Lambda_d^{(\epsilon)}(E),$$

where $\Lambda_d^{(\epsilon)}(E)$ is defined by (i) with the additional requirement that the radii of the covering balls satisfy $0 < r_j \le \epsilon$.

(iii) The d-dimensional dyadic Hausdorff capacity of E is determined by

$$\tilde{\Lambda}_d^{(\infty)}(E) = \inf\left\{\sum_j \ell(I_j)^d : E \subseteq \left(\bigcup_{j=1}^{\infty} I_j\right)^{\circ}\right\},$$

where the infimum ranges only over covers of E by $\{I_j\} \subset \mathcal{D}(\mathbb{R}^n)$ with the interior

$$\left(\bigcup_{j=1}^{\infty} I_j\right)^{\circ} \quad \text{of} \quad \bigcup_{j=1}^{\infty} I_j.$$

(iv) The Choquet integral of $f : \mathbb{R}^n \to [0, \infty]$ with respect to $\Lambda_d^{(\infty)}$ is decided by

$$\int_{\mathbb{R}^n} f \, d\Lambda_d^{(\infty)} = \int_0^{\infty} \Lambda_d^{(\infty)}(\{x \in \mathbb{R}^n : f(x) > t\}) \, dt.$$

Proposition 3.2. *Below is a list of basic properties on the pair $\{\Lambda_d^{(\infty)}, \tilde{\Lambda}_d^{(\infty)}\}$ and the Choquet integral of an arbitrary nonnegative function f on $\mathbb{R}^n$ under $d \in (0, n]$.*

https://doi.org/10.1515/9783110600285-003

(i) *Both capacities are nonnegative, monotone, countably subadditive set functions on all subsets of* $\mathbb{R}^n$.

(ii) $\Lambda_d^{(\infty)}$ *is finite on any bounded subset of* $\mathbb{R}^n$, *while having the same null sets as* Λ_d, *that is,*

$$\Lambda_d^{(\infty)}(E) = 0 \Leftrightarrow \Lambda_d(E) = 0.$$

(iii) *There is*

$$6^{-n}\tilde{\Lambda}_d^{(\infty)}(E) \le \Lambda_d^{(\infty)}(E) \le n^n \tilde{\Lambda}_d^{(\infty)}(E) \quad \forall E \subseteq \mathbb{R}^n.$$

(iv) $\tilde{\Lambda}_d^{(\infty)}(\cdot)$ *satisfies:*

(iv)-a *the outer regularity—if* $E \subseteq \mathbb{R}^n$ *then*

$$\tilde{\Lambda}_d^{(\infty)}(E) = \inf\{\tilde{\Lambda}_d^{(\infty)}(O) : \text{open } O \supseteq E\};$$

(iv)-b *the strong subadditivity—if* E_1, E_2 *are two subsets of* $\mathbb{R}^n$ *then*

$$\tilde{\Lambda}_d^{(\infty)}(E_1 \cup E_2) + \tilde{\Lambda}_d^{(\infty)}(E_1 \cap E_2) \le \tilde{\Lambda}_d^{(\infty)}(E_1) + \tilde{\Lambda}_d^{(\infty)}(E_2);$$

(iv)-c *the upward continuity—if* $\{E_i\}$ *is an increasing sequence of subsets of* $\mathbb{R}^n$ *then*

$$\tilde{\Lambda}_d^{(\infty)}\Big(\bigcup_i E_i\Big) = \lim_{i\to\infty} \tilde{\Lambda}_d^{(\infty)}(E_i);$$

(iv)-d *the downward continuity—if* $\{E_i\}$ *is a decreasing sequence of compact subsets of* $\mathbb{R}^n$ *then*

$$\tilde{\Lambda}_d^{(\infty)}\Big(\bigcap_i E_i\Big) = \lim_{i\to\infty} \tilde{\Lambda}_d^{(\infty)}(E_i).$$

(v)

$$\int_{\mathbb{R}^n} \mathbb{1}_E \, d\Lambda_d^{(\infty)} = \Lambda_d^{(\infty)}(E)$$

holds for any $E \subseteq \mathbb{R}^n$.

(vi)

$$\int_{\mathbb{R}^n} cf \, d\Lambda_d^{(\infty)} = c \int_{\mathbb{R}^n} f \, d\Lambda_d^{(\infty)}$$

holds for any constant c.

(vii)

$$\int_{\mathbb{R}^n} f \, d\Lambda_d^{(\infty)} = 0 \Leftrightarrow f = 0$$

holds $\Lambda_d^{(\infty)}$*-a. e. on* $\mathbb{R}^n$, *that is,* $f = 0$ *on* $\mathbb{R}^n$ *holds except on* E *with* $\Lambda_d^{(\infty)}(E) = 0$.

(viii)
$$\int_{\mathbb{R}^n} f^{\frac{d}{n}}\,\mathrm{d}\Lambda_n^{(\infty)} \le \Big(\int_{\mathbb{R}^n} f\,\mathrm{d}\Lambda_d^{(\infty)}\Big)^{\frac{n}{d}}$$

holds for any constant $d \in (0, n]$.

(ix)
$$\int_{\mathbb{R}^n} \sum_{j=1}^{\infty} f_j\,\mathrm{d}\tilde{\Lambda}_d^{(\infty)} \le \sum_{j=1}^{\infty} \int_{\mathbb{R}^n} f_j\,\mathrm{d}\tilde{\Lambda}_d^{(\infty)}$$

holds for any $f_j \ge 0$ on $\mathbb{R}^n$.

Proof. The first seven properties are well known; see, for example, [1].

To prove (viii), we utilize the inequality

$$\Lambda_n^{(\infty)}(E) \le (\Lambda_d^{(\infty)}(E))^{\frac{n}{d}} \quad \forall E \subseteq \mathbb{R}^n$$

to produce

$$\begin{aligned}\int_{\mathbb{R}^n} f^{\frac{n}{d}}\,\mathrm{d}\Lambda_n^{(\infty)} &= \frac{n}{d}\int_0^\infty (\Lambda_n^{(\infty)}(\{x \in \mathbb{R}^n : f(x) > t\}))t^{\frac{n-d}{d}}\,\mathrm{d}t\\ &\le \int_0^\infty \frac{\mathrm{d}}{\mathrm{d}t}\Big(\int_0^t (\Lambda_n^{(\infty)}(\{x \in \mathbb{R}^n : f(x) > s\}))^{\frac{d}{n}}\,\mathrm{d}s\Big)^{\frac{n}{d}}\,\mathrm{d}t\\ &\le \Big(\int_0^\infty \Lambda_d^{(\infty)}(\{x \in \mathbb{R}^n : f(x) > t\})\,\mathrm{d}t\Big)^{\frac{n}{d}}.\end{aligned}$$

Of course, (ix) follows from (iv)-b. □

Theorem 3.3. *For a locally integrable function f on $\mathbb{R}^n$, denoted by $f \in L^1_{\mathrm{loc}}(\mathbb{R}^n)$, and $x \in \mathbb{R}^n$, let*

$$\mathcal{M}f(x) = \sup_{x \in I \in \mathscr{C}(\mathbb{R}^n)} ⨍_I |f(y)|\,\mathrm{d}y \quad \& \quad \mathcal{M}_d f(x) = \sup_{x \in I \in \mathscr{D}(\mathbb{R}^n)} ⨍_I |f(y)|\,\mathrm{d}y$$

be the Hardy–Littlewood maximal function of f and its dyadic version. If $d \in (0, n]$, then:

(i)
$$\begin{aligned}\Lambda_d^{(\infty)}(\{x \in \mathbb{R}^n : \mathcal{M}_d f(x) > t\}) &\le \Lambda_d^{(\infty)}(\{x \in \mathbb{R}^n : \mathcal{M}f(x) > t\})\\ &\le 3^d \Lambda_d^{(\infty)}(\{x \in \mathbb{R}^n : \mathcal{M}_d f(x) > 4^{-n}t\}) \quad \forall t > 0;\end{aligned}$$

(ii)
$$\int_{\mathbb{R}^n} |f|^p\,\mathrm{d}\Lambda_d^{(\infty)} \gtrsim \begin{cases} \int_{\mathbb{R}^n} (\mathcal{M}f)^p\,\mathrm{d}\Lambda_d^{(\infty)} & \text{as } p \in (\frac{d}{n}, \infty)\\ \sup_{t \in (0,\infty)} t^{\frac{d}{n}} \Lambda_d^{(\infty)}(\{y \in \mathbb{R}^n : \mathcal{M}f(y) > t\}) & \text{as } p = \frac{d}{n}.\end{cases}$$

Proof. (i) This follows from [27, p. 136] and

$$\mathcal{M}_d f(x) \le \mathcal{M} f(x) \quad \forall x \in \mathbb{R}^n.$$

(ii) Two cases are handled below.

Case $p \in (\frac{d}{n}, \infty)$. If x_I is the center of $I \in \mathscr{C}(\mathbb{R}^n)$, then

$$\mathbb{1}_{\mathcal{I}}(x) \lesssim \min\left\{1, \left(\frac{\ell(I)}{|x - x_I|}\right)^n\right\} \quad \forall x \in \mathbb{R}^n,$$

and hence an application of $\frac{d}{n} < p$ gives

$$\int_{\mathbb{R}^n} (\mathcal{M}\mathbb{1}_I)^p \, d\Lambda_d^{(\infty)} \lesssim \ell(I)^d \left(1 + \int_0^1 t^{-\frac{d}{pn}} \, dt\right).$$

Next, for each integer k let

$$\{I_{j,k}\} \subseteq \mathscr{D}(\mathbb{R}^n)$$

be a nonoverlapping family such that

$$\begin{cases} \{x \in \mathbb{R}^n : 2^k < f(x) \le 2^{k+1}\} \subseteq \bigcup_j I_{j,k} \\ \sum_j \ell(I_{j,k})^d \le 2\Lambda_d^{(\infty)}(\{x \in \mathbb{R}^n : 2^k < f(x) \le 2^{k+1}\}). \end{cases}$$

Then

$$J_k = \bigcup_j I_{j,k} \quad \& \quad g = \sum_k 2^{(k+1)p} \mathbb{1}_{J_k} \Rightarrow f^p \le g.$$

▷ In case of $p \ge 1$, we have

$$(\mathcal{M}f)^p \le \mathcal{M}g \le \sum_k 2^{(k+1)p} \sum_j \mathcal{M}\mathbb{1}_{I_{j,k}},$$

thereby getting via the above maximal estimate and Propositions 3.2(iii) & (ix),

$$\begin{aligned} \int_{\mathbb{R}^n} (\mathcal{M}f)^p \, d\Lambda_d^{(\infty)} &\lesssim \int_{\mathbb{R}^n} \sum_k 2^{(k+1)p} \sum_j \mathcal{M}\mathbb{1}_{I_{j,k}} \, d\Lambda_d^{(\infty)} \\ &\lesssim \sum_k 2^{(k+1)p} \sum_j \int_{\mathbb{R}^n} \mathcal{M}\mathbb{1}_{I_{j,k}} \, d\Lambda_d^{(\infty)} \\ &\lesssim \sum_k 2^{(k+1)p} \sum_j \ell(I_{j,k})^d \\ &\lesssim \sum_k 2^{(k+1)p} \Lambda_d^{(\infty)}(\{x \in \mathbb{R}^n : 2^k < f(x) \le 2^{k+1}\}) \\ &\lesssim \int_{\mathbb{R}^n} f^p \, d\Lambda_d^{(\infty)}. \end{aligned}$$

▷ In case of $\frac{d}{n} < p < 1$, we use

$$f \leq \sum_k 2^{k+1} \mathbb{1}_{J_k}$$

to get

$$(\mathcal{M}f)^p \leq \sum_k 2^{(k+1)p} \sum_j (\mathcal{M}\mathbb{1}_{I_{j,k}})^p,$$

whence

$$\int_{\mathbb{R}^n} (\mathcal{M}f)^p \, d\Lambda_d^{(\infty)} \lesssim \sum_k 2^{(k+1)p} \sum_j \ell(I_{j,k})^d \lesssim \int_{\mathbb{R}^n} f^p \, d\Lambda_d^{(\infty)}.$$

Case $p = \frac{d}{n}$. We make the following consideration.

▷ Observe first that if

$$\{I_j\} \subseteq \mathscr{C}(\mathbb{R}^n)$$

is a sequence of nonoverlapping cubes then there is a subsequence $\{I_{j_k}\}$ obeying

$$\begin{cases} \sum_{I_{j_k} \subseteq I} \ell(I_{j_k})^d \leq 2\ell(I)^d \quad \forall I \in \mathscr{D}(\mathbb{R}^n) \\ \Lambda_d^{(\infty)}\left(\bigcup_j I_j\right) \leq 2 \sum_k \ell(I_{j_k})^d. \end{cases}$$

In fact, if

$$j_1 = 1 \quad \& \quad j_1, \ldots, j_k$$

are selected for the first inequality, then j_{k+1} is defined as the first index such that

$$\{I_{j_1}, \ldots, I_{j_{k+1}}\}$$

obeys the first inequality. Upon taking $j \in (j_l, j_{l+1})$ for some $l \in \mathbb{N}$, we can find

$$I_j^* \in \mathscr{D}(\mathbb{R}^n)$$

such that

$$I_j^* \supseteq I_j \quad \& \quad \sum_{I_{j_k} \subseteq I_l^* \ \& \ k \leq l} \ell(I_{j_k})^d + \ell(I_j)^d > 2\ell(I_j^*)^d,$$

whence

$$\sum_{I_{j_k} \subseteq I_l^* \ \& \ k \leq l} \ell(I_{j_k})^d \geq \ell(I_j^*)^d.$$

Without loss of generality, we may assume

$$\sum_k \ell(I_{j_k})^d < \infty,$$

thereby seeing that $\{\ell(I_j^*)\}$ is bounded. Upon considering $\{\tilde{I}_k\}$ as a sequence of maximal cubes taken from $\{I_j^*\}$, we get

$$\bigcup_j I_j \subseteq \left(\bigcup_k I_{j_k}\right) \cup \left(\bigcup_k \tilde{I}_k\right),$$

thereby reaching the second inequality

$$\Lambda_d^{(\infty)}\left(\bigcup_j I_j\right) \le 2\sum_k \ell(I_{j_k})^d.$$

▷ Next, for $t > 0$ let

$$\{I_j\} \subseteq \mathscr{D}(\mathbb{R}^n)$$

be the sequence of maximal cubes with

$$\fint_{I_j} |f(x)|\,\mathrm{d}x > t.$$

Then

$$\{x \in \mathbb{R}^n : \mathcal{M}_{\mathscr{D}} f(x) > t\} = \bigcup_j I_j.$$

Now, an application of Proposition 3.2(viii) derives

$$\ell(I_j)^d \le \left(t^{-1}\int_{I_j} f(x)\,\mathrm{d}x\right)^{\frac{d}{n}} \lesssim t^{-\frac{d}{n}}\int_{I_j} f^{\frac{d}{n}}\,\mathrm{d}\Lambda_d^{(\infty)}.$$

By (i) and the first observation above, we obtain a subsequence $\{I_{j_k}\}$ of $\{I_j\}$ such that

$$\begin{aligned}
\Lambda_d^{(\infty)}(\{x \in \mathbb{R}^n : \mathcal{M}f(x) > 4^n t\}) &\le 3^d\Lambda_d^{(\infty)}(\{x \in \mathbb{R}^n : \mathcal{M}_{\mathscr{D}} f(x) > t\}) \\
&\le 3^{1+d}\sum_k \ell(I_{j_k})^d \\
&\lesssim t^{-\frac{n}{d}}\sum_k \int_{I_{j_k}} |f|^{\frac{d}{n}}\,\mathrm{d}\Lambda_d^{(\infty)} \\
&\lesssim t^{-\frac{n}{d}}\int_{\mathbb{R}^n} |f|^{\frac{d}{n}}\,\mathrm{d}\Lambda_d^{(\infty)}
\end{aligned}$$

holds, thereby reaching the desired conclusion in (ii). □

Below is the Adams preduality.

Theorem 3.4. *For $d \in (0, n]$, let $\mathsf{L}^{1,d}(\mathbb{R}^n)$ be Morrey space of all Radon measures μ on $\mathbb{R}^n$ (denoted by $\mu \in \mathsf{RM}(\mathbb{R}^n)$) with*

$$|||\mu|||_{\mathsf{L}^{1,d}(\mathbb{R}^n)} = \sup_{I \in \mathscr{C}(\mathbb{R}^n)} \frac{|\mu|(I)}{\ell(I)^d} < \infty,$$

where

$$|\mu| = \mu_+ + \mu_- \quad \& \quad \mu_\pm = \max\{0, \pm\mu\}.$$

(i) *If $\mathsf{L}^1(\Lambda_d^{(\infty)})$ is the completion of $\mathsf{C}_0(\mathbb{R}^n)$ of all continuous functions with compact support in $\mathbb{R}^n$ with respect to*

$$\int_{\mathbb{R}^n} |f| \, d\Lambda_d^{(\infty)},$$

then $\mathsf{L}^{1,d}(\mathbb{R}^n)$ is isomorphic to the dual of $\mathsf{L}^1(\Lambda_d^{(\infty)})$ under the pairing $\int_{\mathbb{R}^n} f \, d\mu$.

(ii) *If f is nonnegative and lower semicontinuous on $\mathbb{R}^n$, then*

$$\int_{\mathbb{R}^n} f \, d\Lambda_d^{(\infty)} \approx \sup\left\{\left|\int_{\mathbb{R}^n} f \, d\mu\right| : 0 \le \mu \in \mathsf{L}^{1,d}(\mathbb{R}^n) \ \& \ \|\mu\|_{\mathsf{L}^{1,d}(\mathbb{R}^n)} \le 1\right\}.$$

Proof. (i) Clearly,

$$\mu \in \mathsf{L}^{1,d}(\mathbb{R}^n)$$

induces a bounded linear functional on $\mathsf{L}^1(\Lambda_d^{(\infty)})$ via

$$\mathcal{L}(f) = \int_{\mathbb{R}^n} f \, d\mu$$

thanks to

$$|\mu|(E) \le |||\mu|||_{\mathsf{L}^{1,d}(\mathbb{R}^n)} \Lambda_d^{(\infty)}(E) \quad \forall E \subseteq \mathbb{R}^n$$

and

$$|\mathcal{L}(f)| \le \int_{\mathbb{R}^n} |f| \, d|\mu| \le |||\mu|||_{\mathsf{L}^{1,d}(\mathbb{R}^n)} \int_{\mathbb{R}^n} |f| \, d\Lambda_d^{(\infty)} \quad \forall f \in \mathsf{L}^1(\Lambda_d^{(\infty)}).$$

Conversely, suppose that $\mathcal{L}$ is a linear functional on $\mathsf{L}^1(\Lambda_d^{(\infty)})$ with its norm $\|\mathcal{L}\| < \infty$, then there exists a $\mu \in \mathsf{RM}(\mathbb{R}^n)$ such that $\mathcal{L}$ is determined by

$$\mathcal{L}(f) = \int_{\mathbb{R}^n} f \, d\mu \quad \forall f \in \mathsf{C}_0(\mathbb{R}^n) \subseteq \mathsf{L}^1(\Lambda_d^{(\infty)}).$$

However, for any $g \in C_0(\mathbb{R}^n)$ we have

$$\left|\int_{\mathbb{R}^n} g \, d|\mu|\right| \le \int_{\mathbb{R}^n} |g| \, d|\mu|$$
$$\le \sup\left\{\int_{\mathbb{R}^n} f \, d\mu : f \in C_0(\mathbb{R}^n) \,\&\, |f| \le |g|\right\}$$
$$\le \|\mathcal{L}\| \sup\left\{\int_{\mathbb{R}^n} |f| \, d\Lambda_d^{(\infty)} : f \in C_0(\mathbb{R}^n) \,\&\, |f| \le |g|\right\}$$
$$\le \|\mathcal{L}\| \int_{\mathbb{R}^n} |g| \, d\Lambda_d^{(\infty)}.$$

For

$$(I, \epsilon) \in \mathscr{C}(\mathbb{R}^n) \times \mathbb{R}_+$$

let $I_\epsilon \in \mathscr{C}(\mathbb{R}^n)$ have the same center as I's and the edge-length $\epsilon + \ell(I)$. Upon choosing a function

$$g \in C_0(\mathbb{R}^n) \quad \text{with} \quad g = \begin{cases} 1 & \text{on } I \\ 0 & \text{on } \mathbb{R}^n \setminus I_\epsilon, \end{cases}$$

we achieve

$$|\mu|(I) \le \|\mathcal{L}\| \Lambda_d^{(\infty)}(I_\epsilon),$$

whence

$$\||\mu\||_{L^{1,d}(\mathbb{R}^n)} \le \|\mathcal{L}\|.$$

(ii) Note that the canonical mapping from $L^1(\Lambda_d^{(\infty)})$ to its second dual has the following norm of $g \in L^1(\Lambda_d^{(\infty)})$

$$\int_{\mathbb{R}^n} |g| \, d\tilde{\Lambda}_d^{(\infty)} = \sup\left\{\left|\int_{\mathbb{R}^n} g \, d\mu\right| : 0 \le \mu \in L^{1,d}(\mathbb{R}^n) \,\&\, \||\mu\||_{L^{1,d}(\mathbb{R}^n)} \le 1\right\}.$$

So, if f is nonnegative and lower semicontinuous on $\mathbb{R}^n$, then it can be approximated from below by a nonnegative and increasing sequence

$$\{f_j\} \subseteq C_0(\mathbb{R}^n),$$

and hence

$$\int_{\mathbb{R}^n} f_j \, d\tilde{\Lambda}_d^{(\infty)} \le \sup\left\{\left|\int_{\mathbb{R}^n} f \, d\mu\right| : 0 \le \mu \in L^{1,d}(\mathbb{R}^n) \,\&\, \|\mu\|_{L^{1,d}(\mathbb{R}^n)} \le 1\right\}.$$

Proposition 3.2(iv) is used to derive

$$\lim_{j\to\infty} \tilde{\Lambda}_d^{(\infty)}(\{x \in \mathbb{R}^n : f_j(x) \geq t\}) = \tilde{\Lambda}_d^{(\infty)}(\{x \in \mathbb{R}^n : f(x) \geq t\}) \quad \forall t > 0,$$

whence

$$\int_{\mathbb{R}^n} f \, d\tilde{\Lambda}_d^{(\infty)} \approx \sup\left\{\left|\int_{\mathbb{R}^n} f \, d\mu\right| : 0 \leq \mu \in \mathsf{L}^{1,d}(\mathbb{R}^n) \;\&\; \|\mu\|_{\mathsf{L}^{1,d}(\mathbb{R}^n)} \leq 1\right\}. \qquad \square$$

3.2 Carleson measure and tent space

A focus on the upper half-space $\mathbb{R}_+^{n+1}$ leads to the concepts of a Carleson measure and a nontangential maximal function.

Definition 3.5. Let $p > 0$.

(i) A Radon measure μ on $\mathbb{R}_+^{n+1}$ (denoted by $\mu \in \mathsf{RM}(\mathbb{R}_+^{n+1})$) is a p-Carleson measure provided

$$|||\mu|||_{\mathsf{CM}_p} = \sup \frac{|\mu|(S(I))}{\ell(I)^{pn}} < \infty,$$

where

$$|\mu| = \mu_+ + \mu_- \quad \& \quad \mu_\pm = \max\{0, \pm\mu\}$$

and the supremum is taken over all Carleson boxes

$$S(I) = \{(x,t) : x \in I, t \in (0, \ell(I))\}$$

based on $I \in \mathscr{C}(\mathbb{R}^n)$.

(ii) The nontangential maximal function $\mathcal{N}(f)$ of a measurable function f on $\mathbb{R}_+^{n+1}$ is defined by

$$\mathcal{N}(f)(x) = \sup_{(y,t)\in\Lambda(x)} |f(y,t)|,$$

where

$$\Lambda(x) = \{(y,t) \in \mathbb{R}_+^{n+1} : |y - x| < t\}$$

is the cone at $x \in \mathbb{R}^n$.

Theorem 3.6. *Let*

$$\begin{cases} d \in (0,n] \\ \dfrac{d}{n} < p < \infty \\ G_t(y) = (4\pi t)^{-\frac{n}{2}} \exp\left(-\dfrac{|y|^2}{4t}\right) & \forall\,(y,t) \in \mathbb{R}^{n+1}_+ \\ P_t(y) = \Gamma\left(\dfrac{n+1}{2}\right)\pi^{-\frac{n+1}{2}} t(t^2+|y|^2)^{-\frac{n+1}{2}} & \forall\,(y,t) \in \mathbb{R}^{n+1}_+ \\ \Gamma(s) = \displaystyle\int_0^\infty t^{s-1}e^{-t}\,\mathrm{d}s & \forall\, s > 0 \end{cases}$$

and μ *be a Radon measure on* $\mathbb{R}^{n+1}_+$.

(i) *The following five statements are equivalent:*

(i)-a

$$|||\mu|||_{\mathrm{CM}_{\frac{d}{n}}} < \infty;$$

(i)-b

$$\sup_{(x,s)\in\mathbb{R}^{n+1}_+} \iint_{\mathbb{R}^{n+1}_+} \left(t(|x-y|^2+(t-s)^2)^{-\frac{n+1}{2}}\right)^{\frac{d}{n}} \mathrm{d}|\mu|(y,t) < \infty;$$

(i)-c

$$\iint_{\mathbb{R}^{n+1}_+} |f(y,t)|\,\mathrm{d}|\mu|(y,t) \lesssim \int_{\mathbb{R}^n} \mathcal{N}(f)\,\mathrm{d}\Lambda_d^{(\infty)}$$

holds for all measurable functions f *on* $\mathbb{R}^{n+1}_+$;

(i)-d

$$\iint_{\mathbb{R}^{n+1}_+} |f(y,t)|^p\,\mathrm{d}|\mu|(y,t) \lesssim \int_{\mathbb{R}^n} |f_0|^p\,\mathrm{d}\Lambda_d^{(\infty)}$$

holds for all measurable functions

$$f(y,t) = f_0 * G_{t^2}(y) \quad \textit{or} \quad f(y,t) = f_0 * P_t(y) \quad \textit{with} \quad f_0 \in L^1_{\mathrm{loc}}(\mathbb{R}^n);$$

(i)-e

$$\sup_{s>0} s^p |\mu|(\{x \in \mathbb{R}^{n+1}_+ : |f(y,t)| > s\}) \lesssim \int_{\mathbb{R}^n} |f_0|^p\,\mathrm{d}\Lambda_d^{(\infty)}$$

holds for all measurable functions

$$f(y,t) = f_0 * G_{t^2}(y) \quad \textit{or} \quad f(y,t) = f_0 * P_t(y) \quad \textit{with} \quad f_0 \in L^1_{\mathrm{loc}}(\mathbb{R}^n).$$

(ii) *If* $\mathsf{CM}_{\frac{d}{n}}(\mathbb{R}^{n+1}_+)$ *is the class of all*

$$\mu \in \mathsf{RM}(\mathbb{R}^{n+1}_+) \quad \textit{with} \quad |||\mu|||_{\mathsf{CM}_{\frac{d}{n}}} < \infty$$

and $\mathsf{LN}^1(\Lambda_d^{(\infty)})$ *is the completion of all continuous functions* f *on* $\mathbb{R}^{n+1}_+$ *with compact support (denoted by* $f \in \mathsf{C}_0(\mathbb{R}^{n+1}_+)$*) with respect to*

$$\|f\|_{\mathsf{LN}^1(\Lambda_d^{(\infty)})} = \int_{\mathbb{R}^n} \mathcal{N}(f)\,\mathrm{d}\Lambda_d^{(\infty)},$$

then $\mathsf{CM}_{\frac{d}{n}}(\mathbb{R}^{n+1}_+)$ *is isomorphic to the dual of* $\mathsf{LN}^1(\Lambda_d^{(\infty)})$ *under the pairing* $\iint_{\mathbb{R}^{n+1}_+} f\,\mathrm{d}\mu$.

Proof. (i) This will be verified according to the following four equivalences.

(i)-a⇔(i)-b. If (i)-a holds, then fixing $(x,s) \in \mathbb{R}^{n+1}_+$, letting $I \in \mathscr{C}(\mathbb{R}^n)$ have the center x and $\ell(I) = s$ and noticing

$$\begin{cases} (|y-x|^2 + |t-s|^2)^{\frac{1}{2}} \ge s & \text{as} \quad (y,t) \in S(2I) \\ (|y-x|^2 + |t-s|^2)^{\frac{1}{2}} \approx 2^m s & \text{as} \quad (y,t) \in S(2^m I), \end{cases}$$

we obtain (i)-b through

$$\begin{aligned} \iint_{\mathbb{R}^{n+1}_+} \left(t(|x-y|^2 + (t-s)^2)^{-\frac{n+1}{2}}\right)^{\frac{d}{n}} \mathrm{d}|\mu|(y,t) &= \left(\int_{S(2I)} + \sum_{m=1}^{\infty} \int_{S(2^{m+1}I)\setminus S(2^m I)} \right)(\cdots) \\ &\lesssim \frac{|\mu|(S(2I))}{s^d} + \sum_{m=1}^{\infty} \frac{|\mu|(S(2^{m+1}I))}{2^{md(1+n^{-1})} s^d} \\ &\lesssim |||\mu|||_{\mathsf{CM}_{\frac{d}{n}}}. \end{aligned}$$

Conversely, if (i)-b is valid, then for $I \in \mathscr{C}(\mathbb{R}^n)$ taking (x,s) to be the center of $S(I)$ and utilizing

$$(y,t) \in S(I) \Rightarrow |y-x|^2 + |t-s|^2 \lesssim \ell(I)^2,$$

we achieve

$$\iint_{\mathbb{R}^{n+1}_+} \left(t(|x-y|^2 + (t-s)^2)^{-\frac{n+1}{2}}\right)^{\frac{d}{n}} \mathrm{d}|\mu|(y,t) \ge \int_{S(I)} (\cdots) \gtrsim \frac{|\mu|(S(I))}{\ell(I)^d},$$

so that (i)-a follows up.

(i)-a⇔(i)-c. If (i)-a is valid, then for $t > 0$ and a measurable function f letting

$$O_t(f) = \{x \in \mathbb{R}^n : \mathcal{N}(f)(x) > t\} \quad \& \quad \Lambda_d^{(\infty)}(O_t(f)) < \infty,$$

we cover $O_t(f)$ by

$$\{I_j\} \subseteq \mathscr{D}(\mathbb{R}^n) \quad \text{with} \sum_j \ell(I_j)^d < \infty,$$

thereby being able to choose

$$\{J_k\} \subseteq \mathscr{D}(\mathbb{R}^n)$$

such that (cf. [17, Lemma 4.1])

$$\begin{cases} \bigcup_j I_j = \bigcup_k J_k \\ \sum_k \ell(J_k) \le \sum_j \ell(I_j)^d \\ \{(y,t) \in \mathbb{R}^{n+1}_+ : |f(y,t)| > t\} \subseteq T(O_t(f)) \subseteq \bigcup_k T(J_k^*) \\ J_k^* = 5\sqrt{n} J_k, \end{cases}$$

where

$$T(O) = \{(y,t) \in \mathbb{R}^{n+1}_+ : B(y,t) \subseteq O\}$$

is the tent based on O. Accordingly,

$$|\mu|(\{(y,t) \in \mathbb{R}^{n+1}_+ : |f(y,t)| > t\}) \le \sum_k |\mu|(T(J_k^*)) \lesssim |||\mu|||_{\mathrm{CM}_{\frac{d}{n}}} \sum_j \ell(I_j)^d.$$

This in turn implies

$$|\mu|(\{(y,t) \in \mathbb{R}^{n+1}_+ : |f(y,t)| > t\}) \lesssim |||\mu|||_{\mathrm{CM}_{\frac{d}{n}}} \Lambda_d^{(\infty)}(O_t(f))$$

and (i)-c.

Conversely, if (i)-c holds, then for $I \in \mathscr{C}(\mathbb{R}^n)$ letting

$$\phi(y,t) = \mathbb{1}_{T(I)}$$

and noticing

$$(y,t) \in T(I) \cap \Lambda(x) \Leftrightarrow x \in B(y,t) \subseteq I,$$

we obtain

$$\mathcal{N}(\phi) = \mathbb{1}_I,$$

thereby getting (i)-a through

$$|\mu|(T(I)) \lesssim \int_0^\infty \Lambda_d^{(\infty)}(O_t(\phi))\,\mathrm{d}t \lesssim \ell(I)^d.$$

(i)-a⇔(i)-d. It is enough to do this for the Gauss kernel $G_t(y)$ since the treatment for the Poisson kernel $P_t(y)$ is completely similar.

Suppose that (i)-a holds. Note that

$$\sup_{(y,t)\in\Lambda(x)} |f_0 * G_t(y)| \lesssim \mathcal{M}f_0(x) \quad \forall x \in \mathbb{R}^n.$$

So, an application of Theorem 3.3(ii) gives (i)-d.

Conversely, if (i)-d is true, then $f_0 = \mathbb{1}_B$ of an open ball $B = B(x,t) \subseteq \mathbb{R}^n$ yields that

$$f_0 * G_{t^2}(x) \gtrsim \int_{B(x,t)} t^{-n} \exp\left(-\frac{|x-y|^2}{4t^2}\right) \mathrm{d}y \gtrsim 1$$

and so that

$$|\nu|(T(B)) \lesssim \Lambda_d^{(\infty)}(B) \lesssim \mathrm{m}(B)^{\frac{d}{n}}$$

holds, thereby reaching (i)-a.

(i)-a⇔(i)-e. This is similar to the argument for (i)-a⇔(i)-d.

(ii) The argument for (i)-a⇒(i)-b ensures that any

$$\mu \in \mathsf{CM}_{\frac{d}{n}}(\mathbb{R}^{n+1}_+)$$

induces a bounded linear functional of

$$\mathsf{LN}^1(\Lambda_d^{(\infty)}) \quad \text{via} \quad \iint_{\mathbb{R}^{n+1}_+} f\,\mathrm{d}\mu.$$

Conversely, if $\mathcal{L}$ is a bounded linear functional of $\mathsf{LN}^1(\Lambda_d^{(\infty)})$ with its norm $\|\mathcal{L}\|$, then an application of

$$\mathsf{C}_0(\mathbb{R}^{n+1}_+) \subseteq \mathsf{LN}^1(\Lambda_d^{(\infty)})$$

and the Riesz representation theorem derives $\mu \in \mathsf{RM}(\mathbb{R}^{n+1}_+)$ such that

$$\begin{cases} \mathcal{L}(f) = \displaystyle\iint_{\mathbb{R}^{n+1}_+} f\,\mathrm{d}\mu; \\ |\mathcal{L}(f)| \le \|\mathcal{L}\|\|f\|_{\mathsf{LN}^1(\Lambda_d^{(\infty)})} \quad \forall f \in \mathsf{C}_0(\mathbb{R}^{n+1}_+). \end{cases}$$

But then for any $g \in \mathsf{C}_0(\mathbb{R}^{n+1}_+)$ we have

$$\left|\iint_{\mathbb{R}^{n+1}_+} g\,\mathrm{d}|\mu|\right| \le \|\mathcal{L}\| \sup\left\{\int_{\mathbb{R}^n} f\,\mathrm{d}\Lambda_d^{(\infty)} : f \in \mathsf{C}_0(\mathbb{R}^n)\ \&\ |f| \le |g|\right\} \le \|\mathcal{L}\|\|g\|_{\mathsf{LN}^1(\Lambda_d^{(\infty)})}.$$

Now, for

$$(I,\epsilon) \in \mathscr{C}(\mathbb{R}^n) \times \mathbb{R}_+$$

let $I_\epsilon \in \mathscr{C}(\mathbb{R}^n)$ have the same center as I's and the edge-length $\epsilon + \ell(I)$. Upon taking a function

$$g \in \mathsf{C}_0(\mathbb{R}^{n+1}_+) \quad \text{with} \quad g = \begin{cases} 1 & \text{on } S(I) \\ 0 & \text{on } \mathbb{R}^{n+1}_+ \setminus S(I_\epsilon), \end{cases}$$

we gain

$$\mathcal{N}(g) \lesssim \mathbb{1}_{I_\epsilon} \quad \& \quad |\mu|(S(I)) \le \|\mathcal{L}\| \Lambda_d^{(\infty)}(I_\epsilon)$$

thereby arriving at

$$|||\mu|||_{\mathsf{CM}_{\frac{d}{n}}} \lesssim \|\mathcal{L}\|. \qquad \square$$

Interestingly and importantly, Theorem 3.4 yields the so-called tent space pair

$$\{\mathsf{T}_d^{\infty}(\mathbb{R}^{n+1}_+), \mathsf{T}_d^1(\mathbb{R}^{n+1}_+)\}$$

as described below.

Definition 3.7. Let $(d, n-1) \in (0,n] \times \mathbb{N}$.

(i) $\mathsf{T}_d^{\infty}(\mathbb{R}^{n+1}_+)$ is the class of all measurable functions f on $\mathbb{R}^{n+1}_+$ obeying

$$\|f\|_{\mathsf{T}_d^{\infty}} = \sup_{B \subseteq \mathbb{R}^n} \left(\iint_{T(B)} |f(y,t)|^2 \left(\frac{t}{\mathrm{m}(B)^{\frac{1}{n}}} \right)^d \frac{dydt}{t^{1+n}} \right)^{\frac{1}{2}} < \infty,$$

where the supremum is taken over all open balls $B \subseteq \mathbb{R}^n$.

(ii) A function a on $\mathbb{R}^{n+1}_+$ is said to be a $\mathsf{T}_d^1(\mathbb{R}^{n+1}_+)$-atom if there exists an open ball

$$B = B(x,r) \subseteq \mathbb{R}^n$$

such that a is supported in the tent $T(B)$ and satisfies

$$\iint_{T(B)} |a(y,t)|^2 \left(\frac{\mathrm{m}(B)^{\frac{1}{n}}}{t} \right)^d \frac{dydt}{t^{1-n}} \le 1.$$

(iii) $\mathsf{T}_d^1(\mathbb{R}^{n+1}_+)$ is the class of all measurable functions f on $\mathbb{R}^{n+1}_+$ satisfying

$$\|f\|_{\mathsf{T}_d^1(\mathbb{R}^{n+1}_+)} = \inf_{\omega} \left(\iint_{\mathbb{R}^{n+1}_+} |f(y,t)|^2 \left(\frac{\omega(t,y)^{-1}}{t^d} \right) \frac{dydt}{t^{1-n}} \right)^{\frac{1}{2}} < \infty,$$

where the infimum is taken over all nonnegative measurable functions

$$\omega \text{ on } \mathbb{R}^{n+1}_+ \text{ with } \int_{\mathbb{R}^n} \mathcal{N}(\omega)\, d\Lambda^{(\infty)}_d \le 1$$

and the restriction that ω is allowed to vanish only where f vanishes.

Theorem 3.8. *Let $(d, n-1) \in (0, n] \times \mathbb{N}$.*

(i) *$f \in T^1_d(\mathbb{R}^{n+1}_+)$ if and only if there is a sequence of $T^1_d(\mathbb{R}^{n+1}_+)$-atoms $\{a_j\}$ and an l^1-sequence $\{\lambda_j\}$ such that*

$$f = \sum_j \lambda_j a_j.$$

Moreover,

$$\|f\|_{T^1_d(\mathbb{R}^{n+1}_+)} \approx \inf\Big\{\sum_j |\lambda_j| : f = \sum_j \lambda_j a_j\Big\},$$

where the infimum is taken over all possible atomic decompositions of $f \in T^1_d(\mathbb{R}^{n+1}_+)$. The right-hand side thus defines a norm on $T^1_d(\mathbb{R}^{n+1}_+)$ which makes it into a Banach space.

(ii) *The inequality*

$$\iint_{\mathbb{R}^{n+1}_+} |f(y,t)g(y,t)| \frac{dy\,dt}{t} \lesssim \|f\|_{T^1_d(\mathbb{R}^{n+1}_+)} \|g\|_{T^\infty_d(\mathbb{R}^{n+1}_+)}$$

holds for all $(f, g) \in T^1_d(\mathbb{R}^{n+1}_+) \times T^\infty_d(\mathbb{R}^{n+1}_+)$.

(iii) *The Banach space dual of $T^1_d(\mathbb{R}^{n+1}_+)$ can be identified with $T^\infty_d(\mathbb{R}^{n+1}_+)$ under the following pairing*

$$\langle f, g\rangle = \iint_{\mathbb{R}^{n+1}} f(y,t)g(y,t)\frac{dydt}{t}.$$

Proof. (i) Suppose first that a is a $T^1_d(\mathbb{R}^{n+1}_+)$-atom. Then there is a ball

$$B = B(x_B, r) \subseteq \mathbb{R}^n$$

such that

$$\operatorname{supp}(a) \subseteq T(B) \quad \& \quad \iint_{T(B)} |a(y,t)|^2 \left(\frac{m(B)^{\frac{1}{n}}}{t}\right)^d \frac{dydt}{t^{1-n}} \le 1.$$

For $\epsilon > 0$ let

$$\omega(x,t) = \kappa\, r^{-d} \min\left\{1, \left(\frac{r}{\sqrt{|x-x_B|^2+t^2}}\right)^{d+\epsilon}\right\},$$

where κ is a constant to be determined later on. Note that in $\mathbb{R}^{n+1}_+$ this function is identically equal to κr^{-d} on the upper half-ball of radius r, and decays radially outside the ball. Since the Euclidean distance in $\mathbb{R}^{n+1}_+$

$$\text{from } \Lambda(x) \text{ to } (x_B, 0) \in \mathbb{R}^{n+1}_+ \text{ is } \sqrt{2}^{-1}|x-x_B|,$$

the nontangential maximal function of ω enjoys

$$\mathcal{N}(\omega)(x) \le \kappa\, r^{-d} \min\left\{1, \left(\frac{r\sqrt{2}}{|x-x_B|}\right)^{d+\epsilon}\right\},$$

and a suitable choice of $\kappa > 0$ ensures

$$\int_{\mathbb{R}^n} \mathcal{N}(\omega)\, \mathrm{d}\Lambda_d^{(\infty)} \le 1$$

via

$$\kappa^{-1} \int_{\mathbb{R}^n} \mathcal{N}(\omega)\, \mathrm{d}\Lambda_d^{(\infty)} \le \int_0^{r^{-d}} \Lambda_d^{(\infty)}(B(x_B, \sqrt{2}(r^{\epsilon}\lambda^{-1})^{1/(d+\epsilon)}))\, \mathrm{d}\lambda \lesssim 1.$$

On the one hand, since

$$\omega^{-1} = r^d \quad \text{on} \quad T(B),$$

we have

$$\iint_{T(B)} |a(x,t)|^2 \left(\frac{\omega(x,t)^{-1}}{t^d}\right) \frac{\mathrm{d}x\mathrm{d}t}{t^{1-n}} = \iint_{T(B)} |a(x,t)|^2 \left(\frac{r}{t}\right)^d \frac{\mathrm{d}x\mathrm{d}t}{t^{1-n}} \lesssim 1,$$

whence

$$a \in \mathrm{T}^1_d(\mathbb{R}^{n+1}_+) \quad \& \quad \|a\|_{\mathrm{T}^1_d(\mathbb{R}^{n+1}_+)} \lesssim 1.$$

Upon taking such $\mathrm{T}^1_d(\mathbb{R}^{n+1}_+)$-atoms $\{a_j\}$ that $\sum_j \lambda_j a_j$ is convergent with

$$\|\{\lambda_j\}\|_1 = \sum_j |\lambda_j| < \infty,$$

we gain

$$\sum_j \lambda_j a_j = f \in \mathrm{T}^1_d(\mathbb{R}^{n+1}_+) \quad \& \quad \|f\|_{\mathrm{T}^1_d(\mathbb{R}^{n+1}_+)} \lesssim \|\{\lambda_j\}\|_1.$$

On the other hand, let $f \in \mathrm{T}^1_d(\mathbb{R}^{n+1}_+)$. Choose a measurable function $\omega \geq 0$ on $\mathbb{R}^{n+1}_+$ satisfying

$$\int_{\mathbb{R}^n} \mathcal{N}(\omega)\, d\Lambda^{(\infty)}_d \leq 1 \quad \& \quad \iint_{\mathbb{R}^{n+1}_+} |f(x,t)|^2 \left(\frac{\omega(x,t)^{-1}}{t^d}\right) \frac{dxdt}{t^{1-n}} \leq 2\|f\|_{\mathrm{T}^1_d(\mathbb{R}^{n+1}_+)}.$$

For each $k \in \mathbb{Z}$, put

$$E_k = \{x \in \mathbb{R}^n : \mathcal{N}(\omega)(x) > 2^k\}.$$

Then there is a sequence

$$\{I_{j,k}\} \subseteq \mathscr{D}(\mathbb{R}^n)$$

with disjoint interiors such that

$$\sum_j \ell(I_{j,k})^d \leq 2\Lambda^{(\infty)}_d(E_k) \quad \& \quad T(E_k) \subseteq \bigcup_j S^*(I_{j,k}).$$

Here, each tent

$$T(I^*_{j,k}) \quad \text{over} \quad I^*_{j,k} = 5\sqrt{n} I_{j,k}$$

has been replaced by

$$S^*(I_{j,k}) = \{(y,t) \in \mathbb{R}^{n+1}_+ : y \in I_{j,k},\ t < 2\,\mathrm{diam}(I_{j,k})\}.$$

Consequently, if

$$T_{j,k} = S^*(I_{j,k}) \setminus \bigcup_{m>k} \bigcup_l S^*(I_{l,m}),$$

then

$$\begin{aligned}\bigcup_{k=-K}^{K} \bigcup_j T_{j,k} &= \bigcup_j S^*(I_{j,-K}) \setminus \bigcup_{m>K} \bigcup_l S^*(I_{l,m}) \\ &\supseteq T(E_{-K}) \setminus \bigcup_{m>K} \bigcup_l S^*(I_{l,m}).\end{aligned}$$

Note that

$$\bigcup_k T(E_k) = \{(x,t) \in \mathbb{R}^{n+1}_+ : \omega(x,t) > 0\}$$

owing to

$$(x,t) \notin T(E_k) \Rightarrow \omega(x,t) \le 2^k.$$

Note also that each $S^*(I_{l,m})$ is contained in an $(n+1)$-dimensional cube of edge-length $2\mathrm{diam}(I_{l,m})$. Using Proposition 3.2(i) and

$$\int_{\mathbb{R}^n} \mathcal{N}(\omega)\,\mathrm{d}\Lambda_d^{(\infty)} \le 1,$$

we achieve

$$\Lambda_d^{(\infty)}\Big(\bigcup_{m>k}\bigcup_{l} S^*(I_{l,m})\Big) \lesssim \sum_{m>k}\sum_{l} \ell(I_{l,m})^d \lesssim \sum_{m>k} \Lambda_d^{(\infty)}(E_m) \to 0 \quad \text{as} \quad k \to \infty,$$

thereby getting

$$\bigcup_k \bigcup_j T_{j,k} \supseteq \bigcup_k T(E_k) \setminus \bigcap_k \bigcup_{m>k} \bigcup_l S^*(I_{l,m}) = \{(x,t) \in \mathbb{R}^{n+1}_+ : w(x,t) > 0\} \setminus T_\infty,$$

where T_∞ is a set whose d-dimensional Hausdorff capacity is zero, and consequently $(n+1)$-dimensional Lebesgue measure is zero. Note that ω is allowed to vanish only where f vanishes. So we obtain

$$f = \sum f \mathbb{1}_{T_{j,k}} \quad \text{a. e. on } \mathbb{R}^{n+1}_+.$$

Upon setting

$$\begin{cases} a_{j,k} = f\mathbb{1}_{T_{j,k}}\Big(\ell(I_{j,k}^*)^d \iint_{T_{j,k}} |f(x,t)|^2 \frac{\mathrm{d}x\mathrm{d}t}{t^{1-n+d}}\Big)^{-\frac{1}{2}} \\ \lambda_{j,k} = \Big(\ell(I_{j,k}^*)^d \iint_{T_{j,k}} |f(x,t)|^2 \frac{\mathrm{d}x\mathrm{d}t}{t^{1-n+d}}\Big)^{\frac{1}{2}}, \end{cases}$$

we have

$$f = \sum_{j,k} \lambda_{j,k} a_{j,k} \quad \text{a.e.} \quad \& \quad S^*(I_{j,k}) \subseteq T(B_{j,k}),$$

where $B_{j,k}$ is the ball with the same center as $I_{j,k}$ and radius $2^{-1}\ell(I_{j,k}^*)$. Thus $a_{j,k}$ is supported in $T(B_{j,k})$ and is normalized according to Definition 3.7, so each $a_{j,k}$ is a $\mathrm{T}^1_d(\mathbb{R}^{n+1}_+)$-atom. It remains to validate that $\{\lambda_{j,k}\}$ is l^1-summable. Utilizing

$$\omega \le 2^{k+1} \quad \text{on } T_{j,k} \subseteq \Big(\bigcup_l S^*(I_{l,k+1})\Big)^c \subseteq (T(E_{k+1}))^c$$

and the Cauchy–Schwarz inequality, we get

$$\begin{aligned}
\sum_{j,k} |\lambda_{j,k}| &\le \sum_{j,k} 2^{(k+1)/2} \ell(I^*_{j,k})^{d/2} \Big(\iint_{T_{j,k}} |f(x,t)|^2 \omega(x,t)^{-1} \frac{\mathrm{d}x\mathrm{d}t}{t^{1-n+d}} \Big)^{\frac{1}{2}} \\
&\le \Big(\sum_{k,j} 2^{k+1} \ell(I^*_{j,k})^d \Big)^{\frac{1}{2}} \Big(\sum_{j,k} \iint_{T_{j,k}} |f(x,t)|^2 \omega(x,t)^{-1} \frac{\mathrm{d}x\mathrm{d}t}{t^{1-n+d}} \Big)^{\frac{1}{2}} \\
&\lesssim \|f\|_{\mathrm{T}^1_d(\mathbb{R}^{n+1}_+)} \Big(\sum_k 2^k \sum_j \ell(I_{j,k})^d \Big)^{\frac{1}{2}} \\
&\lesssim \|f\|_{\mathrm{T}^1_d(\mathbb{R}^{n+1}_+)} \Big(\sum_k 2^k \Lambda^{(\infty)}_d (E_k) \Big)^{\frac{1}{2}} \\
&\lesssim \|f\|_{\mathrm{T}^1_d(\mathbb{R}^{n+1}_+)} \Big(\int_{\mathbb{R}^n} \mathcal{N}(\omega)\, \mathrm{d}\Lambda^{(\infty)}_d \Big)^{\frac{1}{2}} \\
&\lesssim \|f\|_{\mathrm{T}^1_d(\mathbb{R}^{n+1}_+)},
\end{aligned}$$

thereby reaching

$$\sum_{j,k} \lambda_{j,k} a_{j,k} = f$$

under the quasi-norm.

Evidently, $\mathrm{T}^\infty_d(\mathbb{R}^{n+1}_+)$ becomes a Banach space under

$$\|f\|_{\mathrm{T}^1_d(\mathbb{R}^{n+1}_+)} \approx |||f|||_{\mathrm{T}^1_d(\mathbb{R}^{n+1}_+)} = \inf\Big\{ \sum_j |\lambda_j| : f = \sum_j \lambda_j a_j \Big\},$$

where the infimum ranges over all possible atomic decompositions of $f \in \mathrm{T}^\infty_d(\mathbb{R}^{n+1}_+)$.

(ii) Suppose that ω is a nonnegative measurable function on $\mathbb{R}^{n+1}_+$. Let $g \in \mathrm{T}^\infty_d(\mathbb{R}^{n+1}_+)$. Then

$$\mathrm{d}\mu_{g,d}(x,t) = |g(x,t)|^2 \frac{\mathrm{d}x\mathrm{d}t}{t^{1+n-d}}$$

is a $\frac{d}{n}$-Carleson measure, and hence

$$\iint_{\mathbb{R}^{n+1}_+} \omega(x,t)|g(x,t)|^2 \frac{\mathrm{d}x\mathrm{d}t}{t^{1+n-d}} \lesssim \|g\|^2_{\mathrm{T}^\infty_d(\mathbb{R}^{n+1}_+)} \int_{\mathbb{R}^n} \mathcal{N}(\omega)\, \mathrm{d}\Lambda^{(\infty)}_d \lesssim \|g\|^2_{\mathrm{T}^\infty_d(\mathbb{R}^{n+1}_+)}.$$

Accordingly, if $f \in \mathrm{T}^\infty_d(\mathbb{R}^{n+1}_+)$, then the Cauchy–Schwarz inequality gives

$$\iint_{\mathbb{R}^{n+1}_+} |f(y,t)g(y,t)| \frac{\mathrm{d}y\mathrm{d}t}{t}$$

$$\leq \left(\iint_{\mathbb{R}^{n+1}_+} |f(y,t)|^2 \omega(y,t)^{-1} \frac{dydt}{t^{1-n+d}} \right)^{\frac{1}{2}} \left(\iint_{\mathbb{R}^{n+1}_+} |g(y,t)|^2 \omega(y,t) \frac{dydt}{t^{1+n-d}} \right)^{\frac{1}{2}}$$

$$\lesssim \left(\iint_{\mathbb{R}^{n+1}_+} |f(y,t)|^2 \omega(y,t)^{-1} \frac{dydt}{t^{1-n+d}} \right)^{\frac{1}{2}} \|g\|_{\mathrm{T}^{\infty}_d(\mathbb{R}^{n+1}_+)}.$$

(iii) By part (ii), every $g \in \mathrm{T}^{\infty}_d(\mathbb{R}^{n+1}_+)$ induces a bounded linear functional on $\mathrm{T}^1_d(\mathbb{R}^{n+1}_+)$. So, it suffices to show the converse. Let $\mathcal{L}$ be a bounded linear functional on $\mathrm{T}^1_d(\mathbb{R}^{n+1}_+)$. Fix a ball $B \subseteq \mathbb{R}^n$ with radius $r > 0$. If f is supported in $T(B)$ with

$$f \in \mathrm{L}^2(T(B), t^{-1}\mathrm{d}x\mathrm{d}t),$$

then

$$\iint_{T(B)} |f(x,t)|^2 \frac{\mathrm{d}x\mathrm{d}t}{t^{1-n+d}} \leq r^{n-d} \iint_{T(B)} |f(x,t)|^2 \frac{\mathrm{d}x\mathrm{d}t}{t},$$

so f is a multiple of a $\mathrm{T}^1_d(\mathbb{R}^{n+1}_+)$-atom with

$$\|f\|^2_{\mathrm{T}^1_d(\mathbb{R}^{n+1}_+)} \lesssim r^n \|f\|_{\mathrm{L}^2(T(B),t^{-1}\mathrm{d}x\mathrm{d}t)}.$$

Hence $\mathcal{L}$ induces a bounded linear functional on the square space on $T(B)$ against the weighted Lebesgue measure $t^{-1}\mathrm{d}x\mathrm{d}t$

$$\mathrm{L}^2(T(B), t^{-1}\mathrm{d}x\mathrm{d}t),$$

and acts via the inner product with some function

$$g_B \in \mathrm{L}^2(T(B), t^{-1}\mathrm{d}x\mathrm{d}t).$$

Upon taking

$$B_j = B(o,j) \quad \forall j \in \mathbb{N},$$

we have

$$g_{B_j} = g_{B_{j+1}} \quad \text{on} \quad T(B_j),$$

so we get a single function g on $\mathbb{R}^{n+1}_+$ which belongs locally to

$$\mathrm{L}^2(\mathbb{R}^{n+1}_+, t^{-1}\mathrm{d}x\mathrm{d}t),$$

and such that

$$\mathcal{L}(f) = \iint_{\mathbb{R}^{n+1}_+} f(x,t)g(x,t)t^{-1}\mathrm{d}x\mathrm{d}t$$

whenever $f \in \mathrm{T}_d^\infty(\mathbb{R}_+^{n+1})$ with support in some finite tent $T(B)$. By the atomic decomposition in part (i), the subspace of such f is dense in $\mathrm{T}_d^\infty(\mathbb{R}_+^{n+1})$. Thus if we can prove

$$g \in \mathrm{T}_d^\infty(\mathbb{R}_+^{n+1}) \quad \text{with} \quad \|g\|_{\mathrm{T}_d^\infty(\mathbb{R}_+^{n+1})} \lesssim \|\mathcal{L}\|,$$

then by taking limits we can get the representation of $\mathcal{L}$ for any

$$f \in \mathrm{T}_d^\infty(\mathbb{R}_+^{n+1}).$$

To see this, fix a ball $B \subseteq \mathbb{R}^n$ with radius $r > 0$, and for every $\epsilon > 0$ let

$$\begin{cases} f_\epsilon(x,t) = t^{d-n}\, g(x,t)\, \mathbb{1}_{T^\epsilon(B)}(x,t) \\ T^\epsilon(B) = T(B) \cap \{(x,t) : t > \epsilon\}. \end{cases}$$

Then

$$\iint\limits_{T(B)} |f_\epsilon(x,t)|^2 \frac{\mathrm{d}x\mathrm{d}t}{t^{1-n+d}} = \iint\limits_{T^\epsilon(B)} |g(x,t)|^2 \frac{\mathrm{d}x\mathrm{d}t}{t^{1+n-d}} < \infty.$$

Accordingly, f_ϵ is a multiple of a $\mathrm{T}_d^1(\mathbb{R}_+^{n+1})$-atom with

$$\|f_\epsilon\|^2_{\mathrm{T}_d^1(\mathbb{R}_+^{n+1})} \lesssim r^d \iint\limits_{T^\epsilon(B)} |g(x,t)|^2 \frac{\mathrm{d}x\mathrm{d}t}{t^{1+n-d}}$$

and the constant being independent of ϵ. However, f_ϵ is used to derive

$$\begin{aligned} \iint\limits_{T^\epsilon(B)} |g(x,t)|^2 \frac{\mathrm{d}x\mathrm{d}t}{t^{1+n-d}} &= |\mathcal{L}(f_\epsilon)| \\ &\le \|\mathcal{L}\| \|f_\epsilon\|_{\mathrm{T}_d^1(\mathbb{R}_+^{n+1})} \\ &\lesssim \|\mathcal{L}\| \Big(r^d \iint\limits_{T^\epsilon(B)} |g(x,t)|^2 \frac{\mathrm{d}x\mathrm{d}t}{t^{1+n-d}} \Big)^{\frac{1}{2}}. \end{aligned}$$

Since this estimation is true for all $\epsilon > 0$ with a constant independent of ϵ, we find the same inequality for the integral over $T(B)$, and since that does not rely on the choice of B, we have actually demonstrated

$$g \in \mathrm{T}_d^\infty(\mathbb{R}_+^{n+1}) \quad \& \quad \|g\|_{\mathrm{T}_d^\infty(\mathbb{R}_+^{n+1})} \lesssim \|\mathcal{L}\|.$$ □

3.3 Notes

Below are some historical remarks on Chapter 3.

Section 3.1. Proposition 3.2 is basically taken from [1, 17, 100]. Theorems 3.3–3.4 are formulated from a deep understanding of [1, 67].

Section 3.2. Theorems 3.6 and 3.8 come from an appropriate modification of [22, Lemma 4.1] and [17, Theorems 4.2–4.4–5.4].

4 Gauss or Poisson lifting

On the basis of Chapter 3, this chapter focuses on lifting $Q_\alpha(\mathbb{R}^n)$ to $\mathbb{R}^{n+1}_+$ via either the Gauss integral or the Poisson integral, characterizing the divergence of $(Q_\alpha(\mathbb{R}^n))^n$ and on linking $Q_\alpha(\mathbb{R}^n)$ up with the square Campanato space $C^{2,2\alpha}(\mathbb{R}^n)$.

4.1 Extension via either Gauss or Poisson integral

Via either Gauss integral $e^{t^2\Delta}f(x)$ or Poisson integral $e^{-t\sqrt{-\Delta}}f(x)$, we can extend each member of $Q_\alpha(\mathbb{R}^n)$ to the upper half-space $\mathbb{R}^{n+1}_+$ and examine its behavior.

Theorem 4.1. *For*

$$\begin{cases}(\alpha, n-1) \in (0,1) \times \mathbb{N} \\ (x, y, t) \in \mathbb{R}^n \times \mathbb{R}^n \times \mathbb{R}_+,\end{cases}$$

let f be a measurable function on $\mathbb{R}^n$ with

$$\begin{cases} e^{t^2\Delta}f(x) = (4\pi t^2)^{-\frac{n}{2}} \int\limits_{\mathbb{R}^n} \left(\exp\left(-\frac{|x-y|^2}{4t^2}\right)\right) f(y)\, dy \\ e^{-t\sqrt{-\Delta}}f(x) = \Gamma\left(\frac{n+1}{2}\right)\pi^{-\frac{n+1}{2}} t(t^2 + |x-y|^2)^{-\frac{n+1}{2}} \\ \Delta = \partial^2_{x_1} + \cdots + \partial^2_{x_n} \\ \nabla = (\partial_{x_1}, \ldots, \partial_{x_n}) \quad or \quad (\partial_t).\end{cases}$$

Then the following four statements are equivalent for $F(x,t)$ being

$$either\ e^{t^2\Delta}f(x)\ or\ e^{-t\sqrt{-\Delta}}f(x):$$

(i) $f \in Q_\alpha(\mathbb{R}^n)$;

(ii)

$$|||f|||_{Q_\alpha(\mathbb{R}^n),1} = \sup_{(x,r)\in\mathbb{R}^{n+1}_+} \left(r^{2\alpha-n} \int_0^r \left(\int_{B(x,r)} |\nabla F(y,t)|^2\, dy\right) \frac{dt}{t^{2\alpha-1}}\right)^{\frac{1}{2}} < \infty;$$

(iii)

$$|||f|||_{Q_\alpha(\mathbb{R}^n),2} = \sup_{(x,s)\in\mathbb{R}^{n+1}_+} \left(\iint\limits_{\mathbb{R}^{n+1}_+} \frac{|\nabla F(y,t)|^2}{(s(|x-y|^2 + (t-s)^2)^{-\frac{n+1}{2}})^{\frac{2\alpha-n}{n}}} \frac{dydt}{t^{2\alpha-1}}\right)^{\frac{1}{2}} < \infty;$$

(iv)

$$|||f|||_{Q_\alpha(\mathbb{R}^n),3} = \sup_{(x,s)\in\mathbb{R}^{n+1}_+} \left(\iint\limits_{\mathbb{R}^{n+1}_+} (G((x,s),(y,t)))^{\frac{n-2\alpha}{n}} |\nabla F(y,t)|^2 \frac{dydt}{t^{2\alpha(1-n^{-1})}}\right)^{\frac{1}{2}} < \infty,$$

https://doi.org/10.1515/9783110600285-004

where

$$G((x,s),(y,t)) = |(x,s)-(y,t)|^{1-n} - |(x,s)-(y,-t)|^{1-n} \geq 0$$

is the Green function of $\mathbb{R}^{n+1}_+$.

Proof. Since the argument for the Poisson integral

$$F(x,t) = e^{-t\sqrt{-\Delta}}f(x)$$

is completely analogous, it suffices to validate the result for

$$F(x,t) = e^{t^2\Delta}f(x).$$

(i)⇔(ii) Suppose that (ii) holds. Then

$$\begin{cases} F(x,0) = f(x) \\ F(x+y,0) = f(x+y) \\ \partial_{x_j}F(x,t) = \int\limits_{\mathbb{R}^n} \partial_{y_j}\Big((4\pi t^2)^{-\frac{n}{2}}\exp\Big(-\frac{|y|^2}{4t^2}\Big)\Big)(f(x+y)-f(x))\,dy \\ \partial_t F(x,t) = \int\limits_{\mathbb{R}^n} \partial_t\Big((4\pi t^2)^{-\frac{n}{2}}\exp\Big(-\frac{|y|^2}{4t^2}\Big)\Big)(f(x+y)-f(x))\,dy \\ \Big|\partial_{y_j}\Big((4\pi t^2)^{-\frac{n}{2}}\exp\Big(-\frac{|y|^2}{4t^2}\Big)\Big)\Big| \lesssim \min\{t^{-1-n}, |y|^{-1-n}\} \\ \Big|\partial_t\Big((4\pi t^2)^{-\frac{n}{2}}\exp\Big(-\frac{|y|^2}{4t^2}\Big)\Big)\Big| \lesssim \min\{t^{-1-n}, |y|^{-1-n}\}. \end{cases}$$

Note that the triangle inequality gives

$$\begin{aligned} |f(x+y)-f(x)| &\leq \Big|F\Big(x+\frac{y}{2},\frac{|y|}{2}\Big) - F(x,0)\Big| + \Big|F\Big(x+y-\frac{y}{2},\frac{|y|}{2}\Big) - F(x+y,0)\Big| \\ &= \mathsf{A}_1 + \mathsf{A}_2. \end{aligned}$$

According to Theorem 1.2, it suffices to prove

$$\mathsf{B}_i = \ell(I)^{2\alpha-n} \int\limits_{|y|<\ell(I)} \Big(\int\limits_I |\mathsf{A}_i|^2\,dx\Big)\frac{dy}{|y|^{n+2\alpha}} \lesssim \|\|f\|\|^2_{\mathcal{Q}_\alpha(\mathbb{R}^n),1} \quad \forall I \in \mathscr{C}(\mathbb{R}^n)\ \&\ i \in \{1,2\}.$$

Due to the symmetry of A_1 and A_2 about $x + 2^{-1}y$, it is enough to estimate B_1. To do so, upon letting

$$\nabla_{z,s} = (\partial_{z_1},\ldots,\partial_{z_n},\partial_s)$$

and noticing

$$\mathsf{A}_1 = \left|\int_0^1 \frac{d}{dt}F\left(x + \frac{ty}{2}, \frac{t|y|}{2}\right) dt\right| \le \int_0^1 \left|\nabla_{z,s}F(z,s)|_{(z,s)=(x+\frac{ty}{2},\frac{t|y|}{2})} \cdot \left(\frac{y}{2}, \frac{|y|}{2}\right)\right| dt,$$

we utilize the Minkowski inequality to deduce

$$\begin{aligned}
&\left(\int_I |\mathsf{A}_1|^2\,dx\right)^{\frac{1}{2}}\\
&\le \left(\int_I \left(\int_0^1 \left|\nabla_{z,s}F(z,s)|_{(z,s)=(x+\frac{ty}{2},\frac{t|y|}{2})} \cdot \left(\frac{y}{2}, \frac{|y|}{2}\right)\right| dt\right)^2 dx\right)^{\frac{1}{2}}\\
&\le \left(\int_I \left(\int_0^1 \left(|\partial_s F(z,s)|_{(z,s)=(x+\frac{ty}{2},\frac{t|y|}{2})}| + |\nabla_z F(z,s)|_{(z,s)=(x+\frac{ty}{2},\frac{t|y|}{2})}|\right)\frac{|y|}{2}\,dt\right)^2 dx\right)^{\frac{1}{2}}\\
&\lesssim \int_0^{\frac{|y|}{2}} \left(\int_I \left|\nabla F\left(x + s\frac{y}{|y|}, s\right)\right|^2 dx\right)^{\frac{1}{2}} ds.
\end{aligned}$$

This, along with $|y| \le \ell(I)$ and the Hardy inequality, yields

$$\begin{aligned}
\mathsf{B}_1 &\lesssim (\ell(I))^{2\alpha-n} \int_{|y|<\ell(I)} \left(\int_0^{\frac{|y|}{2}} \left(\int_{4I} |\nabla F(x,t)|^2\,dx\right)^{\frac{1}{2}} dt\right)^2 \frac{dy}{|y|^{n+2\alpha}}\\
&\approx \int_0^{\ell(I)} \left(\int_0^s \|\nabla F(\cdot,t)\|_{\mathsf{L}^2(4I)}\,dt\right)^2 (\ell(I))^{2\alpha-n} \frac{ds}{s^{1+2\alpha}}\\
&\lesssim \int_0^{\ell(I)} \|\nabla F(\cdot,t)\|^2_{\mathsf{L}^2(4I)} \left(\frac{t}{\ell(I)}\right)^{n-2\alpha} \frac{dt}{t^{n-1}}\\
&\lesssim |||f|||^2_{\mathsf{Q}_\alpha(\mathbb{R}^n),1}.
\end{aligned}$$

Next, let us check (i)⟹(ii). It is enough to validate

$$\int_0^{\ell(I)}\int_I |\partial_{x_j}F(x,t)|^2 \left(\frac{t}{\ell(I)}\right)^{n-2\alpha} \frac{dx\,dt}{t^{n-1}} \lesssim \|f\|^2_{\mathsf{Q}_\alpha(\mathbb{R}^n)} \quad \forall I \in \mathscr{C}(\mathbb{R}^n)\ \&\ j \in \{1,\dots,n\}.$$

According to the first half of the argument for [22, Lemma 3.1], we have

$$\|\partial_{x_j}F(\cdot,t)\|^2_{\mathsf{L}^2(\mathbb{R}^n)} \lesssim t^{-2n-2}\left(\int_0^t \left(\int_{|\zeta|=1} \|f(\cdot + r\zeta) - f\|_{\mathsf{L}^2(\mathbb{R}^n)}\,d\zeta\right) r^{n-1}\,dr\right)^2$$

$$+\left(\int_t^\infty\left(\int_{|\zeta|=1}\|f(\cdot+r\zeta)-f\|_{L^2(\mathbb{R}^n)}\,d\zeta\right)r^{-2}\,dr\right)^2.$$

On the one hand, via the Hardy inequality we get

$$\int_0^\infty\int_{\mathbb{R}^n}|\partial_{x_j}F(x,t)|^2\frac{dxdt}{t^{2\alpha-1}}\lesssim\int_{\mathbb{R}^n}\int_{\mathbb{R}^n}|f(x+y)-f(x)|^2\frac{dxdy}{|y|^{n+2\alpha}}.$$

Now, for $J\in\mathscr{C}(\mathbb{R}^n)$ with $\ell(J)\geq 3\ell(I)$ and the same center as I's let φ be a function with

$$\begin{cases}0\leq\varphi\leq 1\\ \varphi=1\quad\text{on }\frac{2}{3}J\\ \operatorname{supp}(\varphi)\subseteq\frac{3}{4}J\\ |\varphi(x)-\varphi(y)|\lesssim(\ell(J))^{-1}|x-y|\\ f=f_J+(f-f_J)\varphi+(f-f_J)(1-\varphi)\\ F(x,t)=e^{t^2\Delta}f_J(x)+e^{t^2\Delta}((f-f_J)\varphi)(x)+e^{t^2\Delta}((f-f_J)(1-\varphi))(x).\end{cases}$$

Following the second half of the argument for [22, Lemma 3.1] and noticing the basic inclusion,

$$Q_\alpha(\mathbb{R}^n)\subseteq \mathrm{BMO}(\mathbb{R}^n),$$

we can find

$$\begin{aligned}&\int_0^{\ell(I)}\int_I|\partial_{x_j}F(x,t)|^2\left(\frac{t}{\ell(I)}\right)^{n-2\alpha}\frac{dx\,dt}{t^{n-1}}\\
&\quad\lesssim\int_0^{\ell(I)}\int_I|\partial_{x_j}e^{t^2\Delta}((f-f_J)\varphi)(x)|^2\left(\frac{t}{\ell(I)}\right)^{n-2\alpha}\frac{dx\,dt}{t^{n-1}}\\
&\qquad+\int_0^{\ell(I)}\int_I|\partial_{x_j}e^{t^2\Delta}((f-f_J)(1-\varphi))(x)|^2\left(\frac{t}{\ell(I)}\right)^{n-2\alpha}\frac{dx\,dt}{t^{n-1}}\\
&\quad\lesssim\|f\|^2_{Q_\alpha(\mathbb{R}^n)}+\|f\|^2_{\mathrm{BMO}(\mathbb{R}^n)}\\
&\quad\lesssim\|f\|^2_{Q_\alpha(\mathbb{R}^n)},\end{aligned}$$

as desired.

(i)⇔(iii) The proof is immediate by the above (i)⇔(ii) and Theorem 3.4:(i)-a⇔(i)-b.

(i)⇔(iv) Note that if

$$(x,s),(y,t)\in\mathbb{R}^{n+1}_+\quad\&\quad n\geq 2$$

then

$$\frac{st}{|(x,s)-(y,-t)|^{n+1}} \lesssim G((x,s),(y,t)) \lesssim \frac{st}{|(x,s)-(y,-t)|^2|(x,s)-(y,t)|^{n-1}}.$$

So, if (iv) holds, then the above equivalence (i)⇔(iii) yields (i).

Conversely, if $f \in Q_\alpha(\mathbb{R}^n)$, then upon writing

$$\int_{\mathbb{R}^{n+1}_+} (G((x,s),(y,t)))^{\frac{n-2\alpha}{n}} |\nabla F(y,t)|^2 \frac{dydt}{t^{2\alpha(1-n^{-1})}} = C_1 + C_2,$$

where

$$\begin{cases} C_1 = \int\limits_{\frac{|(x,s)-(y,t)|}{|(x,s)-(y,-t)|}\le\frac{1}{2}} (G((x,s),(y,t)))^{\frac{n-2\alpha}{n}} |\nabla F(y,t)|^2 \frac{dydt}{t^{2\alpha(1-n^{-1})}} \\ C_2 = \int\limits_{\frac{|(x,s)-(y,t)|}{|(x,s)-(y,-t)|}>\frac{1}{2}} (G((x,s),(y,t)))^{\frac{n-2\alpha}{n}} |\nabla F(y,t)|^2 \frac{dydt}{t^{2\alpha(1-n^{-1})}}, \end{cases}$$

we not only use

$$Q_\alpha(\mathbb{R}^n) \subseteq \mathrm{BMO}(\mathbb{R}^n)$$

to get

$$\begin{aligned} C_1 &\lesssim \int\limits_{|(x,s)-(y,t)|\le 2t} \left(\frac{|(x,s)-(y,t)|}{t}\right)^{\frac{2\alpha(n-1)}{n}} G((x,s),(y,t))|\nabla F(y,t)|^2\, dydt \\ &\lesssim \int_{\mathbb{R}^{n+1}_+} G((x,s),(y,t))|\nabla F(y,t)|^2\, dydt \\ &\lesssim \|f\|^2_{Q_\alpha(\mathbb{R}^n)}, \end{aligned}$$

but also utilize the above equivalence (i)⇔(iii) to estimate

$$\begin{aligned} C_2 &\lesssim \int\limits_{\frac{|(x,s)-(y,t)|}{|(x,s)-(y,-t)|}\ge\frac{1}{2}} t^{2\alpha(n^{-1}-1)}\left(\frac{st}{|(x,s)-(y,-t)|^{n+1}}\right)^{1-\frac{2\alpha}{n}} |\nabla F(y,t)|^2\, dydt \\ &\lesssim \int_{\mathbb{R}^{n+1}_+} t^{1-2\alpha}\left(\frac{s}{|(x,s)-(y,-t)|^{n+1}}\right)^{1-\frac{2\alpha}{n}} |\nabla F(y,t)|^2\, dydt \\ &\lesssim \|f\|^2_{Q_\alpha(\mathbb{R}^n)}. \end{aligned}$$

□

4.2 Divergence of Q^n

Using the Gauss integral situation of Theorem 4.1, we characterize the divergence of $(Q_\alpha(\mathbb{R}^n))^n$ as seen below.

Proposition 4.2. *For $(\alpha, n-1) \in (0,1) \times \mathbb{N}$, we say that*

$$g \in Q_\alpha^{-1}(\mathbb{R}^n) = \operatorname{div}(Q_\alpha(\mathbb{R}^n))^n$$

provided

$$g = \sum_{j=1} \partial_{x_j} g_j$$

for some

$$g_j \in Q_\alpha(\mathbb{R}^n) \quad \text{with} \quad \|g\|_{Q_\alpha^{-1}(\mathbb{R}^n)} = \sum_{j=1}^{n} \|g_j\|_{Q_\alpha(\mathbb{R}^n)}.$$

Let f be a tempered distribution on $\mathbb{R}^n$. Then

$$f \in Q_\alpha^{-1}(\mathbb{R}^n) \Leftrightarrow \|f\|_{Q_\alpha^{-1}(\mathbb{R}^n),*} = \sup_{(x,r)\in\mathbb{R}_+^{n+1}} \left(\int_0^{r^2} \int_{B(x,r)} |e^{t\Delta} f(y)|^2 \left(\frac{\sqrt{t}}{r}\right)^{n-2\alpha} \frac{dxdt}{t^{\frac{n}{2}}} \right)^{\frac{1}{2}} < \infty.$$

Proof. On the one hand, let $f \in Q_\alpha^{-1}(\mathbb{R}^n)$. Then there exists

$$f_1, \ldots, f_n \in Q_\alpha(\mathbb{R}^n)$$

such that

$$\begin{cases} f(x) = \sum_{j=1}^{n} \partial_{x_j} f_j(x) \\ \|f\|_{Q_\alpha^{-1}(\mathbb{R}^n)} = \sum_{j=1}^{n} \|f_j\|_{Q_\alpha(\mathbb{R}^n)}. \end{cases}$$

Via the Minkowski inequality, we get

$$\|f\|_{Q_\alpha^{-1}(\mathbb{R}^n),*} = \left\| \sum_{j=1}^{n} \partial_{x_j} f_j \right\|_{Q_\alpha^{-1}(\mathbb{R}^n),*} \lesssim \|f\|_{Q_\alpha^{-1}(\mathbb{R}^n)}.$$

On the other hand, let

$$\|f\|_{Q_\alpha^{-1}(\mathbb{R}^n),*} < \infty.$$

For simplicity, set

$$\partial_j = \partial_{x_j} \quad \& \quad f_{j,k}(x) = \partial_j \partial_k (-\Delta)^{-1} f(x) = \partial_{x_j} \partial_{x_k} (-\Delta)^{-1} f(x) \quad \forall j,k = 1, \ldots, n.$$

As long as

$$\|f_{j,k}\|_{Q_\alpha^{-1}(\mathbb{R}^n),*} < \infty$$

is proved, we will achieve $f \in Q_\alpha^{-1}(\mathbb{R}^n)$ thanks to

$$\|f_{j,k}\|_{Q_\alpha^{-1}(\mathbb{R}^n),*} < \infty \Rightarrow f_k = -\partial_k(-\Delta)^{-1}f \in Q_\alpha(\mathbb{R}^n),$$

and there exists the representation

$$f = \sum_{k=1}^n \partial_k f_k$$

via

$$\widehat{\left(\sum_{k=1}^n \partial_k f_k\right)}(\xi) = -\sum_{k=1}^n \sqrt{-1}\xi_k \hat{f}_k(\xi) = \sum_{k=1}^n (\sqrt{-1}\xi_k)(\sqrt{-1}\xi_k)|\xi|^{-2}\hat{f}(\xi) = \hat{f}(\xi).$$

Now, it remains to validate

$$\|f_{j,k}\|_{Q_\alpha^{-1}(\mathbb{R}^n),*} < \infty.$$

In doing so, take a function $\phi \in C_0^\infty(\mathbb{R}^n)$ such that

$$\begin{cases} \operatorname{supp}(\phi) \subseteq B(o,1) = \{x \in \mathbb{R}^n : |x| < 1\} \\ \int_{\mathbb{R}^n} \phi(x)\,\mathrm{d}x = 1 \\ \phi_r(x) = r^{-n}\phi\left(\frac{x}{r}\right) \\ g_r(x,t) = \phi_r * \partial_j\partial_k(-\Delta)^{-1}e^{t\Delta}f(x). \end{cases}$$

Then

$$e^{t\Delta}f_{j,k}(x) = \partial_j\partial_k(-\Delta)^{-1}e^{t\Delta}f(x) = f_r(x,t) + g_r(x,t).$$

Let

$$\mathrm{BMO}^{-1}(\mathbb{R}^n) = \operatorname{div}(\mathrm{BMO}(\mathbb{R}^n))^n.$$

If

$$\dot{B}^1_{1,1}(\mathbb{R}^n) = \lim_{\beta\to 1} \dot{B}^{1,1}_\beta(\mathbb{R}^n)$$

stands for the predual of the homogeneous anti-Besov space

$$\dot{B}^{-1}_{\infty,\infty}(\mathbb{R}^n) = \operatorname{div}(\dot{B}^1_{\infty,\infty}(\mathbb{R}^n))^n = \operatorname{div}\Big(\lim_{\beta\to 1} \dot{B}^{\infty,\infty}_\beta(\mathbb{R}^n)\Big)^n = \Big\{f : \sup_{t>0} t\|e^{t^2\Delta}f\|_{L^\infty(\mathbb{R}^n)} < \infty\Big\},$$

then

$$\begin{aligned}\|g_r(\cdot,t)\|_{L^\infty(\mathbb{R}^n)} &\lesssim \|\phi\|_{\dot{B}^1_{1,1}(\mathbb{R}^n)}\|\partial_j\partial_k(-\Delta)^{-1}e^{t\Delta}f\|_{\dot{B}^{-1}_{\infty,\infty}(\mathbb{R}^n)}\\ &\lesssim r^{-1}\|f\|_{\dot{B}^{-1}_{\infty,\infty}(\mathbb{R}^n)}\\ &\lesssim r^{-1}\|f\|_{\mathrm{BMO}^{-1}(\mathbb{R}^n)}\\ &\lesssim r^{-1}\|f\|_{Q^{-1}_\alpha(\mathbb{R}^n),*},\end{aligned}$$

and hence

$$\int_0^{r^2}\int_{B(x,r)}|g_r(x,t)|^2\left(\frac{\sqrt{t}}{r}\right)^{n-2\alpha}\frac{dydt}{t^{\frac{n}{2}}}\lesssim\|f\|^2_{Q^{-1}_\alpha(\mathbb{R}^n),*}.$$

Next, we estimate f_r. Upon taking $\varphi\in C_0^\infty(\mathbb{R}^n)$ with

$$\begin{cases}\varphi(y)=1 \quad \forall\, y\in B(o,10)=\{x\in\mathbb{R}^n:|x|<10\}\\ \varphi_{x,r}(y)=\varphi\left(\frac{y-x}{r}\right)\\ f_r(y)=F_{x,r}(y,t)+G_{x,r}(y,t)\\ G_{x,r}(y)=\partial_j\partial_k(-\Delta)^{-1}\varphi_{x,r}e^{t\Delta}f(y)-\phi_r*\partial_j\partial_k(-\Delta)^{-1}\varphi_{x,r}e^{t\Delta}f(y),\end{cases}$$

and using the Plancherel formula for the space variable, we obtain

$$\begin{aligned}&\int_0^{r^2}\|\partial_j\partial_k(-\Delta)^{-1}\varphi_{x,r}e^{t\Delta}f\|^2_{L^2(\mathbb{R}^n)}\left(\frac{\sqrt{t}}{r}\right)^{n-2\alpha}\frac{dt}{t^{\frac{n}{2}}}\\ &\quad\lesssim\int_0^{r^2}\|\widehat{\partial_j\partial_k(-\Delta)^{-1}\varphi_{x,r}e^{t\Delta}f}\|^2_{L^2(\mathbb{R}^n)}\left(\frac{\sqrt{t}}{r}\right)^{n-2\alpha}\frac{dt}{t^{\frac{n}{2}}}\\ &\quad\lesssim\int_0^{r^2}\left(\int_{\mathbb{R}^n}\left|y_jy_k|y|^{-2}(\widehat{\varphi_{x,r}e^{t\Delta}f})(y)\right|^2dy\right)\left(\frac{\sqrt{t}}{r}\right)^{n-2\alpha}\frac{dt}{t^{\frac{n}{2}}}\\ &\quad\lesssim\int_0^{r^2}\|\widehat{\varphi_{x,r}e^{t\Delta}f}\|^2_{L^2(\mathbb{R}^n)}\left(\frac{\sqrt{t}}{r}\right)^{n-2\alpha}\frac{dt}{t^{\frac{n}{2}}}\\ &\quad\lesssim\int_0^{r^2}\|\varphi_{x,r}e^{t\Delta}f\|^2_{L^2(\mathbb{R}^n)}\left(\frac{\sqrt{t}}{r}\right)^{n-2\alpha}\frac{dt}{t^{\frac{n}{2}}}.\end{aligned}$$

Similarly, we have

$$\int_0^{r^2}\|\phi_r*\partial_j\partial_k(-\Delta)^{-1}\varphi_{x,r}e^{t\Delta}f\|^2_{L^2(\mathbb{R}^n)}\left(\frac{\sqrt{t}}{r}\right)^{n-2\alpha}\frac{dt}{t^{\frac{n}{2}}}$$

$$\lesssim \int_0^{r^2} \|\varphi_{x,r} e^{t\Delta} f\|^2_{L^2(\mathbb{R}^n)} \left(\frac{\sqrt{t}}{r}\right)^{n-2\alpha} \frac{dt}{t^{\frac{n}{2}}}.$$

Thus we can obtain the following inequality for $G_{x,r}$

$$\int_0^{r^2} \|G_{x,r}(\cdot,t)\|^2_{L^2(\mathbb{R}^n)} \left(\frac{\sqrt{t}}{r}\right)^{n-2\alpha} \frac{dt}{t^{\frac{n}{2}}} \lesssim \int_0^{r^2} \|\varphi_{x,r} e^{t\Delta} f\|^2_{L^2(\mathbb{R}^n)} \left(\frac{\sqrt{t}}{r}\right)^{n-2\alpha} \frac{dt}{t^{\frac{n}{2}}}.$$

Meanwhile, in order to bound $F_{x,r}$, upon noticing

$$\int_{B(x,r)} |F_{x,r}(y,t)|^2 dy \lesssim r^{n+1} \int_{|w-x|\geq 10r} |e^{t\Delta} f(w)|^2 |x-w|^{-(n+1)} dw,$$

we get

$$\begin{aligned}
&\int_0^{r^2} \left(\int_{B(x,r)} |F_{x,r}(y,t)|^2 dy\right) \left(\frac{\sqrt{t}}{r}\right)^{n-2\alpha} \frac{dt}{t^{\frac{n}{2}}} \\
&\lesssim r^{n+1} \int_{|w-x|\geq 10r} |x-w|^{-(n+1)} \left(\int_0^{r^2} |e^{t\Delta} f(w)|^2 \left(\frac{\sqrt{t}}{r}\right)^{n-2\alpha} \frac{dt}{t^{\frac{n}{2}}}\right) dw \\
&\lesssim r^{n+1} \sum_{l=1}^{\infty} \int_{10^l r \leq |w-x| \leq 10^{l+1} r} |x-w|^{-(n+1)} \left(\int_0^{r^2} |e^{t\Delta} f(w)|^2 \left(\frac{\sqrt{t}}{r}\right)^{n-2\alpha} \frac{dt}{t^{\frac{n}{2}}}\right) dw \\
&\lesssim \sum_{l=1}^{\infty} 10^{-l(n+1)} \int_{|w-x|\leq 10^{l+1} r} \left(\int_{\alpha}^{r^2} |e^{t\Delta} f(w)|^2 \left(\frac{\sqrt{t}}{r}\right)^{n-2\alpha} \frac{dt}{t^{\frac{n}{2}}}\right) dw \\
&\lesssim \|f\|^2_{Q_\alpha^{-1}(\mathbb{R}^n),*}.
\end{aligned}$$

The estimates for $G_{x,r}$ and $F_{x,r}$ give

$$\int_0^{r^2} \int_{B(x,r)} |f_r(y,t)|^2 \left(\frac{\sqrt{t}}{r}\right)^{n-2\alpha} \frac{dydt}{t^{\frac{n}{2}}} \lesssim \|f\|^2_{Q_\alpha^{-1}(\mathbb{R}^n),*}.$$

The foregoing argument actually reveals

$$\|f_{j,k}\|_{Q_\alpha^{-1}(\mathbb{R}^n),*} < \infty. \qquad \square$$

4.3 Connection with square Campanato space

The analysis of Proposition 4.2 suggests a consideration of the so-called Campanato spaces.

Definition 4.3. Let $(\alpha,\lambda,p,n) \in (-\infty,\infty)\times(-\infty,\infty)\times[1,\infty)\times\mathbb{N}$.

(i) $(-\Delta)^{\frac{\alpha}{2}}$ is the $\frac{\alpha}{2}$-th power of the standard Laplacian operator $-\Delta$. In other words, the following Fourier transform:

$$\widehat{(-\Delta)^{\frac{\alpha}{2}}f}(\xi) = |\xi|^{\alpha}\hat{f}(\xi) = |\xi|^{\alpha}\int_{\mathbb{R}^n} f(y)e^{-i\xi\cdot y}\,dy$$

holds for any function or tempered distribution for which the right-hand side is meaningful. In particular, $(-\Delta)^0$ is defined as the identity; and if $0<\alpha<2$ then there are two constants $c_{1,\alpha}$, $c_{2,\alpha}$ such that

$$\begin{cases} (-\Delta)^{\frac{\alpha}{2}}f(x) = c_{1,\alpha}\displaystyle\int_{\mathbb{R}^n}\frac{f(x)-f(y)}{|x-y|^{n+\alpha}} \\ (-\Delta)^{-\frac{\alpha}{2}}f(x) = c_{2,\alpha}\displaystyle\int_{\mathbb{R}^n} f(y)(|y-x|^{\alpha-n}-|y|^{\alpha-n})\,dy, \end{cases}$$

holds for any smooth function f that decays rapidly at infinity.

(ii) $C^{p,\lambda}(\mathbb{R}^n)$ and $M^{p,\lambda}(\mathbb{R}^n)$ express respectively the (p,λ)-Campanato space and the (p,λ)-Morrey space whose norms are decided respectively by (cf. [29, 65, 68])

$$\begin{cases} \|f\|_{C^{p,\lambda}(\mathbb{R}^n)} = \sup\limits_{(x,r)\in\mathbb{R}^{n+1}_+}\left(r^{\lambda}\displaystyle\fint_{B(x,r)}|f(y)-f_{B(x,r)}|^p\,dy\right)^{\frac{1}{p}} \\ \|f\|_{M^{p,\lambda}(\mathbb{R}^n)} = \sup\limits_{(x,r)\in\mathbb{R}^{n+1}_+}\left(r^{\lambda}\displaystyle\fint_{B(x,r)}|f(y)|^p\,dy\right)^{\frac{1}{p}}. \end{cases}$$

Theorem 4.4. *Given a function $\phi\in C^1(\mathbb{R}^n)$ that satisfies*

$$\begin{cases} \phi\in L^1(\mathbb{R}^n) \\ \displaystyle\int_{\mathbb{R}^n}\phi(x)\,dx = 0 \\ |\phi(x)| \lesssim (1+|x|)^{-1-n} & \forall\, x\in\mathbb{R}^n \\ |\nabla\phi(x)| \lesssim (1+|x|)^{-2-n} & \forall\, x\in\mathbb{R}^n \\ \phi_t(x) = t^{-n}\phi(t^{-1}x) & \forall\,(x,t)\in\mathbb{R}^{n+1}_+. \end{cases}$$

For $f\in L^1_{\mathrm{loc}}(\mathbb{R}^n)$ let its Littlewood–Paley $\mathcal{G}$-function be

$$\mathcal{G}_{\phi,j}(f)(x) = \begin{cases} (\int_0^\infty |f*\phi_t(x)|^2\frac{dt}{t})^{\frac{1}{2}} & as\ \ j=1 \\ (\int_0^\infty\int_{B(x,t)}|f*\phi_t(y)|^2\frac{dydt}{t^{n+1}})^{\frac{1}{2}} & as\ \ j=2 \\ (\int_0^\infty\int_{\mathbb{R}^n}|f*\phi_t(x-y)|^2(1+\frac{|y|}{t})^{-\kappa n}\frac{dydt}{t^{n+1}})^{\frac{1}{2}} & as\ \ j=3. \end{cases}$$

If

$$\begin{cases} 1 < p < \infty \\ -p < \lambda \le n \\ 3 + \dfrac{2}{n} < \kappa < \infty \\ j \in \{1,2,3\} \\ \mathcal{G}_{\phi,j}(x_0) < \infty \quad \textit{for some} \quad x_0 \in \mathbb{R}^n, \end{cases}$$

then

$$\|\mathcal{G}_{\phi,j}(f)\|_{\mathcal{C}^{p,\lambda}(\mathbb{R}^n)} \lesssim \|f\|_{\mathcal{C}^{p,\lambda}(\mathbb{R}^n)}.$$

Proof. It is enough to prove the last inequality for $j = 1$. For $(x,r) \in \mathbb{R}^{n+1}_+$ let

$$\begin{cases} f \in \mathcal{C}^{p,\lambda}(\mathbb{R}^n) \\ f_1 = (f - f_{B(x,2r)})1_{B(x,2r)} \\ f_2 = (f - f_{B(x,2r)})1_{\mathbb{R}^n \setminus B(x,2r)} \\ f_3 = f_{B(x,2r)}. \end{cases}$$

Since $\mathcal{G}_{\phi,1}(f_3) = 0$, it remains to handle $\mathcal{G}_{\phi,1}(f_k) = 0$ for $k = 1,2$. If

$$\begin{cases} \mathcal{G}_{\phi,1,0}(f_k)(y) = \left(\displaystyle\int_0^r |f * \phi_t(y)|^2 \frac{dt}{t}\right)^{\frac{1}{2}} \\ \mathcal{G}_{\phi,1,\infty}(f_k)(y) = \left(\displaystyle\int_r^\infty |f * \phi_t(y)|^2 \frac{dt}{t}\right)^{\frac{1}{2}}, \end{cases}$$

then

$$\max\{\mathcal{G}_{\phi,1,0}(f_k)(y), \mathcal{G}_{\phi,1,\infty}(f_k)(y)\} \le \mathcal{G}_{\phi,1}(f_k)(y) \le \mathcal{G}_{\phi,1,0}(f_k)(y) + \mathcal{G}_{\phi,1,\infty}(f_k)(y).$$

First, since

$$f \in \mathcal{C}^{p,\lambda}(\mathbb{R}^n),$$

for

$$\beta > 0 \quad \& \quad -\frac{\lambda}{p} < \min\{\beta, 1\}$$

one has (cf. [23])

$$\int_{\mathbb{R}^n} \frac{|f(y) - f_B(x,r)|}{r^{n+\beta} + |x-y|^{n+\beta}}\, dy \lesssim r^{-\frac{\lambda}{p}-\beta} \|f\|_{\mathcal{C}^{p,\lambda}(\mathbb{R}^n)}.$$

Moreover,

$$\mathcal{G}_{\phi,1}(f_1) \in L^p(\mathbb{R}^n)$$

implies that $\mathcal{G}_{\phi,1}(f_1)$ exists a. e. in $\mathbb{R}^n$ and

$$\|\mathcal{G}_{\phi,1}(f_1)\|_{L^p(B(x,r))} \lesssim \|f_1\|_{L^p(\mathbb{R}^n)} \lesssim r^{\frac{n-\lambda}{p}} \|f\|_{C^{p,\lambda}(\mathbb{R}^n)},$$

as well as

$$\begin{aligned}
\mathcal{G}_{\phi,1,\infty}(f_1)(y) &\le \left(\int_r^\infty \left(\int_{B(x,2r)} |\phi_t(y-z)||f(z) - f_{B(x,2r)}| \, dz \right)^2 \frac{dt}{t} \right)^{\frac{1}{2}} \\
&\lesssim \left(\int_r^\infty \left| \int_{B(x,2r)} t^{-n} \left(1 + \frac{|y-z|}{t}\right)^{-n-1} |f(z - f_{B(x,2r)}| \, dz) \right|^2 \frac{dt}{t} \right)^{\frac{1}{2}} \\
&\lesssim r^{-\frac{\lambda}{p}} \|f\|_{C^{p,\lambda}(\mathbb{R}^n)}.
\end{aligned}$$

Second, let us prove that

$$y \in B(x,2r) \Rightarrow \begin{cases} \mathcal{G}_{\phi,1,0}(f_2)(y) \lesssim r^{-\frac{\lambda}{p}} \|f\|_{C^{p,\lambda}(\mathbb{R}^n)} \\ |\mathcal{G}_{\phi,1,\infty}(f_2)(y) - \mathcal{G}_{\phi,1,\infty}(f)(x)| \lesssim r^{-\frac{\lambda}{p}} \|f\|_{C^{p,\lambda}(\mathbb{R}^n)}. \end{cases}$$

In fact, if

$$\begin{cases} t < r \\ |y - x| < 2r \\ |z - x| \ge 2r, \end{cases}$$

then

$$t + |y - z| \approx r + |x - y|,$$

and hence

$$\mathcal{G}_{\phi,1,0}(f_2) \lesssim \left(\int_0^r \left| \int_{\mathbb{R}^n} \frac{t|f_2(z)|}{t^{n+1} + |y-z|^{n+1}} \, dz \right|^2 \frac{dt}{t} \right)^{\frac{1}{2}} \lesssim r^{-\frac{\lambda}{p}} \|f\|_{C^{p,\lambda}(\mathbb{R}^n)}.$$

At the same time, we have

$$\begin{aligned}
&|\mathcal{G}_{\phi,1,\infty}(f_2)(y) - \mathcal{G}_{\phi,1,\infty}(f)(x)| \\
&\lesssim \left(\int_r^\infty \left| \int_{\mathbb{R}^n} |\phi_t(y-z)f_2(z) - \phi_t(x-z)f(z)| \, dz \right|^2 \frac{dt}{t} \right)^{\frac{1}{2}}
\end{aligned}$$

$$\lesssim \mathcal{G}_{\phi,1,\infty}(f_1)(x) + \left(\int_r^\infty \left|\int_{\mathbb{R}^n} |\phi_t(y-z) - \phi_t(x-z)||f_2(z)|\,\mathrm{d}z\right|^2 \frac{\mathrm{d}t}{t}\right)^{\frac{1}{2}}$$

$$\lesssim \mathcal{G}_{\phi,1,\infty}(f_1)(x) + \int_{|z-x|\ge 2r} |f(z) - f_{B(x,2r)}| \left(\int_r^\infty |\phi_t(y-z) - \phi_t(x-z)|^2 \frac{\mathrm{d}t}{t}\right)^{\frac{1}{2}} \mathrm{d}z$$

$$\lesssim \mathcal{G}_{\phi,1,\infty}(f_1)(x) + \int_{|z-x|\ge 2r} |f(z) - f_{B(x,2r)}| \left(\int_r^\infty \left(\frac{t^{-1}|y-x|}{(1+t^{-1}|x-z|)^{n+2}}\right)^2 \frac{\mathrm{d}t}{t^{2n+1}}\right)^{\frac{1}{2}} \mathrm{d}z$$

$$\lesssim \mathcal{G}_{\phi,1,\infty}(f_1)(x) + r \int_{|z-x|\ge 2r} \frac{|f(z) - f_{B(x,2r)}|}{|x-z|^{n+1}}\,\mathrm{d}z$$

$$\lesssim r^{-\frac{\lambda}{p}} \|f\|_{\mathrm{C}^{p,\lambda}(\mathbb{R}^n)}.$$

Accordingly, $\mathcal{G}_{\phi,1}(f_2)(y)$ is finite for any $y \in B(x,2r)$—this, plus the a. e.-existence of $\mathcal{G}_{\phi,1}(f_1)$, implies that $\mathcal{G}_{\phi,1}$ is finite a. e. in $B(x,2r)$ and so in $\mathbb{R}^n$.

Finally, upon letting

$$E = \{x \in \mathbb{R}^n : \mathcal{G}_{\phi,1}(f)(x) < \infty\},$$

we are about to verify the boundedness of

$$\mathcal{G}_{\phi,1} \quad \text{on} \quad \mathrm{C}^{p,\lambda}(\mathbb{R}^n).$$

To do so, for $x_0 \in E$ and $r_0 > 0$ set

$$\begin{cases} f = f_{1,0} + f_{2,0} + f_{3,0} \in \mathrm{C}^{p,\lambda}(\mathbb{R}^n) \\ f_{1,0} = (f - f_{B(x_0,2r_0)})1_{B(x_0,2r_0)} \\ f_{2,0} = (f - f_{B(x_0,2r_0)})1_{\mathbb{R}^n \setminus B(x_0,2r_0)} \\ f_{3,0} = f_{B(x_0,2r_0)}. \end{cases}$$

Then

$$\begin{cases} \mathcal{G}_{\phi,1}(f_{3,0}) = 0 \\ \|\mathcal{G}_{\phi,1}(f_{1,0})\|_{L^p(B(x_0,r_0))} \lesssim \|f_{1,0}\|_{L^p(B(x_0,r_0))} \lesssim r_0^{\frac{n-\lambda}{p}} \|f\|_{\mathrm{C}^{p,\lambda}(\mathbb{R}^n)} \\ \sup_{x \in B(x_0,r_0)} \mathcal{G}_{\phi,1,0}(f_2)(x) \lesssim r_0^{-\frac{\lambda}{p}} \|f\|_{\mathrm{C}^{p,\lambda}(\mathbb{R}^n)} \\ \sup_{x \in B(x_0,r_0)} |\mathcal{G}_{\phi,1,\infty}(f_2)(x) - \mathcal{G}_{\phi,1,\infty}(f)(x)| \lesssim r_0^{-\frac{\lambda}{p}} \|f\|_{\mathrm{C}^{p,\lambda}(\mathbb{R}^n)}. \end{cases}$$

Consequently, we arrive at

$$\left(\int_{B(x_0,r_0)} |\mathcal{G}_{\phi,1}(f)(x) - (\mathcal{G}_{\phi,1}(f))_{B(x_0,r_0)}|^p\,\mathrm{d}x\right)^{\frac{1}{p}}$$

$$
\begin{aligned}
&\lesssim \Big(\int\limits_{B(x_0,r_0)} |\mathcal{G}_{\phi,1}(f)(x) - (\mathcal{G}_{\phi,1}(f_{2,0}))_{B(x_0,r_0)}|^p \,dx \Big)^{\frac{1}{p}} \\
&\quad + \Big(\int\limits_{B(x_0,r_0)} |\mathcal{G}_{\phi,1}(f_{2,0})(x) - (\mathcal{G}_{\phi,1,\infty}(f))_{B(x_0,r_0)}|^p \,dx \Big)^{\frac{1}{p}} \\
&\lesssim \|\mathcal{G}_{\phi,1}(f_{1,0})\|_{L^p(B(x_0,r_0))} + \Big(\int\limits_{B(x_0,r_0)} |\mathcal{G}_{\phi,1}(f_{2,0})(x)|^p \,dx \Big)^{\frac{1}{p}} \\
&\quad + r_0^{\frac{n}{p}} \sup_{x\in B(x_0,r_0)} |\mathcal{G}_{\phi,1}(f_{2,0})(x) - \mathcal{G}_{\phi,1}(f)(x_0)| \\
&\lesssim r_0^{\frac{n-\lambda}{p}} \|f\|_{C^{p,\lambda}(\mathbb{R}^n)}.
\end{aligned}
$$

□

Utilizing the Poisson integral situation of Theorem 4.1 and the Littlewood–Paley $\mathcal{G}$-function used in Theorem 4.4 we can discover a relationship between $Q_\alpha(\mathbb{R}^n)$ and the square Campanato space $C^{2,2\alpha}(\mathbb{R}^n)$.

Theorem 4.5. *Let* $(\alpha, n-1) \in (0,1) \times \mathbb{N}$. *Then:*

(i)

$$Q_\alpha(\mathbb{R}^n) = (-\Delta)^{-\frac{\alpha}{2}} C^{2,2\alpha}(\mathbb{R}^n);$$

(ii) *Consequently, if* $\psi \in S(\mathbb{R}^n)$ *satisfies*

$$
\begin{cases}
\operatorname{supp}(\hat{\psi}) \subseteq \{x \in \mathbb{R}^n : 2^{-1} \le |x| \le 2\} \\
|\hat{\psi}(x)| \gtrsim 1 \quad \forall \frac{3}{5} \le |x| \le \frac{5}{3} \\
\psi_j(x) = 2^{jn}\psi(2^j x) \quad \forall (j,x) \in \mathbb{Z} \times \mathbb{R}^n \\
\sum_{j\in\mathbb{Z}} \hat{\psi}_j = 1,
\end{cases}
$$

then

$$
\begin{cases}
Q_\alpha(\mathbb{R}^n) \\
C^{2,2\alpha}(\mathbb{R}^n)
\end{cases}
$$

has the Littlewood–Paley-type characterization

$$
\begin{cases}
\|f\|_{Q_\alpha(\mathbb{R}^n)} \approx \sup_{I\in\mathscr{C}(\mathbb{R}^n)} \Big(\ell(I)^{2\alpha-n} \sum_{j\ge -\log_2 \ell(I)} 2^{2\alpha j} \|f * \psi_j\|_{L^2(I)}^2 \Big)^{\frac{1}{2}} \\
\|f\|_{C^{2,2\alpha}(\mathbb{R}^n)} \approx \sup_{I\in\mathscr{C}(\mathbb{R}^n)} \Big(\ell(I)^{2\alpha-n} \sum_{j\ge -\log_2 \ell(I)} \|f * \psi_j\|_{L^2(I)}^2 \Big)^{\frac{1}{2}}.
\end{cases}
$$

Proof. (i) The argument consists of two steps.

Step 1. We prove that

$$f \in \mathrm{C}^{2,2\alpha}(\mathbb{R}^n) \Leftrightarrow |||f|||_{\mathrm{C}^{2,2\alpha}(\mathbb{R}^n)} = \sup_{(x,r)\in\mathbb{R}^{n+1}_+} \left(r^{2\alpha-n} \int_0^r \int_{B(x,r)} |t\partial_t e^{-t\sqrt{-\Delta}} f(y)|^2 \frac{dy\,dt}{t} \right)^{\frac{1}{2}} < \infty.$$

On the one hand, assume that

$$\begin{cases} f \in \mathrm{C}^{2,2\alpha}(\mathbb{R}^n) \\ f_1 = (f - f_{B(x,2r)})1_{B(x,2r)} \\ f_2 = (f - f_{B(x,2r)})1_{\mathbb{R}^n\setminus B(x,2r)} \\ f_3 = f_{B(x,2r)}. \end{cases}$$

Note that

$$\partial_t e^{-t\sqrt{-\Delta}} f(y) = \partial_t e^{-t\sqrt{-\Delta}} f_1(y) + \partial_t e^{-t\sqrt{-\Delta}} f_2(y).$$

So, using the well-known $\mathrm{L}^2(\mathbb{R}^n)$-boundedness

$$\|\mathcal{G}(f_1)\|_{\mathrm{L}^2(\mathbb{R}^n)} \lesssim \|f_1\|_{\mathrm{L}^2(\mathbb{R}^n)}$$

for the Littlewood–Paley $\mathcal{G}$-function of f_1 determined by

$$\mathcal{G}(f_1)(y) = \left(\int_0^\infty |t\partial_t e^{-t\sqrt{-\Delta}} f_1(y)|^2 \frac{dt}{t} \right)^{\frac{1}{2}},$$

we get

$$\begin{aligned} \int_0^r \left(\int_{B(x,r)} |t\partial_t e^{-t\sqrt{-\Delta}} f_1(y)|^2 \, dy \right) \frac{dt}{t} &\le \int_{\mathbb{R}^n} \left(\int_0^\infty |t\partial_t e^{-t\sqrt{-\Delta}} f_1(y)|^2 \frac{dt}{t} \right) dy \\ &\lesssim \|f\|^2_{\mathrm{C}^{2,2\alpha}(\mathbb{R}^n)} r^{n-2\alpha}. \end{aligned}$$

Meanwhile, using the Cauchy–Schwarz inequality, we obtain that if $(y,t) \in B(x,r) \times (0,r)$ then

$$|\partial_t e^{-t\sqrt{-\Delta}} f_2(y)| \lesssim \sum_{k=1}^\infty \int_{B(x,2^k r)\setminus B(x,2^{k-1}r)} \frac{|f(z) - f_{B(x,2r)}|}{|x-z|^{n+1}} \, dz \lesssim r^{-(1+\alpha)} \|f\|_{\mathrm{C}^{2,2\alpha}(\mathbb{R}^n)}.$$

This in turn implies

$$\int_0^r \int_{B(x,r)} |t\partial_t e^{t^2\Delta} f_1(y)|^2 t^{-1} \, dy\,dt \lesssim r^{n-2\alpha} \|f\|^2_{\mathrm{C}^{2,2\alpha}(\mathbb{R}^n)}.$$

The above estimates give

$$|||f|||_{C^{2,2\alpha}(\mathbb{R}^n)} \lesssim \|f\|_{C^{2,2\alpha}(\mathbb{R}^n)}.$$

On the other hand, suppose

$$|||f|||_{C^{2,2\alpha}(\mathbb{R}^n)} < \infty.$$

Then

$$|||f|||_{C^{2,2\alpha}(\mathbb{R}^n),1} = \sup_{I\in\mathscr{C}(\mathbb{R}^n)}\left(\ell(I)^{2\alpha-n}\int_0^{\ell(I)}\int_I \left|\partial_t e^{-t\sqrt{-\Delta}}f(y)\right|^2 t\,\mathrm{d}y\,\mathrm{d}t\right)^{\frac{1}{2}} \approx |||f|||_{C^{2,2\alpha}(\mathbb{R}^n)} < \infty.$$

Set

$$\begin{cases} \psi(x) = \dfrac{1+|x|^2-(n+1)\Gamma(\frac{n+1}{2})\pi^{-\frac{n+1}{2}}}{(1+|x|^2)^{\frac{n+3}{2}}} \\ \psi_t(x) = t^{-n}\psi\left(\dfrac{x}{t}\right) \\ \Upsilon_\psi F(x) = \displaystyle\int_{\mathbb{R}^{n+1}_+} F(y,t)\psi_t(x-y)\,\frac{\mathrm{d}y\,\mathrm{d}t}{t}. \end{cases}$$

Upon demonstrating that

$$\sup_{I\in\mathscr{C}(\mathbb{R}^n)}\left(\ell(I)^{2\alpha-n}\int_\alpha^{\ell(I)}\int_I |F(y,t)|^2\frac{\mathrm{d}y\,\mathrm{d}t}{t}\right)^{\frac{1}{2}} < \infty$$

implies

$$\sup_{J\in\mathscr{C}(\mathbb{R}^n)}\left(\ell(J)^{2\alpha-n}\int_J |\Upsilon_\psi F(x)-(\Upsilon_\psi F)_J|^2\,\mathrm{d}x\right)^{\frac{1}{2}} < \infty,$$

we can use

$$F(x,t) = \psi_t * f(x) = t\partial_t e^{-t\sqrt{-\Delta}}f(y) \quad \& \quad \Upsilon_\psi(\psi_t * f)(x) = f(x)$$

to find

$$\|f\|_{C^{2,2\alpha}(\mathbb{R}^n)} \lesssim |||f|||_{C^{2,2\alpha}(\mathbb{R}^n)}.$$

To do so, given $J \in \mathscr{C}(\mathbb{R}^n)$ let

$$F_1(y,t) = F(y,t)|_{S(2J)} \quad \& \quad F_2(y,t) = F(y,t) - F_1(y,t).$$

Then, from a result of Coifman–Mayer–Stein [16, pp. 328–329] it follows that

$$\int_J |\Upsilon_\psi F_1(x)|^2\, dx \lesssim \ell(J)^{n-2\alpha} \sup_{I\in\mathscr{C}(\mathbb{R}^n)} \left(\ell(I)^{2\alpha-n} \int_0^{\ell(I)} \int_I |F(y,t)|^2 \frac{dy\, dt}{t} \right).$$

Also, from the Cauchy–Schwarz inequality it follows that

$$\begin{aligned}\int_J |\Upsilon_\psi F_2(x)|^2\, dx &\le \int_J \left(\sum_{k=1}^{\infty} \iint_{S(2^{k+1}J)\setminus S(2^kJ)} |\psi_t(x-y)||F_2(y,t)| \frac{dy\, dt}{t} \right)^2 dx \\ &\lesssim \int_J \left(\sum_{k=1}^{\infty} (2^k\ell(J))^{-(n+1)} \iint_{S(2^{k+1}J)\setminus S(2^kJ)} |F_2(y,t)|\, dy\, dt \right)^2 dx \\ &\lesssim \int_J \left(\sum_{k=1}^{\infty} (2^k\ell(J))^{-\frac{n}{2}} \left(\iint_{S(2^{k+1}J)\setminus S(2^kJ)} |F_2(y,t)|^2 t^{-1}\, dy\, dt \right)^{\frac12} \right)^2 dx \\ &\lesssim \ell(J)^{n-2\alpha} \sup_{I\in\mathscr{C}(\mathbb{R}^n)} \left(\ell(I)^{2\alpha-n} \int_0^{\ell(I)} \int_I |F(y,t)|^2 \frac{dy\, dt}{t} \right).\end{aligned}$$

Accordingly, we get the desired estimate

$$\begin{aligned}\int_J |\Upsilon_\psi F(x) - (\Upsilon_\psi F)_J|^2\, dx &\lesssim \int_J (|\Upsilon_\psi F_1|^2 + |\Upsilon_\psi F_2|^2)(x)\, dx \\ &\lesssim \ell(J)^{n-2\alpha} \sup_{I\in\mathscr{C}(\mathbb{R}^n)} \left(\ell(I)^{2\alpha-n} \int_0^{\ell(I)} \int_I |F(y,t)|^2 \frac{dy\, dt}{t} \right).\end{aligned}$$

Step 2. We are about to verify

$$Q_\alpha(\mathbb{R}^n) = (-\Delta)^{-\frac{\alpha}{2}} C^{2,2\alpha}(\mathbb{R}^n).$$

For $f \in C^{2,2\alpha}(\mathbb{R}^n)$ let

$$F(y,t) = t^{1+\alpha} \partial_t e^{-t\sqrt{-\Delta}} f(y).$$

Then

$$\sup_{(x,r)\in\mathbb{R}^{n+1}_+} r^{2\alpha-n} \int_0^r \left(\int_{B(x,r)} |F(y,t)|^2\, dy \right) \frac{dt}{t^{1+2\alpha}} \lesssim \|f\|^2_{C^{2,2\alpha}(\mathbb{R}^n)},$$

and so Theorem 4.1 yields

$$\Upsilon_\psi F \in Q_\alpha(\mathbb{R}^n).$$

Since

$$\widehat{F(\cdot,t)}(x) = -t^{\alpha+2}|x|\hat{f}(x)\exp(-t|x|),$$

a computation derives

$$\widehat{\Upsilon_\psi F}(x) = \left(\frac{\Gamma(2+\alpha)}{2^{2+\alpha}}\right)\frac{\hat{f}(x)}{|x|^\alpha} = 2^{-2}\pi^\alpha\Gamma(2+\alpha)\widehat{(-\Delta)^{\frac{\alpha}{2}}f}(x),$$

whence

$$(-\Delta)^{-\frac{\alpha}{2}}f \in \mathrm{Q}_\alpha(\mathbb{R}^n).$$

Conversely, if $g \in \mathrm{Q}_\alpha(\mathbb{R}^n)$, then from Theorem 4.1 and the argument of *Step 1* it follows that if

$$G(y,t) = t^{1-\alpha}\partial_t e^{-t\sqrt{-\Delta}}g(y)$$

then

$$\Upsilon_\psi G \in \mathrm{C}^{2,2\alpha}(\mathbb{R}^n).$$

Due to

$$\widehat{\Upsilon_\psi G} = 2^{-2}\pi^{-\alpha}\Gamma(2-\alpha)\widehat{(-\Delta)^{\frac{\alpha}{2}}g},$$

we get

$$f = (-\Delta)^{\frac{\alpha}{2}}g \in \mathrm{C}^{2,2\alpha}(\mathbb{R}^n),$$

equivalently

$$g = (-\Delta)^{-\frac{\alpha}{2}}f \in (-\Delta)^{-\frac{\alpha}{2}}\mathrm{C}^{2,2\alpha}(\mathbb{R}^n).$$

(ii) The equivalence

$$f \in \mathrm{C}^{2,2\alpha}(\mathbb{R}^n) \Leftrightarrow \sup_{I\in\mathscr{C}(\mathbb{R}^n)}\left(\ell(I)^{2\alpha-n}\sum_{j\ge-\log_2\ell(I)}\|f*\psi_j\|^2_{\mathrm{L}^2(I)}\right)^{\frac{1}{2}} < \infty$$

follows from the foregoing *Step 1* or the forthcoming demonstration. In accordance with [4, Theorem 10(i)], we have

$$f \in (-\Delta)^{-\frac{\alpha}{2}}\mathrm{C}^{2,2\alpha}(\mathbb{R}^n) \Leftrightarrow \sup_{I\in\mathscr{C}(\mathbb{R}^n)}\left(\ell(I)^{2\alpha-n}\sum_{j\ge-\log_2\ell(I)}2^{2\alpha j}\|f*\psi_j\|^2_{\mathrm{L}^2(I)}\right)^{\frac{1}{2}} < \infty,$$

thereby reaching the first equivalence in (ii) via using (i). Nevertheless, below is a direct demonstration.

On the one hand, if

$$(f,I,J) \in \mathrm{Q}_\alpha(\mathbb{R}^n)\times\mathscr{C}(\mathbb{R}^n)\times\mathscr{D}_k(I)\quad \forall k+1\in\mathbb{N},$$

then

$$f=(f-f_{2J})\mathbb{1}_{2J}+(f-f_{2J})\mathbb{1}_{\mathbb{R}^n\setminus 2J}+f_{2J},$$

and hence

$$\sum_{k\ge 0}2^{(2\alpha-n)k}\sum_{J\in\mathscr{D}_k(I)}\ell(J)^{-n}\sum_{j\ge -\log_2\ell(J)}\|f*\psi_j\|^2_{\mathrm{L}^2(J)}\lesssim\sum_{(1)}+\sum_{(2)}+\sum_{(3)},$$

where

$$\begin{cases}\sum_{(1)}=\sum_{k\ge 0}2^{(2\alpha-n)k}\sum_{J\in\mathscr{D}_k(I)}\ell(J)^{-n}\sum_{j\ge -\log_2\ell(J)}\|((f-f_{2J})\mathbb{1}_{2J})*\psi_j\|^2_{\mathrm{L}^2(J)}\\ \sum_{(2)}=\sum_{k\ge 0}2^{(2\alpha-n)k}\sum_{J\in\mathscr{D}_k(I)}\ell(J)^{-n}\sum_{j\ge -\log_2\ell(J)}\|((f-f_{2J})\mathbb{1}_{\mathbb{R}^n\setminus 2J})*\psi_j\|^2_{\mathrm{L}^2(J)}\\ \sum_{(3)}=\sum_{k\ge 0}2^{(2\alpha-n)k}\sum_{J\in\mathscr{D}_k(I)}\ell(J)^{-n}\sum_{j\ge -\log_2\ell(J)}\|f_{2J}*\psi_j\|^2_{\mathrm{L}^2(J)}.\end{cases}$$

Since

$$\sum_{(3)}=0,$$

it remains to handle

$$\sum_{(1)}\ \&\ \sum_{(2)}.$$

Using the Plancherel formula, we get

$$\begin{aligned}\sum_{(1)}&\le\sum_{k\ge 0}2^{(2\alpha-n)k}\sum_{J\in\mathscr{D}_k(I)}\ell(J)^{-n}\|(f-f_{2J})\mathbb{1}_{2J}\|^2_{\mathrm{L}^2(\mathbb{R}^n)}\\&\lesssim\sum_{k\ge 0}2^{(2\alpha-n)k}\sum_{J\in\mathscr{D}_k(I)}\ell(J)^{-n}\|(f-f_{2J})\mathbb{1}_{2J}\|^2_{\mathrm{L}^2(\mathbb{R}^n)}\\&\lesssim\|f\|^2_{\mathrm{Q}_\alpha(\mathbb{R}^n)}.\end{aligned}$$

Note also that there is a constant $N>n=2\alpha$ such that

$$|\psi_j(x-y)|\lesssim\frac{2^{jn}}{(1+|x-y|2^j)^{N+n}}.$$

So for $x\in J$, one has

$$|(f-f_{2J})\mathbb{1}_{\mathbb{R}^n\setminus 2J}*\psi_j(x)|\lesssim\sum_{l\ge 1}\int_{2^{l+1}J\cap(\mathbb{R}^n\setminus 2^lJ)}|\psi_j(x-y)||f(y)-f_{2J}|\,\mathrm{d}y$$

$$\lesssim 2^{-jN}\ell(J)^{-N}\sum_{l\geq 1}2^{-lN}|f-f_{2J}|_{2^{l+1}J}$$
$$\lesssim 2^{-jN}\ell(J)^{-N}\Big(\sum_{l\geq 1}2^{-lN}\Big)^{\frac{1}{2}}\Big(\sum_{l\geq 1}2^{-lN}|f-f_{2J}|^2_{2^{l+1}J}\Big)^{\frac{1}{2}}.$$

This, plus the argument for Theorem 1.6(i) (or [22, Lemma 5.3]), implies

$$\begin{aligned}\sum_{(2)} &\lesssim \sum_{k\geq 0}2^{(2\alpha-n)k}\sum_{J\in\mathscr{D}_k(I)}\Big(\sum_{l\geq 1}2^{-lN}(2^{l+1}\ell(J))^{-n}\int_{2^{l+1}J}|f-f_{2J}|\,dy\Big)^2\\ &\lesssim \sum_{l\geq 1}2^{-l(N+n)}\sum_{k\geq 0}\sum_{J\in\mathscr{D}_k(I)}\ell(J)^{-2n}\int_{2^{l+1}J}\int_{2^{l+1}J}|f(x)-f(y)|^2\,dxdy\\ &\lesssim \sum_{l\geq 1}2^{-l(N-2\alpha-n)}\|f\|^2_{Q_\alpha(\mathbb{R}^n)}\\ &\lesssim \|f\|^2_{Q_\alpha(\mathbb{R}^n)}.\end{aligned}$$

Consequently,

$$\ell(I)^{2\alpha-n}\sum_{j\geq -\log_2\ell(I)}2^{2\alpha j}\|f*\psi_j\|^2_{L^2(I)}\lesssim \sum_{(1)}+\sum_{(2)}+\sum_{(3)}\lesssim \|f\|^2_{Q_\alpha(\mathbb{R}^n)}.$$

On the other hand, suppose

$$\|f\|_{Q_\alpha(\mathbb{R}^n),\dagger}=\sup_{I\in\mathscr{C}(\mathbb{R}^n)}\Big(\ell(I)^{2\alpha-n}\sum_{j\geq -\log_2\ell(I)}2^{2\alpha j}\|f*\psi_j\|^2_{L^2(I)}\Big)^{\frac{1}{2}}<\infty.$$

Note that

$$f\in L^2_{\mathrm{loc}}(\mathbb{R}^n)$$

can be written as

$$f=\sum_{j\in\mathbb{Z}}f*\psi_f=\sum_{j>-\log_2\ell(I)}f*\psi_f+\sum_{j\leq -\log_2\ell(I)}f*\psi_f\quad\forall I\in\mathscr{C}(\mathbb{R}^n).$$

So for $I\in\mathscr{C}(\mathbb{R}^n)$, one has

$$\Theta_{f,\alpha}(I)\lesssim\Theta^{(1)}_{f,\alpha}(I)+\Theta^{(2)}_{f,\alpha}(I),$$

where

$$\begin{cases}\Theta^{(1)}_{f,\alpha}(I)=\ell(I)^{2\alpha-n}\displaystyle\int_I\int_I\Big|\sum_{j<-\log_2\ell(I)}(f*\psi_j(x)-f*\psi_j(y))\Big|^2|x-y|^{-n-2\alpha}\,dxdy\\ \Theta^{(2)}_{f,\alpha}(I)=\ell(I)^{2\alpha-n}\displaystyle\int_I\int_I\Big|\sum_{j\geq -\log_2\ell(I)}(f*\psi_j(x)-f*\psi_j(y))\Big|^2|x-y|^{-n-2\alpha}\,dxdy.\end{cases}$$

Since

$$\sum_{j<-\log_2 \ell(I)} \max_{x\in I} |\nabla f * \psi_j(x)| \lesssim \|f\|_{\mathrm{BMO}(\mathbb{R}^n)} \ell(I)^{-1},$$

one gets

$$\Theta^{(1)}_{f,\alpha}(I) \lesssim \|f\|^2_{\mathrm{BMO}(\mathbb{R}^n)} \ell(I)^{2\alpha-n-2} \int_I \int_I |x-y|^{2-2\alpha-n}\, dxdy \lesssim \|f\|^2_{Q_\alpha(\mathbb{R}^n),\dagger}.$$

To control $\Theta^{(2)}_{f,\alpha}(I)$, let us choose a nonnegative radial function $\phi \in C_0^\infty(\mathbb{R}^n)$ enjoying

$$\begin{cases} \check{\phi}(x) = \mathcal{F}^{-1}\phi(x) \geq 1 \quad \forall\, |x| \leq \frac{1}{2} \\ \mathrm{supp}(\phi) \subseteq \left\{x \in \mathbb{R}^n : |x| \leq \frac{1}{2}\right\} \\ \sigma_I(x) = \ell(I)^{\alpha-\frac{n}{2}} \check{\phi}\left(\frac{\pi(x-c_I)}{\ell(I)}\right), \end{cases}$$

where c_I is the center of I. Clearly, σ_I satisfies

$$\begin{cases} |\sigma_I(x)|^2 \gtrsim \ell(I)^{2\alpha-n} \quad \forall\, x \in I \\ \mathrm{supp}(\hat{\sigma}_I) \subseteq \left\{x \in \mathbb{R}^n : |x| \leq \frac{1}{2\ell(I)}\right\}. \end{cases}$$

Since

$$|1 - e^{-2\pi i y\cdot\xi}| \lesssim \min\{1, (|y||\xi|)^{\alpha_0}\}$$

holds for arbitrarily given $\alpha_0 \in (\alpha, 1)$, it follows that

$$\int_{|y|\leq \ell(I)} \frac{|1 - e^{-2\pi i y\cdot\xi}|}{|y|^{n+2\alpha}}\, dy \lesssim |\xi|^{2\alpha}.$$

Upon defining $\varsigma_j(f)$ via

$$\widehat{\varsigma_j(f)}(\xi) = |2^{-j}\xi|^\alpha \hat{\psi}(2^{-j}\xi)\hat{f}(\xi),$$

we have

$$\left| \sum_{j\geq -\log_2 \ell(I)} \hat{\psi}_I * \widehat{\varsigma_j(f)}(\xi)\right|^2 \lesssim \sum_{j\geq -\log_2 \ell(I)} |\hat{\psi}_I * \widehat{\varsigma_j(f)}(\xi)|^2,$$

thereby finding

$$\Theta^{(2)}_{f,\alpha}(I) \lesssim \ell(I)^{2\alpha-n} \int_{|y|\leq \ell(I)} \int_I \left| \sum_{j\geq -\log_2 \ell(I)} (f * \psi_j(x) - f * \psi_j(x+y))\right|^2 |y|^{-n-2\alpha}\, dxdy$$

$$\lesssim \int_{|y|\le \ell(I)} \int_{\mathbb{R}^n} |\phi_I(x)|^2 \left| \sum_{j\ge -\log_2 \ell(I)} (f * \psi_j(x) - f * \psi_j(x+y)) \right|^2 |y|^{-n-2\alpha}\, dxdy$$

$$\lesssim \sum_{j\ge -\log_2 \ell(I)} 2^{2\alpha j} \int_{\mathbb{R}^n} |\widehat{\phi_I} * \widehat{\varsigma_j(f)}(\xi)|^2\, d\xi$$

$$\lesssim \sum_{j\ge -\log_2 \ell(I)} 2^{2\alpha j} \int_{\mathbb{R}^n} |\phi_I(x)\widehat{\varsigma_j(f)}(x)|^2\, dx$$

$$\lesssim \sup_{I\in\mathscr{C}(\mathbb{R}^n)} \ell(I)^{2\alpha-n} \sum_{j\ge -\log_2 \ell(I)} 2^{2\alpha j} \|\varsigma_j(f)\|^2_{L^2(I)}$$

$$\lesssim \sup_{I\in\mathscr{C}(\mathbb{R}^n)} \ell(I)^{2\alpha-n} \sum_{j\ge -\log_2 \ell(I)} 2^{2\alpha j} \|f * \psi_j\|^2_{L^2(I)}.$$

Putting

$$\Theta^{(1)}_{f,\alpha}(I) \quad \& \quad \Theta^{(2)}_{f,\alpha}(I)$$

together deduces

$$\Theta_{f,\alpha}(I) \lesssim \|f\|^2_{Q_\alpha(\mathbb{R}^n),\dagger}. \qquad \square$$

4.4 Notes

Below are some historical remarks on Chapter 4.

Section 4.1. Theorem 4.1 comes from [22, Lemmas 3.1 and 4.1 and Theorems 3.2, 4.2 and 4.5]. Although the case $\alpha \in (-\infty, 0)$ in Theorem 4.1 corresponds to the well-known Carleson measure characterization of $\mathrm{BMO}(\mathbb{R}^n)$, according to [96, Proposition 2.1] we have that if

$$F(y,t) = \text{either } e^{t^2\Delta} f(y) \text{ or } e^{-t\sqrt{-\Delta}} f(y)$$

then

$$\|f\|_{Q_\alpha(\mathbb{R}^n)} \lesssim \sup_{I\in\mathscr{C}(\mathbb{R}^n)} \left(\int_0^{\ell(I)} \left(\int_I |\nabla F(y,t)|^2 \left(\ln \frac{e\ell(I)}{t} \right)^2 dy \right) \frac{t\,dt}{\ell(I)^n} \right)^{\frac{1}{2}}$$

$$\lesssim \sup_{I\in\mathscr{C}(\mathbb{R}^n)} \left(\int_{B(0,\ell(I))} \left(\int_I |f(x+y) - f(x)|^2 dx \right) \left| \ln \frac{e\ell(I)}{|y|} \right|^3 \frac{dy}{(\ell(I)|y|)^n} \right)^{\frac{1}{2}}.$$

The foregoing inequality can be established via an application of the following Hardy inequality (cf. [3, Lemma 11] or [52]) – if $(p,b) \in (1,\infty) \times (0,\infty]$ and $\{\mu, \nu\}$ is a pair of two measurable nonnegative functions on $(0,b)$, then:

▷

$$\int_0^b \left(\int_0^s f(t)\,\mathrm{d}t\right)^p \mu(s)\,\mathrm{d}s \lesssim \int_0^b (f(s))^p v(s)\,\mathrm{d}s$$

holds for any measurable nonnegative function f on $(0,b)$ when and only when

$$\sup_{s\in(0,b)} \left(\int_s^b \mu(t)\,\mathrm{d}t\right)^{\frac{1}{p}} \left(\int_0^s (v(t))^{\frac{1}{1-p}}\,\mathrm{d}t\right)^{\frac{p-1}{p}} < \infty;$$

▷

$$\int_0^b \left(\int_s^b f(t)\,\mathrm{d}t\right)^p \mu(s)\,\mathrm{d}s \lesssim \int_0^b (f(s))^p v(s)\,\mathrm{d}s$$

holds for any measurable and nonnegative function f on $(0,b)$ when and only when

$$\sup_{s\in(0,b)} \left(\int_0^s \mu(t)\,\mathrm{d}t\right)^{\frac{1}{p}} \left(\int_s^b (v(t))^{\frac{1}{1-p}}\,\mathrm{d}t\right)^{\frac{p-1}{p}} < \infty.$$

Section 4.2. The case $\alpha \in (-\infty, 0)$ of Proposition 4.2 is well known; see [46]. The case $\alpha \in (0,1)$ of this proposition comes from [92, Theorem 1.2(iii)] (cf. [93]). Of course, this can be pushed to the endpoint case $\alpha = 0$ through bringing

$$\sup_{(x,r)\in\mathbb{R}^{n+1}_+} \left(\int_0^{r^2} \left(\int_{B(x,r)} |e^{t\Delta}f(y)|^2 \left(\ln \frac{er}{\sqrt{t}}\right)^2 dy\right) \frac{dt}{r^n}\right)^{\frac{1}{2}}$$

into play; see also [96, Proposition 2.2].

Section 4.3. Theorem 4.4 comes from shortening [81, Theorems 1, 2 and 3]; Theorem 4.5(i) is a combination of Lemma 2.1 and Theorem 1.2(i) in [92] (cf. [93]), and yet Theorem 4.5(ii) exists as a consequence of Theorem 4.5(i) and [4, Proposition 8(i) and Theorem 10(i)] and its current proof is suitably taken from [59]. Three more comments on Theorem 4.5 are in order.

▷ According to [37, Theorem 1.2(i) and Theorem 1.3(ii)] and Theorem 4.5, we have that if

$$\mathrm{HQ}_\alpha(\mathbb{R}^{n+1}_+) \quad \text{or} \quad \mathrm{CQ}_\alpha(\mathbb{R}^{n+1}_+)$$

stands for the class of all harmonic or caloric functions u on $\mathbb{R}^{n+1}_+$ with

$$\|u\|_{\mathrm{HQ}_\alpha(\mathbb{R}^{n+1}_+)} = \sup_{(x_0,r_0)\in\mathbb{R}^{n+1}_+} \left(r_0^{2\alpha-n} \int_0^{r_0} \int_{B(x_0,r_0)} |\nabla u(x,t)|^2 t^{1-2\alpha} \mathrm{d}x\,\mathrm{d}t\right)^{\frac{1}{2}} < \infty$$

or

$$\|u\|_{\mathrm{CQ}_\alpha(\mathbb{R}^{n+1}_+)} = \sup_{(x_0,r_0)\in\mathbb{R}^{n+1}_+} \left(r_0^{2\alpha-n} \int_0^{r_0} \int_{B(x_0,r_0)} |\nabla u(x,t)|^2 t^{1-2\alpha} dx\, dt \right)^{\frac{1}{2}} < \infty$$

then

$$u \in \mathrm{HQ}_\alpha(\mathbb{R}^{n+1}_+) \quad \text{or} \quad u \in \mathrm{CQ}_\alpha(\mathbb{R}^{n+1}_+)$$

if and only if there is some $f \in \mathrm{Q}_\alpha(\mathbb{R}^n)$ such that

$$u(x,t) = e^{-t\sqrt{-\Delta}} f(x) \quad \text{or} \quad u(x,t^2) = e^{t^2\Delta} f(x).$$

▷ According to [80, Theorems 3.2, 3.3 and 3.4], if the BMO-Sobolev space

$$(-\Delta)^{-\frac{\alpha}{2}} \mathrm{BMO}(\mathbb{R}^n)$$

expresses the image of $\mathrm{BMO}(\mathbb{R}^n)$ under the $[0,1] \ni \alpha$-Riesz potential

$$(-\Delta)^{-\frac{\alpha}{2}} f = \mathcal{F}^{-1}(|\cdot|^{-\alpha} \hat{f}(\cdot))$$

which is defined on tempered distributions module linear functions and where $\mathcal{F}^{-1}$ stands for the inverse Fourier transform, and if the homogeneous Besov space $\dot{\mathrm{B}}^\alpha_{\infty,2}(\mathbb{R}^n)$ is defined for $\alpha \in (0,1)$ modulo $\mathrm{K}(\mathbb{R}^n)$ by the following condition (cf. [26, Chapter 5])

$$\|f\|_{\dot{\mathrm{B}}^\alpha_{\infty,2}(\mathbb{R}^n)} = \left(\int_{\mathbb{R}^n} \frac{\|f(\cdot+y) - f(\cdot)\|^2_{\mathrm{L}^\infty(\mathbb{R}^n)}}{|y|^{n+2\alpha}} dy \right)^{\frac{1}{2}} < \infty,$$

then:

(i)

$$(-\Delta)^{-\frac{0}{2}} \mathrm{BMO}(\mathbb{R}^n) = \mathrm{BMO}(\mathbb{R}^n);$$

(ii)

$$f \in (-\Delta)^{-\frac{\alpha}{2}} \mathrm{BMO}(\mathbb{R}^n) \Leftrightarrow \sup_{I\in\mathscr{C}(\mathbb{R}^n)} \left(\ell(I)^{-n} \int_I \int_I \frac{|f(x)-f(y)|^2}{|x-y|^{n+2\alpha}} dx dy \right)^{\frac{1}{2}} < \infty;$$

(iii)

$$f \in (-\Delta)^{-\frac{1}{2}} \mathrm{BMO}(\mathbb{R}^n) \Leftrightarrow \sup_{|y|\neq 0} \left\| \frac{f(\cdot+y) - f(\cdot)}{|y|} \right\|_{\mathrm{BMO}(\mathbb{R}^n)} < \infty;$$

(iv)

$$\dot{\mathrm{B}}^\alpha_{\infty,2}(\mathbb{R}^n) \subseteq (-\Delta)^{-\frac{\alpha}{2}} \mathrm{BMO}(\mathbb{R}^n) \subseteq \dot{\mathrm{B}}^\alpha_{\infty,\infty}(\mathbb{R}^n).$$

▷ According to [88, Theorem 4], if

$$(\alpha,\beta) \in (0,1) \times (-1,1) \quad \& \quad \alpha - \beta < 1$$

and $Q_{\beta,\alpha}(\mathbb{R}^n)$ is the class of all real-valued functions f on $\mathbb{R}^n$ obeying

$$\sup_{(x,r)\in\mathbb{R}^{n+1}_+} r^{2\beta-n} \int_{B(x,r)} \int_{B(x,r)} \frac{|f(y) - f(z)|^2}{|y - z|^{n+2\alpha}} dydz < \infty,$$

then

$$Q_{\beta,\alpha}(\mathbb{R}^n) = (-\Delta)^{-\frac{\alpha}{2}} C^{2,2\beta}(\mathbb{R}^n),$$

which is the square Campanato–Sobolev space.

5 Preduality and decomposition

With the help of Chapters 3–4, we can solve not only [22, problem 8.2] on finding a predual of $\mathrm{Q}_\alpha(\mathbb{R}^n)$ via the so-called Hardy–Hausdorff space but also [22, problem 8.3] about giving a Fefferman–Stein type decomposition of $\mathrm{Q}_\alpha(\mathbb{R}^n)$ via

$$\mathrm{Q}_\alpha(\mathbb{R}^n) \cap \mathrm{L}^\infty(\mathbb{R}^n)$$

and its Riesz transforms.

5.1 Hardy–Hausdorff space as a predual

Such a Hardy–Hausdorff space is determined below.

Definition 5.1. Let $(\alpha, n-1) \in (-\infty, \infty) \times \mathbb{N}$ and $\mathrm{S}(\mathbb{R}^n)$ be the Schwarz class.

(i) For $E \subseteq \mathbb{R}^n$ let $\dot{\mathrm{L}}^2_\alpha(E)$ be the homogeneous of functions f on E with

$$\|f\|_{\dot{\mathrm{L}}^2_\alpha(E)} = \Big(\int_E |\hat{f}(x)|^2 |x|^{2\alpha}\,dx\Big)^{\frac{1}{2}} < \infty.$$

In particular, $\dot{\mathrm{L}}^2_0(\mathbb{R}^n)$ is identified with $\mathrm{L}^2(\mathbb{R}^n)$.

(ii) $\mathrm{HH}^1_{-\alpha}(\mathbb{R}^n)$ with $0 < \alpha < 1$ is the so-called Hardy–Hausdorff class of all distributions $f \in \dot{\mathrm{L}}^2_{-\frac{n}{2}}(\mathbb{R}^n)$ with

$$\|f\|_{\mathrm{HH}^1_{-\alpha}(\mathbb{R}^n)} = \|t\nabla(e^{-t\sqrt{-\Delta}} f)(x)\|_{\mathrm{T}^1_{n-2\alpha}(\mathbb{R}^{n+1}_+)} < \infty.$$

In particular, $\mathrm{HH}^1_0(\mathbb{R}^n)$ is an alternative of the classical Hardy space $\mathrm{H}^1(\mathbb{R}^n)$.

(iii) A tempered distribution a is called an $\mathrm{HH}^1_{-\alpha}$-atom provided that there is an $I \in \mathscr{C}(\mathbb{R}^n)$ such that:

▷
$$\operatorname{supp}(a) \subseteq I;$$

▷
$$|\langle a, \phi\rangle| \le \int_0^{\ell(I)} \int_I |\nabla(e^{-t\sqrt{-\Delta}}\phi)(x)|^2 \Big(\frac{t}{\ell(I)}\Big)^{n-2\alpha} \frac{dx\,dt}{t^{n-1}} \quad \forall \phi \in \mathrm{S}(\mathbb{R}^n);$$

▷
$$\langle a, \phi\rangle = 0 \quad \forall \phi \in \mathrm{S}(\mathbb{R}^n)$$

with its restriction to a neighborhood of I being a polynomial of degree up to $1 + \frac{n}{2}$. In particular, an HH^1_{-0}-atom is essentially a classical H^1-atom.

The following is the atomic decomposition of $\mathrm{HH}^1_{-\alpha}(\mathbb{R}^n)$.

https://doi.org/10.1515/9783110600285-005

Theorem 5.2. *Let $(\alpha, n-1) \in (0,1) \times \mathbb{N}$. Then:*

(i) $\mathrm{HH}^1_{-\alpha}(\mathbb{R}^n)$ *is complete under* $\|\cdot\|_{\mathrm{HH}^1_{-\alpha}(\mathbb{R}^n)}$;

(ii) *A tempered distribution f on $\mathbb{R}^n$ belongs to $\mathrm{HH}^1_{-\alpha}(\mathbb{R}^n)$ if and only if there exist an $\mathrm{HH}^1_{-\alpha}(\mathbb{R}^n)$-atomic sequence $\{a_j\}$ and an l^1-summable sequence $\{\lambda_j\}$ such that*

$$f = \sum_j \lambda_j a_j$$

holds in the sense of distributions. Moreover,

$$\|f\|_{\mathrm{HH}^1_{-\alpha}(\mathbb{R}^n)} \approx \inf\Big\{\sum_j |\lambda_j| : f = \sum_j \lambda_j a_j\Big\}.$$

Proof. In what follows, we will always assume

$$\begin{cases} \psi(x) = \dfrac{1+|x|^2-(n+1)\Gamma(\frac{n+1}{2})\pi^{-\frac{n+1}{2}}}{(1+|x|^2)^{\frac{n+3}{2}}} \\ \psi_t(x) = t^{-n}\psi\left(\dfrac{x}{t}\right) \\ F(y,t) = f * \psi_t(y). \end{cases}$$

(i) Suppose that $\{f_j\}$ is a Cauchy sequence in $\mathrm{HH}^1_{-\alpha}(\mathbb{R}^n)$. An application of the Calderón reproducing formula gives that

$$f_j = \int_0^\infty f_j * \psi_t * \psi_t \frac{dt}{t}$$

holds in the sense of distributions. So, for any $\phi \in \mathrm{S}(\mathbb{R}^n)$ we use Theorems 3.8, 4.1 and 1.4 to derive

$$\begin{aligned} |\langle f_j - f_k, \phi\rangle| &\lesssim \|(f_j - f_k) * \psi_t(x)\|_{\mathrm{T}^1_{n-2\alpha}(\mathbb{R}^{n+1}_+)} \|\psi_t * \phi\|_{\mathrm{T}^\infty_{n-2\alpha}(\mathbb{R}^{n+1}_+)} \\ &\lesssim \|(f_j - f_k) * \psi_t(x)\|_{\mathrm{T}^1_{n-2\alpha}(\mathbb{R}^{n+1}_+)} \|\phi\|_{\mathrm{Q}_\alpha(\mathbb{R}^n)} \\ &\lesssim \|(f_j - f_k) * \psi_t(x)\|_{\mathrm{T}^1_{n-2\alpha}(\mathbb{R}^{n+1}_+)} \|\phi\|_{\dot{\mathrm{L}}^2_{\frac{n}{2}}(\mathbb{R}^n)}. \end{aligned}$$

Accordingly, $\{f_j\}$ is a Cauchy sequence in $\dot{\mathrm{L}}^2_{-\frac{n}{2}}(\mathbb{R}^n)$ which is complete, and so

$$f = \lim_{j\to\infty} f_j$$

exists in $\dot{\mathrm{L}}^2_{-\frac{n}{2}}(\mathbb{R}^n)$. Upon passing to a subsequence, we find that

$$f = f_1 + \sum_{j=2}^\infty (f_{j+1} - f_j)$$

holds in the sense of distributions, and the subsequence is taken such that

$$\sum_{j=2}^{\infty} \|f_{j+1} - f_j\|_{\mathrm{HH}^1_{-\alpha}(\mathbb{R}^n)} < \infty.$$

Note that

$$\|f * \psi_t(x)\|_{\mathrm{T}^1_{n-2\alpha}(\mathbb{R}^{n+1}_+)} \lesssim \|f_1 * \psi_t(x)\|_{\mathrm{T}^1_{n-2\alpha}(\mathbb{R}^{n+1}_+)} + \sum_{j=2}^{\infty} \|(f_{j+1} - f_j) * \psi_t(x)\|_{\mathrm{T}^1_{n-2\alpha}(\mathbb{R}^{n+1}_+)} < \infty.$$

So

$$f \in \mathrm{HH}^1_{-\alpha}(\mathbb{R}^n) \quad \& \quad \lim_{j\to\infty} \|f_j - f\|_{\mathrm{HH}^1_{-\alpha}(\mathbb{R}^n)} = 0.$$

(ii) For the if-part, by (i) it is enough to prove that if a is an $\mathrm{HH}^1_{-\alpha}(\mathbb{R}^n)$-atom then a is in $\mathrm{HH}^1_{-\alpha}(\mathbb{R}^n)$. To do so, fix $\epsilon \in (0, 2)$ and let

$$\omega(x,t) = \kappa \ell(I)^{2\alpha-n} \min\left\{1, \left(\frac{\ell(I)}{\sqrt{|x - x_I|^2 + t^2}}\right)^{n-2\alpha+\epsilon}\right\},$$

where a is supported in a given $I \in \mathscr{C}(\mathbb{R}^n)$, x_I denotes the center of I, and κ is a constant. As in the argument for Theorem 3.3 we can choose a suitable κ such that

$$\begin{cases} \mathcal{N}(\omega)(x) \le \kappa \ell(I)^{2\alpha-n} \min\left\{1, \left(\frac{\sqrt{2}\ell(I)}{|x - x_I|}\right)^{n-2\alpha+\epsilon}\right\} \\ \int_{\mathbb{R}^n} \mathcal{N}(\omega) d\Lambda^{(\infty)}_{n-2\alpha} \le 1. \end{cases}$$

If

$$\begin{cases} B_I = B(x_I, \mathrm{diam}(I)) \\ E_I = B_I \times (0, \mathrm{diam}(I)) \\ E_I^c = \mathbb{R}^{n+1}_+ \setminus E_I \\ S_a = \mathrm{supp}\,(a * \psi_t(x)), \end{cases}$$

then

$$\iint_{\mathbb{R}^{n+1}_+} |a * \psi_t(x)|^2 \omega(x,t)^{-1} \frac{dx\,dt}{t^{1-2\alpha}} = \left(\iint_{E_I} + \iint_{E_I^c \cap S_a}\right) |a * \psi_t(x)|^2 \omega(x,t)^{-1} \frac{dx\,dt}{t^{1-2\alpha}}.$$

Thanks to the following facts:

▷ E_I is a subset of some half ball in $\mathbb{R}^{n+1}_+$ centered at $(x_I, 0)$ so

$$\omega^{-1} \gtrsim \ell(I)^{n-2\alpha} \quad \text{on} \quad E_I;$$

▷ a is a distribution with compact support and the convolution $a * \psi_t$ has Fourier transform $\hat{a}\hat{\psi}_t$ in $\mathsf{L}^2(\mathbb{R}^n)$,

we use Plancherel's formula and the size condition of a to estimate

$$\begin{aligned}\iint\limits_{E_I} |a * \psi_t(x)|^2 \omega(x,t)^{-1} \frac{dx\,dt}{t^{1-2\alpha}} &\lesssim \ell(I)^{n-2\alpha} \int\limits_0^{\operatorname{diam}(I)} \int\limits_{\mathbb{R}^n} |a * \psi_t(x)|^2 dx \frac{dt}{t^{1-2\alpha}} \\ &\lesssim \ell(I)^{n-2\alpha} \int\limits_0^{\infty} \int\limits_{\mathbb{R}^n} |\hat{a}(\xi)|^2 |\hat{\psi}_t(\xi)|^2 d\xi \frac{dt}{t^{1-2\alpha}} \\ &\lesssim \ell(I)^{n-2\alpha} \|a\|^2_{\mathsf{L}^2_{-\alpha}(\mathbb{R}^n)} \\ &\lesssim 1.\end{aligned}$$

Meanwhile, in order to handle the integral on $E_I^c \cap S_a$, we observe

$$|(a * \psi_t)(x)| \lesssim \begin{cases} 0 & \text{as } x \notin B_I \ \&\ t \le 2^{-1}|x - x_I| \\ \operatorname{diam}(I)\, t^{-(n+1)} & \text{otherwise,} \end{cases}$$

which means

$$(x,t) \in E_I^c \cap S_a \Rightarrow x \notin B_I \ \&\ 2t > |x - x_I| \quad \text{or} \quad x \in B_I \ \&\ t > \operatorname{diam}(I),$$

and consequently

$$t \approx \sqrt{|x - x_I|^2 + t^2} \ge \operatorname{diam}(I) \quad \& \quad \omega^{-1} \approx \ell(I)^{-\epsilon} t^{n-2\alpha+\epsilon}.$$

With the help of this last estimation, we have

$$\iint\limits_{E_I^c \cap S_a} |a * \psi_t(x)|^2 \omega(x,t)^{-1} \frac{dx\,dt}{t^{1-2\alpha}} \lesssim 1.$$

Putting

$$\iint\limits_{E_I} (\cdots) \quad \& \quad \iint\limits_{E_I^c \cap S_a} (\cdots)$$

together gives

$$\|a\|_{\mathsf{HH}^1_{-\alpha}(\mathbb{R}^{n+1}_+)} \le \iint\limits_{\mathbb{R}^{n+1}_+} |a * \psi_t(x)|^2 \omega(x,t)^{-1} \frac{dx\,dt}{t^{1-2\alpha}} \lesssim 1.$$

For the only-if-part, let

$$\begin{cases} f \in \mathrm{HH}^1_{-\alpha}(\mathbb{R}^n) \\ 0 < \epsilon < N < \infty \\ f^{\epsilon,N}(x) = \int_\epsilon^N f * \psi_t * \psi_t(x) \frac{\mathrm{d}t}{t} \\ S^{\epsilon,N} = \{(x,t) \in \mathbb{R}^{n+1}_+ : \epsilon \le t \le N\}. \end{cases}$$

Then by the Calderón reproducing formula, we have

$$\begin{aligned} \lim_{\epsilon\to 0, N\to\infty} f^{\epsilon,N}(x) &= \int_0^\infty f * \psi_t * \psi_t(x) \frac{\mathrm{d}t}{t} \\ &= \iint_{\mathbb{R}^{n+1}_+} F(y,t)\psi_t(x-y)\frac{\mathrm{d}y\,\mathrm{d}t}{t} \\ &= \lim_{\epsilon\to 0, N\to\infty} \iint_{S^{\epsilon,N}} F(y,t)\psi_t(x-y)\frac{\mathrm{d}y\,\mathrm{d}t}{t}. \end{aligned}$$

By Definition 5.1(ii), there is a measurable function $\omega \ge 0$ on $\mathbb{R}^{n+1}_+$ such that

$$\begin{cases} \iint_{\mathbb{R}^{n+1}_+} |F(x,t)|^2 \omega(x,t)^{-1} \frac{\mathrm{d}x\,\mathrm{d}t}{t^{1-2\alpha}} \le 2\|F\|_{T^1_{n-2\alpha}(\mathbb{R}^{n+1}_+)} \\ \int_{\mathbb{R}^n} \mathcal{N}(\omega)\mathrm{d}\Lambda^{(\infty)}_{n-2\alpha} \le 1. \end{cases}$$

Now, for $k \in \mathbb{Z}$ let $T_{j,k}$ be the tent (constructed in the proof of Theorem 3.8) over

$$E_k = \{x \in \mathbb{R}^n : \mathcal{N}(\omega)(x) > 2^k\}.$$

Recall that the tents $T_{j,k}$ have mutually disjoint interiors and

$$F = \sum_{j,k} F\mathbb{1}_{T_{j,k}} \quad \text{a.e on } \mathbb{R}^{n+1}_+.$$

Let

$$g^{\epsilon,N}_{j,k}(x) = \iint_{S^{\epsilon,N}\cap T_{j,k}} F(y,t)\psi_t(x-y)\frac{\mathrm{d}y\,\mathrm{d}t}{t}.$$

Note that there exist smooth functions with support in

$$\{x : \Lambda(x) \cap T_{j,k} \ne \emptyset\} \subseteq I^*_{j,k} = 5\sqrt{n}I_{j,k}$$

and the same number of vanishing moments as ψ. So we are required to show that $g_{j,k}^{\epsilon,N}$ converge, as

$$\epsilon \to 0 \quad \& \quad N \to \infty,$$

to distributions $g_{j,k}$ with

$$f = \sum_{j,k} g_{j,k} \quad \text{in } S'(\mathbb{R}^n).$$

Thanks to

$$\omega \le 2^{k+1} \quad \text{on } T_{j,k} \quad \forall \tau \in S(\mathbb{R}^n),$$

we can utilize Fubini's theorem, the Cauchy–Schwarz inequality and Plancherel's formula to estimate

$$\begin{aligned}
|\langle g_{j,k}^{\epsilon,N}, \tau\rangle| &= \left| \int_{\mathbb{R}^n} \left(\iint_{S^{\epsilon,N} \cap T_{j,k}} F(y,t)\psi_t(x-y) \frac{dy\,dt}{t} \right) \tau(x)\,dx \right| \\
&\le \iint_{S^{\epsilon,N} \cap T_{j,k}} |F(y,t)| \left| \int_{\mathbb{R}^n} \psi_t(x-y)\tau(x)\,dx \right| \frac{dydt}{t} \\
&\le 2^{\frac{k+1}{2}} \left(\iint_{S^{\epsilon,N} \cap T_{j,k}} \left(\frac{|F(y,t)|}{\sqrt{\omega(y,t)}} \right)^2 \frac{dy\,dt}{t^{1-2\alpha}} \right)^{\frac{1}{2}} \left(\iint_{T_{j,k}} |\tau * \psi_t(y)|^2 \frac{dy\,dt}{t^{1+2\alpha}} \right)^{\frac{1}{2}} \\
&\lesssim 2^{\frac{k+1}{2}} \left(\iint_{S^{\epsilon,N} \cap T_{j,k}} \left(\frac{|F(y,t)}{\sqrt{\omega(y,t)}} \right)^2 \frac{dy\,dt}{t^{1-2\alpha}} \right)^{\frac{1}{2}} \|\tau\|_{\dot{L}^2_\alpha(\mathbb{R}^n)}.
\end{aligned}$$

Similarly, if

$$\epsilon_1 < \epsilon_2 \quad \& \quad N_1 > N_2,$$

then

$$|\langle g_{j,k}^{\epsilon_1,N_1} - g_{j,k}^{\epsilon_2,N_2}, \tau\rangle| \lesssim 2^{\frac{k+1}{2}} \left(\iint_{(S^{\epsilon_1,N_1} \setminus S^{\epsilon_2,N_2}) \cap T_{j,k}} \left(\frac{|F(y,t)}{\sqrt{\omega(y,t)}} \right)^2 \frac{dy\,dt}{t^{1-2\alpha}} \right)^{\frac{1}{2}} \|\tau\|_{\dot{L}^2_\alpha(\mathbb{R}^n)},$$

and hence

$$\|g_{j,k}^{\epsilon_1,N_1} - g_{j,k}^{\epsilon_2,N_2}\|_{\dot{L}^2_{-\alpha}(\mathbb{R}^n)} \to 0 \quad \text{as} \quad \epsilon_1, \epsilon_2 \to 0 \,\&\, N_1, N_2 \to \infty,$$

and so $g_{j,k}^{\epsilon,N}$ converge to $g_{j,k}$ in the sense of distributions with

$$\begin{cases} \|g_{j,k}\|_{\dot{L}^2_{-\alpha}(3I^*_{j,k})} \lesssim 2^{\frac{k+1}{2}} \left(\iint\limits_{T_{j,k}} \left(\frac{|F(y,t)|}{\sqrt{\omega(y,t)}} \right)^2 \frac{dy\,dt}{t^{1-2\alpha}} \right)^{\frac{1}{2}} \\ |\langle g_{j,k}, \tau \rangle| \lesssim \left(\iint\limits_{T_{j,k}} |\tau * \psi_t(y)|^2 \frac{dy\,dt}{t^{1+2\alpha}} \right)^{\frac{1}{2}}. \end{cases}$$

Moreover, each $g_{j,k}$ satisfies the cancellation condition in Definition 5.1(iii) when N is suitably chosen.

If

$$\begin{cases} a_{j,k} = g_{j,k} \|g_{j,k}\|^{-1}_{\dot{L}^2_{-\alpha}(3I^*_{j,k})} \ell(3I^*_{j,k})^{\frac{2\alpha-n}{2}} \\ \lambda_{j,k} = \|g_{j,k}\|_{\dot{L}^2_{-\alpha}(3I^*_{j,k})} \ell(3I^*_{j,k})^{\frac{n-2\alpha}{2}}, \end{cases}$$

then each $a_{j,k}$ is $\mathsf{HH}^1_{-\alpha}(\mathbb{R}^n)$-atom, and from the Cauchy–Schwarz inequality it follows that

$$\begin{aligned} \sum_{j,k} |\lambda_{j,k}| &\le \left(\sum_{j,k} \ell(3I^*_{j,k})^{n-2\alpha} 2^{k+1} \right)^{1/2} \left(\sum_{j,k} \iint\limits_{T_{j,k}} \left(\frac{|F(y,t)|}{\sqrt{\omega(y,t)}} \right)^2 \frac{dydt}{t^{1-2\alpha}} \right)^{\frac{1}{2}} \\ &\lesssim \left(\int\limits_{\mathbb{R}^n} \mathcal{N}(\omega) d\Lambda^{(\infty)}_{n-2\alpha} \right)^{\frac{1}{2}} \left(\iint\limits_{\mathbb{R}^{n+1}_+} \left(\frac{|F(y,t)|}{\sqrt{\omega(y,t)}} \right)^2 \frac{dydt}{t^{1-2\alpha}} \right)^{\frac{1}{2}} \\ &\lesssim \|f\|_{\mathsf{HH}^1_{-\alpha}(\mathbb{R}^n)}, \end{aligned}$$

and hence

$$\sum_{j,k} g_{j,k} = \sum_{j,k} \lambda_{j,k} a_{j,k}$$

is equal to a distribution g in $\mathsf{HH}^1_{-\alpha}(\mathbb{R}^n)$.

It remains to check $g = f$. To do so, for

$$\tau \in \mathsf{S}(\mathbb{R}^n) \quad \& \quad 0 < \epsilon < N$$

we use Theorem 4.1 to obtain

$$\begin{aligned} |\langle g^{\epsilon,N}_{j,k}, \tau \rangle| &\lesssim 2^{\frac{k+1}{2}} \left(\iint\limits_{T_{j,k}} \left(\frac{|F(y,t)|}{\sqrt{\omega(y,t)}} \right)^2 \frac{dydt}{t^{1-2\alpha}} \right)^{\frac{1}{2}} \left(\iint\limits_{T_{j,k}} |\tau * \psi_t(y)|^2 \frac{dy\,dt}{t^{1+2\alpha}} \right)^{\frac{1}{2}} \\ &\lesssim \ell(3I^*_{j,k})^{\frac{n}{2}-\alpha} 2^{\frac{k+1}{2}} \left(\iint\limits_{T_{j,k}} \left(\frac{|F(y,t)|}{\sqrt{\omega(y,t)}} \right)^2 \frac{dydt}{t^{1-2\alpha}} \right)^{\frac{1}{2}} \|\tau\|_{\mathsf{Q}_\alpha(\mathbb{R}^n)}. \end{aligned}$$

Consequently, we achieve

$$\lim_{\epsilon\to 0, N\to\infty} \sum_{j,k} \langle g_{j,k}^{\epsilon,N}, \tau\rangle = \sum_{j,k} \lim_{\epsilon\to 0, N\to\infty} \langle g_{j,k}^{\epsilon,N}, \tau\rangle = \sum_{j,k} \langle g_{j,k}, \tau\rangle,$$

whence

$$f = \lim_{\epsilon\to 0, N\to\infty} f^{\epsilon,N} = \lim_{\epsilon\to 0, N\to\infty} \sum_{j,k} g_{j,k}^{\epsilon,N} = \sum_{j,k} g_{j,k} = g \quad \text{in } \mathsf{S}'(\mathbb{R}^n) = \text{the dual of } \mathsf{S}(\mathbb{R}^n). \quad \square$$

The duality between

$$\mathsf{HH}^1_{-\alpha}(\mathbb{R}^n) \quad \text{and} \quad \mathsf{Q}_\alpha(\mathbb{R}^n)$$

is presented below.

Theorem 5.3. *Let $(\alpha, n-1) \in (0,1) \times \mathbb{N}$. Then $\mathsf{Q}_\alpha(\mathbb{R}^n)$ exists as the dual of $\mathsf{HH}^1_{-\alpha}(\mathbb{R}^n)$ in the following sense.*

(i) *If $f \in \mathsf{Q}_\alpha(\mathbb{R}^n)$, then the functional*

$$g \mapsto \int_{\mathbb{R}^n} f(x)g(x)\,\mathrm{d}x \quad \textit{initially defined} \quad \forall g \in \mathsf{HH}^1_{-\alpha}(\mathbb{R}^n) \cap \mathsf{C}_0^\infty(\mathbb{R}^n),$$

has a bounded linear extension to all $g \in \mathsf{HH}^1_{-\alpha}(\mathbb{R}^n)$.

(ii) *Conversely, if $\mathcal{L}$ is a bounded linear functional on $\mathsf{HH}^1_{-\alpha}(\mathbb{R}^n)$ then there is $f \in \mathsf{Q}_\alpha(\mathbb{R}^n)$ such that*

$$\mathcal{L}(g) = \int_{\mathbb{R}^n} f(x)g(x)\,\mathrm{d}x \quad \forall g \in \mathsf{HH}^1_{-\alpha}(\mathbb{R}^n).$$

Proof. (i) If

$$f \in \mathsf{HH}^1_{-\alpha}(\mathbb{R}^n) \cap C_0^\infty(\mathbb{R}^n) \quad \& \quad g \in \mathsf{Q}_\alpha(\mathbb{R}^n),$$

then both Theorem 5.2 and Calderón's reproducing formula

$$f(x) = \int_0^\infty f * \psi_t * \psi_t(x) \frac{\mathrm{d}t}{t}$$

are used to deduce

$$\left|\int_{\mathbb{R}^n} f(x)g(x)\,\mathrm{d}x\right| \approx \left|\int_{\mathbb{R}^{1+n}_+} (f * \psi_t)(t,y) \cdot (g * \psi_t)(t,y) \frac{\mathrm{d}x\mathrm{d}t}{t}\right| \lesssim \|f\|_{\mathsf{HH}^1_{-\alpha}(\mathbb{R}^n)} \|g\|_{\mathsf{Q}_\alpha(\mathbb{R}^n)},$$

and hence $g \in \mathsf{Q}_\alpha(\mathbb{R}^n)$ induces a bounded linear functional on

$$\mathsf{HH}^1_{-\alpha}(\mathbb{R}^n) \cap C_0^\infty(\mathbb{R}^n).$$

Since

$$\mathrm{HH}^1_{-\alpha}(\mathbb{R}^n) \cap C_0^\infty(\mathbb{R}^n)$$

is dense in $\mathrm{HH}^1_{-\alpha}(\mathbb{R}^n)$ (cf. [17, Lemma 6.4(ii)]), we obtain that

$$f \mapsto \int_{\mathbb{R}^n} f(x)g(x)\,\mathrm{d}x$$

can be extended to a bounded linear functional L on $\mathrm{HH}^1_{-\alpha}(\mathbb{R}^n)$ with its norm

$$\|\mathcal{L}\| \lesssim \|g\|_{\mathrm{Q}_\alpha(\mathbb{R}^n)}.$$

(ii) If $\mathcal{L}$ is a bounded linear functional on $\mathrm{HH}^1_{-\alpha}(\mathbb{R}^n)$ and

$$\varrho_\psi(f)(x) = f * \psi_t(x) = F(x,t) \quad \forall\,(x,t) \in \mathbb{R}^{n+1}_+,$$

then

$$\varrho_\psi(\mathrm{HH}^1_{-\alpha}(\mathbb{R}^n)) \subseteq \mathrm{T}^1_{n-2\alpha}(\mathbb{R}^{n+1}_+)$$

thanks to Definitions 3.7 and 5.1 and Theorem 5.2, and hence a combination of Theorem 3.8 and the Hahn–Banach theorem yields an extension $\tilde{\mathcal{L}}$ which is bounded linear functional on $\mathrm{T}^1_{n-2\alpha}(\mathbb{R}^{n+1}_+)$ and induces

$$G(\cdot,\cdot) \in \mathrm{T}^\infty_{n-2\alpha}(\mathbb{R}^{n+1}_+)$$

such that

$$\tilde{\mathcal{L}}(F) = \iint_{\mathbb{R}^{n+1}_+} F(x,t)G(x,t)\,\frac{\mathrm{d}x\mathrm{d}t}{t} \quad \forall\,F \in \mathrm{T}^1_{n-2\alpha}(\mathbb{R}^{n+1}_+).$$

Accordingly, if

$$f \in \mathrm{HH}^1_{-\alpha}(\mathbb{R}^n) \cap C_0^\infty(\mathbb{R}^n) \quad \& \quad g(\cdot) = \int_0^\infty G(\cdot,t) * \psi_t(\cdot)\,\frac{\mathrm{d}t}{t},$$

then an application of

$$\varrho_\psi(f) \in \mathrm{T}^1_{n-2\alpha}(\mathbb{R}^{n+1}_+)$$

gives

$$\mathcal{L}(f) = \tilde{\mathcal{L}}(\varrho_\psi(f))$$

$$= \iint_{\mathbb{R}^{n+1}_+} f * \psi_t(y)\, G(y,t)\, \frac{\mathrm{d}y\mathrm{d}t}{t}$$

$$= \iint_{\mathbb{R}^{n+1}_+} f(x)(G(\cdot,t) * \psi_t)(x)\, \frac{\mathrm{d}x\mathrm{d}t}{t}$$

$$= \int_{\mathbb{R}^n} f(x)g(x)\,\mathrm{d}x.$$

Since the proof of Theorem 4.1(i)⇔(ii) (cf. [17, Theorem 7.0(i)]) ensures

$$\|g\|_{\mathrm{Q}_\alpha(\mathbb{R}^n)} \lesssim \|G\|_{\mathrm{T}^\infty_{n-2\alpha}(\mathbb{R}^{n+1}_+)} \lesssim \|\mathcal{L}\|,$$

we arrive at the desired result. □

5.2 Hardy–Hausdorff–Sobolev space as a predual

Recall that

$$\mathcal{R}_j(f)(x) = \frac{\Gamma(\frac{n+1}{2})}{\pi^{\frac{n+1}{2}}} \mathrm{p.v.} \int_{\mathbb{R}^n} \frac{(x_j - y_j)f(y)}{|x-y|^{n+1}}\, dy \quad \forall\, (j,f) \in \{1,\ldots,n\} \times \mathrm{C}^\infty_0(\mathbb{R}^n)$$

is the j-th Riesz transform, and the vector-valued Riesz transform $(\mathcal{R}_j(f))_{j=1}^n$ exists as the gradient of the 1-order Riesz potential $(-\Delta)^{-\frac{1}{2}}f$, up to a dimensional constant (cf. [74] and Definition 4.3(i)), that is,

$$\mathcal{R}_j(f)(x) = \left(\frac{\Gamma(\frac{n+1}{2})}{(1-n)\pi^{\frac{n+1}{2}} c_{2,1}}\right)\partial_{x_j}(-\Delta)^{-\frac{1}{2}}f(x).$$

So, in accordance with [91, Theorem 4.1] we have

$$\|f\|_{\mathrm{Q}_\alpha(\mathbb{R}^n)} \lesssim \|\nabla f\|_{\mathrm{L}^n(\mathbb{R}^n)}$$

thereby obtaining

$$\|(-\Delta)^{-\frac{1}{2}}f\|_{\mathrm{Q}_\alpha(\mathbb{R}^n)} \lesssim \|\nabla(-\Delta)^{-\frac{1}{2}}f\|_{\mathrm{L}^n(\mathbb{R}^n)} \lesssim \sum_{j=1}^n \|\mathcal{R}_j(f)\|_{\mathrm{L}^n(\mathbb{R}^n)} \lesssim \|f\|_{\mathrm{L}^n(\mathbb{R}^n)},$$

via the boundedness of the Riesz transform

$$(\mathcal{R}_1,\ldots,\mathcal{R}_n) \quad \text{on } \mathrm{L}^n(\mathbb{R}^n),$$

in other words (cf. Proposition 2.4(i)),

$$(-\Delta)^{-\frac{1}{2}}(\mathrm{L}^n(\mathbb{R}^n)) \subseteq \mathrm{Q}_\alpha(\mathbb{R}^n).$$

This, along with Theorem 5.3, the fact that the dual of $\mathsf{L}^n(\mathbb{R}^n)$ is $\mathsf{L}^{\frac{n}{n-1}}(\mathbb{R}^n)$, and the equation

$$\int_{\mathbb{R}^n} ((-\Delta)^{-\frac{1}{2}} f(x)) g(x)\, dx = \int_{\mathbb{R}^n} f(x)((-\Delta)^{-\frac{1}{2}} g(x))\, dx,$$

reveals

$$(-\Delta)^{-\frac{1}{2}}(\mathsf{HH}^1_{-\alpha}(\mathbb{R}^n)) \subseteq \mathsf{L}^{\frac{n}{n-1}}(\mathbb{R}^n)$$

whose limiting case $\alpha \to 0$ can be found in Stein–Weiss [79]. Evidently, the foregoing space-inclusion implies the subsequent Sobolev-type inequality

$$\|f\|_{\mathsf{L}^{\frac{n}{n-1}}(\mathbb{R}^n)} \lesssim \|(-\Delta)^{-\frac{1}{2}}|\nabla f|\|_{\mathsf{L}^{\frac{n}{n-1}}(\mathbb{R}^n)} \lesssim \|\nabla f\|_{\mathsf{HH}^1_{-\alpha}(\mathbb{R}^n)} \quad \forall f \in \mathsf{C}^\infty_0(\mathbb{R}^n),$$

which not only presents a new perspective of the classical Sobolev inequality

$$\|f\|_{\mathsf{L}^{\frac{n}{n-1}}(\mathbb{R}^n)} \lesssim \|\nabla f\|_{\mathsf{L}^1(\mathbb{R}^n)} \quad \forall f \in \mathsf{C}^\infty_0(\mathbb{R}^n),$$

but also leads to the following assertion.

Theorem 5.4. *For $(\alpha, n-1) \in (0,1) \times \mathbb{N}$, let $\mathsf{HHS}^1_{-\alpha}(\mathbb{R}^n)$ be the Hardy–Hausdorff–Sobolev space of all functions f on $\mathbb{R}^n$ satisfying*

$$\nabla f \in \underbrace{\mathsf{HH}^1_{-\alpha}(\mathbb{R}^n) \times \cdots \times \mathsf{HH}^1_{-\alpha}(\mathbb{R}^n)}_{n \text{ copies}} = (\mathsf{HH}^1_{-\alpha}(\mathbb{R}^n))^n.$$

Then

$$\mathsf{Q}^{-1}_\alpha(\mathbb{R}^n) = \operatorname{div} \underbrace{(\mathsf{Q}_\alpha(\mathbb{R}^n), \ldots, \mathsf{Q}_\alpha(\mathbb{R}^n))}_{n \text{ copies}} = \nabla \cdot (\mathsf{Q}_\alpha(\mathbb{R}^n))^n$$

exists as the dual of $\mathsf{HHS}^1_{-\alpha}(\mathbb{R}^n)$.

Proof. Note that $\mathsf{HHS}^1_{-\alpha}(\mathbb{R}^n)$ can be also identified with the closure of $C^\infty_0(\mathbb{R}^n)$ in the norm

$$\||\nabla f|\|_{\mathsf{HH}^1_{-\alpha}(\mathbb{R}^n)}.$$

Thus, the coming-up-next argument is two-fold.

On the one hand, if $f \in \mathsf{Q}^{-1}_\alpha(\mathbb{R}^n)$ then there is

$$(f_1, \ldots, f_n) \in (\mathsf{Q}_\alpha(\mathbb{R}^n))^n$$

such that

$$f(x) = \nabla \cdot (f_1, \ldots, f_n)(x) = \sum_{j=1}^n \partial_{x_j} f_j(x)$$

and hence f induces a bounded linear functional on $\mathrm{HHS}^1_{-\alpha}(\mathbb{R}^n)$ thanks to

$$C_0^\infty(\mathbb{R}^n) \ni g \mapsto \int_{\mathbb{R}^n} f(x)g(x)\,dx = -\int_{\mathbb{R}^n} (f_1(x),\ldots,f_n(x)) \cdot \nabla g(x)\,dx.$$

This in turn implies

$$Q_\alpha^{-1}(\mathbb{R}^n) \subseteq (\mathrm{HHS}^1_{-\alpha}(\mathbb{R}^n))^*.$$

On the other hand, suppose that $\mathcal{L}$ is a bounded linear functional on $\mathrm{HHS}^1_{-\alpha}(\mathbb{R}^n)$, that is,

$$\mathcal{L} \in (\mathrm{HHS}^1_{-\alpha}(\mathbb{R}^n))^*.$$

Since ∇ exists as a bounded linear operator

$$\text{from } \mathrm{HHS}^1_{-\alpha}(\mathbb{R}^n) \text{ to } (\mathrm{HH}^1_{-\alpha}(\mathbb{R}^n))^n,$$

the completeness of $\mathrm{HHS}^1_{-\alpha}(\mathbb{R}^n)$ ensures that the range

$$\mathscr{R}(\nabla\, \mathrm{HHS}^1_{-\alpha}(\mathbb{R}^n)) \quad \text{of } \mathrm{HHS}^1_{-\alpha}(\mathbb{R}^n)$$

under ∇ is a closed subspace of

$$(\mathrm{HH}^1_{-\alpha}(\mathbb{R}^n))^n.$$

Upon defining

$$\mathcal{L}_1(\nabla f) = \mathcal{L}(f) \quad \forall\, \nabla f \in \mathscr{R}(\nabla\, \mathrm{HHS}^1_{-\alpha}(\mathbb{R}^n)),$$

we achieve

$$\|\mathcal{L}_1\|_{[\mathscr{R}(\nabla\, \mathrm{HHS}^1_{-\alpha}(\mathbb{R}^n))]^*} = \|\mathcal{L}\|_{[\mathrm{HHS}^1_{-\alpha}(\mathbb{R}^n)]^*}.$$

This, plus the Hahn–Banach theorem, yields a norm-preserving extension

$$\mathcal{L}_2 \text{ of } \mathcal{L}_1 \text{ to } (\mathrm{HH}^1_{-\alpha}(\mathbb{R}^n))^n.$$

Meanwhile, the vector-valued form of Theorem 5.3 is used to produce a vector

$$(f_1,\ldots,f_n) \in (Q_\alpha(\mathbb{R}^n))^n$$

such that

$$\mathcal{L}_2((g_1,\ldots,g_n)) = \sum_{j=1}^n \int_{\mathbb{R}^n} f_j(x)g_j(x)\,dx \quad \forall\,(g_1,\ldots,g_n) \in (\mathrm{HH}^1_{-\alpha}(\mathbb{R}^n))^n$$

and

$$\|(f_1,\dots,f_n)\|_{(\mathrm{Q}_\alpha^{-1}(\mathbb{R}^n))^n} = \|\mathcal{L}_2\|_{(\mathrm{HH}^1_{-\alpha}(\mathbb{R}^n))^n} = \|\mathcal{L}_1\|_{[\mathscr{R}(\nabla\,\mathrm{HHS}^1_{-\alpha}(\mathbb{R}^n))]^*} = \|\mathcal{L}\|_{[\mathrm{HHS}^1_{-\alpha}(\mathbb{R}^n)]^*}.$$

Consequently, for each $\phi \in C_0^\infty(\mathbb{R}^n)$ we obtain

$$\mathcal{L}(\phi) = \mathcal{L}_1(\nabla\phi) = \mathcal{L}_2(\nabla\phi) = \int_{\mathbb{R}^n} (f_1(x),\dots,f_n(x))\cdot\nabla\phi(x)\,dx,$$

thereby reaching

$$\begin{cases} \mathcal{L} = -\operatorname{div}(f_1,\dots,f_n) \\ \|(f_1,\dots,f_n)\|_{(\mathrm{Q}_\alpha^{-1}(\mathbb{R}^n))^n} = \|\mathcal{L}\|_{[\mathrm{HHS}^1_{-\alpha}(\mathbb{R}^n)]^*}, \end{cases}$$

and so

$$\mathrm{Q}_\alpha^{-1}(\mathbb{R}^n) \supseteq [\mathrm{HHS}^1_{-\alpha}(\mathbb{R}^n)]^*.$$

Therefore, the above-established set-inclusions derive the desired duality. □

5.3 Fefferman–Stein type decomposition

According to Theorem 1.3(ii) and its proof, we see

$$\mathrm{L}^\infty(\mathbb{R}^n) \subseteq \mathrm{Q}_\alpha(\mathbb{R}^n) \Leftrightarrow \alpha \in (-\infty, 0).$$

This equivalence suggests the following decomposition of Fefferman–Stein type.

Theorem 5.5. *Let $(\alpha, n-1) \in (0,1)\times\mathbb{N}$. Then*

$$\mathrm{Q}_\alpha(\mathbb{R}^n) = \mathrm{L}^\infty(\mathbb{R}^n)\cap\mathrm{Q}_\alpha(\mathbb{R}^n) + \sum_{j=1}^n \mathcal{R}_j(\mathrm{L}^\infty(\mathbb{R}^n)\cap\mathrm{Q}_\alpha(\mathbb{R}^n))$$

holds in the sense that $f \in \mathrm{Q}_\alpha(\mathbb{R}^n)$ if and only if

$$\exists\, g_0,\dots,g_n \in \mathrm{L}^\infty(\mathbb{R}^n)\cap\mathrm{Q}_\alpha(\mathbb{R}^n) \quad \text{such that} \quad f = g_0 + \sum_{j=1}^n \mathcal{R}_j g_j.$$

Proof. On the one hand, let us prove

$$\mathrm{Q}_\alpha(\mathbb{R}^n) \supseteq \mathrm{Q}_\alpha(\mathbb{R}^n)\cap\mathrm{L}^\infty(\mathbb{R}^n) + \sum_{j=1}^n \mathcal{R}_j(\mathrm{Q}_\alpha(\mathbb{R}^n)\cap\mathrm{L}^\infty(\mathbb{R}^n)).$$

To do so, let

$$\begin{cases} f = g_0 + \sum_{j=1}^{n} R_j g_j \\ g_j \in \mathrm{L}^\infty(\mathbb{R}^n) \cap \mathrm{Q}_\alpha(\mathbb{R}^n) \\ g_j(x,t) = g_j * P_t(x). \end{cases}$$

According to [32, p. 263, 4.1.8], we have

$$(\mathcal{R}_j P_t)(x) \approx x_j(t^2 + |x|^2)^{-\frac{n+1}{2}} \quad \forall\,(j,x,t) \in \{1,\dots,n\} \times \mathbb{R}^n \times \mathbb{R}_+,$$

thereby obtaining

$$\begin{aligned}
\|\mathcal{R}_j g_j\|_{\mathrm{Q}_\alpha(\mathbb{R}^n)} &\approx \sup_{I\in\mathscr{C}(\mathbb{R}^n)} \left(\int_0^{\ell(I)} \int_I |\nabla(\mathcal{R}_j g_j(x,t))|^2 \left(\frac{t}{\ell(I)}\right)^{n-2\alpha} \frac{\mathrm{d}x\mathrm{d}t}{t^{n-1}} \right)^{2^{-1}} \\
&= \sup_{I\in\mathscr{C}(\mathbb{R}^n)} \left(\int_0^{\ell(I)} \int_I |\nabla((\mathcal{R}_j g_j) * P_t(x))|^2 \left(\frac{t}{\ell(I)}\right)^{n-2\alpha} \frac{\mathrm{d}x\mathrm{d}t}{t^{n-1}} \right)^{2^{-1}} \\
&= \sup_{I\in\mathscr{C}(\mathbb{R}^n)} \left(\int_0^{\ell(I)} \int_I |\nabla(g_j * (\mathcal{R}_j P_t)(x))|^2 \left(\frac{t}{\ell(I)}\right)^{n-2\alpha} \frac{\mathrm{d}x\mathrm{d}t}{t^{n-1}} \right)^{2^{-1}} \\
&\lesssim \sup_{I\in\mathscr{C}(\mathbb{R}^n)} \left(\int_0^{\ell(I)} \int_I |\nabla(g_j * P_t(x))|^2 \left(\frac{t}{\ell(I)}\right)^{n-2\alpha} \frac{\mathrm{d}x\mathrm{d}t}{t^{n-1}} \right)^{2^{-1}} \\
&\lesssim \|g_j\|_{\mathrm{L}^\infty(\mathbb{R}^n)\cap\mathrm{Q}_\alpha(\mathbb{R}^n)}.
\end{aligned}$$

Hence

$$\|f\|_{\mathrm{Q}_\alpha(\mathbb{R}^n)} \le \|g_0\|_{\mathrm{Q}_\alpha(\mathbb{R}^n)} + \sum_{j=1}^{n} \|\mathcal{R}_j g_j\|_{\mathrm{Q}_\alpha(\mathbb{R}^n)} \lesssim \sum_{j=0}^{n} \|g_j\|_{\mathrm{L}^\infty(\mathbb{R}^n)\cap\mathrm{Q}_\alpha(\mathbb{R}^n)}.$$

On the other hand, let us verify

$$\mathrm{Q}_\alpha(\mathbb{R}^n) \subseteq \mathrm{L}^\infty(\mathbb{R}^n) \cap \mathrm{Q}_\alpha(\mathbb{R}^n) + \sum_{j=1}^{n} \mathcal{R}_j(\mathrm{L}^\infty(\mathbb{R}^n) \cap \mathrm{Q}_\alpha(\mathbb{R}^n)).$$

In doing so, we define Ξ_α as the space of all functions f on $\mathbb{R}^n$ with its norm

$$\|f\|_{\Xi_\alpha} = \sup\left\{ \left| \int_{\mathbb{R}^n} f(x)h(x)\,\mathrm{d}x \right| : \ \|h\|_{\mathrm{L}^\infty(\mathbb{R}^n)\cap\mathrm{Q}_\alpha(\mathbb{R}^n)} = \|h\|_{\mathrm{L}^\infty(\mathbb{R}^n)} + \|h\|_{\mathrm{Q}_\alpha(\mathbb{R}^n)} \le 1 \right\}.$$

Let

$$\begin{cases} \mathsf{U} = \{(g_0, g_1, \dots, g_n) : \ g_0, g_1, \dots, g_n \in \Xi_\alpha\} = \underbrace{\Xi_\alpha \oplus \cdots \oplus \Xi_\alpha}_{n+1 \text{ copies}}; \\ \|(g_0, g_1, \dots, g_n)\|_{\mathsf{U}} = \sum_{j=0}^{n} \|g_j\|_{\Xi_\alpha}. \end{cases}$$

If

$$\mathsf{V} = \{(g_0, g_1, \dots, g_n) \in \mathsf{U} : \ g_j = \mathcal{R}_j g_0 \ \forall j \in \{1, \dots, n\}\},$$

then V is a closed subset of U, and hence the mapping

$$g_0 \mapsto (g_0, g_1, \dots, g_n)$$

defines a norm-preserving map from $\mathsf{HH}^1_{-\alpha}(\mathbb{R}^n)$ to V. In order to see the last statement, we are only required to prove

$$g_0 \in \mathsf{HH}^1_{-\alpha}(\mathbb{R}^n) \Rightarrow \mathcal{R}_j g_0 \in \Xi_\alpha \quad \forall j \in \{1, \dots, n\}.$$

In fact, an application of Theorem 5.3 gives

$$\begin{aligned} \|\mathcal{R}_j g_0\|_{\Xi_\alpha} &= \sup_{\|h\|_{\mathsf{L}^\infty(\mathbb{R}^n)\cap \mathsf{Q}_\alpha(\mathbb{R}^n)} \le 1} \left| \int_{\mathbb{R}^n} (\mathcal{R}_j g_0)(x) h(x)\, \mathrm{d}x \right| \\ &= \sup_{\|h\|_{\mathsf{L}^\infty(\mathbb{R}^n)\cap \mathsf{Q}_\alpha(\mathbb{R}^n)} \le 1} \left| \int_{\mathbb{R}^n} g_0(x) (\mathcal{R}_j h)(x)\, \mathrm{d}x \right| \\ &\le \sup_{\|h\|_{\mathsf{L}^\infty(\mathbb{R}^n)\cap \mathsf{Q}_\alpha(\mathbb{R}^n)} \le 1} \|g_0\|_{\mathsf{HH}^1_{-\alpha}(\mathbb{R}^n)} \|\mathcal{R}_j h\|_{\mathsf{Q}_\alpha(\mathbb{R}^n)} \\ &\lesssim \sup_{\|h\|_{\mathsf{L}^\infty(\mathbb{R}^n)\cap \mathsf{Q}_\alpha(\mathbb{R}^n)} \le 1} \|g_0\|_{\mathsf{HH}^1_{-\alpha}(\mathbb{R}^n)} \|h\|_{\mathsf{L}^\infty(\mathbb{R}^n)\cap \mathsf{Q}_\alpha(\mathbb{R}^n)} \\ &\lesssim \|g_0\|_{\mathsf{HH}^1_{-\alpha}(\mathbb{R}^n)}, \end{aligned}$$

where we have used the above-established inequality

$$\|\mathcal{R}_j h\|_{\mathsf{Q}_\alpha(\mathbb{R}^n)} \lesssim \|h\|_{\mathsf{L}^\infty(\mathbb{R}^n)\cap \mathsf{Q}_\alpha(\mathbb{R}^n)}.$$

Now, any bounded linear functional on $\mathsf{HH}^1_{-\alpha}(\mathbb{R}^n)$ can be identified with a corresponding functional on V, and so from the Hahn–Banach theorem it follows that this functional is extendable to a bounded linear functional on U. Note that the dual space of U is

$$\underbrace{(\mathsf{L}^\infty(\mathbb{R}^n) \cap \mathsf{Q}_\alpha(\mathbb{R}^n)) \oplus \cdots \oplus (\mathsf{L}^\infty(\mathbb{R}^n) \cap \mathsf{Q}_\alpha(\mathbb{R}^n))}_{n+1 \text{ copies}}$$

Thus, upon restricting to

$$\mathsf{V}\ \&\ \mathsf{HH}^1_{-\alpha}(\mathbb{R}^n),$$

we get the forthcoming demonstration.

Suppose that $f \in Q_\alpha(\mathbb{R}^n)$ and $\mathcal{L}$ is a bounded linear functional on $\mathsf{HH}^1_{-\alpha}(\mathbb{R}^n)$. Then, for each $j \in \{0,\dots,n\}$ there is

$$\tilde{f}_j \in L^\infty(\mathbb{R}^n) \cap Q_\alpha(\mathbb{R}^n)$$

such that if

$$g_0 \in \mathsf{HH}^1_{-\alpha}(\mathbb{R}^n)$$

then

$$\begin{aligned}\mathcal{L}(f) &= \int_{\mathbb{R}^n} f(x) g_0(x)\,dx\\ &= \int_{\mathbb{R}^n} \tilde{f}_0(x) g_0(x)\,dx + \sum_{j=1}^n \int_{\mathbb{R}^n} \tilde{f}_j(x)(\mathcal{R}_j g_0)(x)\,dx\\ &= \int_{\mathbb{R}^n} \tilde{f}_0(x) g_0(x)\,dx - \sum_{j=1}^n \int_{\mathbb{R}^n} \mathcal{R}_j\tilde{f}_j(x) g_0(x)\,dx\\ &= \int_{\mathbb{R}^n} \left(\tilde{f}_0(x) - \sum_{j=1}^n (\mathcal{R}_j\tilde{f}_j)(x)\right) g_0(x)\,dx.\end{aligned}$$

Accordingly, we gain the desired representation

$$f(x) = \tilde{f}_0(x) - \sum_{j=1}^n (\mathcal{R}_j\tilde{f}_j)(x). \qquad \square$$

5.4 Notes

Below are some historical remarks on Chapter 5.

Section 5.1. Theorem 5.2 and its proof come from a slight modification of [17, Theorem 6.3] and its proof. Theorem 5.3 is the same as [17, Theorem 7.1]. Of course, the case $\alpha \in (-\infty, 0)$ of this proposition is the classical atomic decomposition of $\mathsf{BMO}(\mathbb{R}^n)$.

Section 5.2. Theorem 5.4 comes from [97, Theorem 1.4(ii)], and its situation $\alpha \in (-\infty, 0)$ is also valid. More precisely, $\mathsf{BMO}^{-1}(\mathbb{R}^n)$ coincides with the dual of the Hardy–Sobolev space

$$\mathsf{HS}^1(\mathbb{R}^n) = \mathsf{HHS}^1_{-0}(\mathbb{R}^n).$$

Section 5.3. Theorem 5.5 comes from [97, Theorem 1.4(i)] (cf. [101] for another treatment). This indicates that $Q_\alpha(\mathbb{R}^n)$ is a probably appropriate resolution of the problem described on pages 395–396 of [7]. "This raises the question what are the function spaces X,

$$\dot{W}^{1,n}(\mathbb{R}^n) \subset X \subset BMO(\mathbb{R}^n),$$

such that every $F \in X$ has a decomposition

$$F = \sum_{j=1}^{n} \mathcal{R}_j Y_j, \tag{1.10}$$

where $Y_j \in L^\infty(\mathbb{R}^n)$ or (assuming the Riesz transforms bounded on X) the stronger property $Y_j \in L^\infty \cap X$." Clearly, the case $\alpha \in (-\infty, 0)$ of Theorem 5.5 is the celebrated Fefferman–Stein decomposition theorem (cf. [25] for the original one and [86] for a constructive approach)

$$BMO(\mathbb{R}^n) = L^\infty(\mathbb{R}^n) + \sum_{j=1}^{n} \mathcal{R}_j(L^\infty(\mathbb{R}^n)).$$

This decomposition can be generalized in [36, Theorem 1] to $BMO_\phi(\mathbb{R}^n)$ which consists of all locally integrable functions f with

$$\begin{cases} \sup_{I \in \mathscr{C}(\mathbb{R}^n)} \fint_I |f(y) - f_I| \dfrac{dy}{\phi(\ell(I))} < \infty \\ \phi \text{ being a positive increasing function on } \mathbb{R}_+ \\ \int_r^\infty t^{-2}\phi(t)\,dt \lesssim r^{-1}\phi(r) \quad \forall r \in (0,\infty). \end{cases}$$

More precisely,

$$BMO_\phi(\mathbb{R}^n) = \Lambda_\phi(\mathbb{R}^n) + \sum_{j=1}^{n} \mathcal{R}_j(\Lambda_\phi(\mathbb{R}^n)),$$

where

$$f \in \Lambda_\phi(\mathbb{R}^n) \Leftrightarrow \sup_{\mathbb{R}^n \ni x \neq y \in \mathbb{R}^n} \frac{|f(x) - f(y)|}{\phi(|x-y|)} < \infty.$$

Moreover, if

$$\begin{cases} j \in \{0, 1, \dots, n\} \\ x_0 = t \\ f_j \in L^2(\mathbb{R}^n) \\ u_j(x,t) = f_j * P_t(x), \end{cases}$$

then

$$f_j = \mathcal{R}_j(f_0) \quad \forall j \in \{0, 1, \ldots, n\}$$

amounts to the generalized Cauchy–Riemann system (cf. [32, p. 263, 4.19])

$$\sum_{j=0}^{n} \partial_{x_j} u_j = 0 \quad \& \quad \partial_{x_k} u_j = \partial_{x_j} u_k \quad \forall j \neq k.$$

6 Incompressible Navier–Stokes system

This chapter is devoted to an application of $Q_\alpha(\mathbb{R}^n)$ and $Q_\alpha^{-1}(\mathbb{R}^n)$ in treating not only the existence and uniqueness of a mild solution u to the evolutionary incompressible Navier–Stokes system with the pressure p on $\mathbb{R}^{n+1}_+$ and the initial data a

$$(\star)\quad \begin{cases} \partial_t u - \Delta u + u\cdot\nabla u + \nabla \mathrm{p} = 0 & \text{in } \mathbb{R}^{n+1}_+ \\ \nabla\cdot u = 0 & \text{in } \mathbb{R}^n \\ u(\cdot,0) = a(\cdot) & \text{in } \mathbb{R}^n, \end{cases}$$

but also the zero solution of the stationary incompressible Navier–Stokes system with the pressure p on $\mathbb{R}^n$ (as the steady state of $(\star)$)

$$(\star\star)\quad \begin{cases} \Delta u - (u\cdot\nabla)u + \nabla \mathrm{p} = 0 & \text{in } \mathbb{R}^n \\ \operatorname{div} u = 0 & \text{in } \mathbb{R}^n \\ u(x) \to 0 & \text{as } \infty \leftarrow x \in \mathbb{R}^n. \end{cases}$$

6.1 Evolutionary N-S system in $(Q^{-1})^n$

Classically, the Cauchy problem for $(\star)$ is to establish the existence of a solution/velocity

$$u = u(x,t) = \big(u_1(x,t),\dots,u_n(x,t)\big)$$

with a pressure $\mathrm{p} = \mathrm{p}(x,t)$ of the fluid at time $t \in [0,\infty)$ and position $x \in \mathbb{R}^n$ assuming the initial data/velocity

$$a = a(x) = \big(a_1(x),\dots,a_n(x)\big).$$

Of particular importance is the invariance of $(\star)$ under the scaling transformation

$$\begin{cases} u(x,t) \mapsto u_\lambda(x,t) = \lambda u(\lambda^2 t, \lambda x) \\ \mathrm{p}(x,t) \mapsto \mathrm{p}_\lambda(x,t) = \lambda^2 \mathrm{p}(\lambda^2 t, \lambda x) \\ a(x) \mapsto a_\lambda(x) = \lambda a(\lambda x). \end{cases}$$

More precisely, if

$$\big(u(x,t), \mathrm{p}(x,t), a(x)\big)$$

satisfies $(\star)$ then

$$\big(u_\lambda(x,t), \mathrm{p}_\lambda(x,t), a_\lambda(x)\big)$$

https://doi.org/10.1515/9783110600285-006

also satisfies $(\star)$ for any $\lambda > 0$. This suggests a consideration of $(\star)$ with an initial data being of the scale invariance. Through the scale-invariance

$$\|a_\lambda\|_{L^n(\mathbb{R}^n)} = \sum_{j=1}^{n} \|(a_j)_\lambda\|_{L^n(\mathbb{R}^n)} = \|a\|_{L^n(\mathbb{R}^n)},$$

Kato proved in [42] that $(\star)$ has a mild solution locally in time if $a \in (L^n(\mathbb{R}^n))^n$ and globally if $\|a\|_{L^n(\mathbb{R}^n)}$ is small enough (for some generalizations of Kato's result see, e. g., [84] and [99]). Here and henceforward, by a mild solution $u(x,t)$ of $(\star)$ we mean that $u(x,t)$ solves the integral equation

$$u(x,t) = e^{t\Delta}a(x) - \int_0^t e^{(t-s)\Delta}\mathcal{P}\nabla\cdot(u\otimes u)\,ds,$$

where

$$\begin{cases} e^{t\Delta}a(x) = (e^{t\Delta}a_1(x),\dots,e^{t\Delta}a_n(x)) \\ \mathcal{P} = \{\mathcal{P}_{jk}\}_{j,k=1,\dots,n} = \{\delta_{jk} + (\partial_j(-\Delta)^{-\frac{1}{2}})(\partial_k(-\Delta)^{-\frac{1}{2}})\}_{j,k=1,\dots,n} \\ \delta_{jk} = \text{Kronecker symbol.} \end{cases}$$

Note that $\|\cdot\|_{Q^{-1}_{\alpha,\infty}(\mathbb{R}^n)}$ is invariant under the scale transformation

$$a(x) \mapsto \lambda a(\lambda x).$$

So it is a natural thing to extend the Kato's result to

$$(Q^{-1}_\alpha(\mathbb{R}^n))^n.$$

To do this, we introduce the following concept.

Definition 6.1. Let $\alpha \in [0,1)$ and $t_0 \in (0,\infty]$.

(i) A distribution f on $\mathbb{R}^n$ is said to be in $Q^{-1}_{\alpha,t_0}(\mathbb{R}^n)$ provided

$$\|f\|_{Q^{-1}_{\alpha,t_0}(\mathbb{R}^n)} = \sup_{(x,r)\in\mathbb{R}^n\times(0,t_0)} \left(r^{2\alpha-n}\int_0^{r^2}\int_{B(x,r)} |e^{t\Delta}f(y)|^2\frac{dy\,dt}{t^\alpha}\right)^{\frac{1}{2}} < \infty.$$

(ii) A distribution f on $\mathbb{R}^n$ is said to be in $\overline{VQ^{-1}_\alpha(\mathbb{R}^n)}$ provided

$$\lim_{t_0\to 0}\|f\|_{Q^{-1}_{\alpha,t_0}(\mathbb{R}^n)} = 0.$$

(iii) A function g on $\mathbb{R}^{n+1}_+$ is said to be in $X_{\alpha,t_0}(\mathbb{R}^{n+1}_+)$ provided

$$\|g\|_{X_{\alpha,t_0}(\mathbb{R}^{n+1}_+)} = \sup_{t\in(0,t_0)}\sqrt{t}\|g(\cdot,t)\|_{L^\infty(\mathbb{R}^n)}$$

$$+ \sup_{(x,r)\in\mathbb{R}^n\times(0,t_0)}\left(r^{2\alpha-n}\int_0^{r^2}\int_{B(x,r)}|g(y,t)|^2\frac{dy\,dt}{t^\alpha}\right)^{\frac{1}{2}} < \infty.$$

Clearly, if

$$\begin{cases} f_\lambda(x) = \lambda f(\lambda x) \\ g_\lambda(x,t) = \lambda g(\lambda^2 t, \lambda x) \\ (\lambda, t, x) \in (0,\infty) \times (0,\infty) \times \mathbb{R}^n, \end{cases}$$

then one has the scale invariance

$$\begin{cases} \|f_\lambda\|_{\mathrm{Q}^{-1}_{\alpha,\infty}(\mathbb{R}^n)} = \|f\|_{\mathrm{Q}^{-1}_{\alpha,\infty}(\mathbb{R}^n)} \\ \|g_\lambda\|_{\mathrm{X}_{\alpha,\infty}(\mathbb{R}^{n+1}_+)} = \|g\|_{\mathrm{X}_{\alpha,\infty}(\mathbb{R}^{n+1}_+)}. \end{cases}$$

Also, recalling (cf. [12])

$$f \in \dot{\mathrm{B}}^{-1+\frac{n}{p}}_{p,\infty}(\mathbb{R}^n) \Leftrightarrow \|e^{t\Delta} f\|_{\mathrm{L}^p(\mathbb{R}^n)} \lesssim t^{\frac{n-p}{2p}} \quad \text{under } p > n,$$

we have

$$p > n > \alpha p \Rightarrow \mathrm{L}^n(\mathbb{R}^n) \subseteq \dot{\mathrm{B}}^{-1+\frac{n}{p}}_{p,\infty}(\mathbb{R}^n) \subseteq \mathrm{Q}^{-1}_{\alpha,1}(\mathbb{R}^n),$$

which follows from Hölder's inequality based calculation for $r \in (0,1)$

$$\int_0^{r^2} \int_{B(x,r)} |e^{t\Delta} f(y)|^2 \frac{dy\, dt}{t^\alpha} \lesssim r^{n(1-\frac{2}{p})} \int_0^{r^2} \|e^{t\Delta} f\|^2_{\mathrm{L}^p(\mathbb{R}^n)} t^{-\alpha} dt \lesssim r^{n-2\alpha}.$$

In order to establish the existence and uniqueness of a mild solution of $(\star)$ with an initial data in

$$(\mathrm{Q}^{-1}_\alpha(\mathbb{R}^n))^n,$$

we need two lemmas.

Lemma 6.2. *Given* $(\alpha, n-1, t_0) \in [0,1) \times \mathbb{N} \times (0,\infty]$ *and a function* $f(\cdot,\cdot)$ *on* $\mathbb{R}^{n+1}_+$, *let*

$$\mathcal{I}(f,x,t) = \int_0^t e^{(t-s)\Delta} \Delta f(x,s)\, ds \quad \forall\, (x,t) \in \mathbb{R}^{n+1}_+.$$

Then

$$\int_0^{t_0} \|\mathcal{I}(f,\cdot,t)\|^2_{\mathrm{L}^2(\mathbb{R}^n)} \frac{dt}{t^\alpha} \lesssim \int_0^{t_0} \|f(\cdot,t)\|^2_{\mathrm{L}^2(\mathbb{R}^n)} \frac{dt}{t^\alpha}.$$

Proof. It suffices to check the last inequality for $t_0 = \infty$ thanks to the following observation:

- ▷ $\mathcal{I}(f,x,t)$ counts only on the values of f on $\mathbb{R}^n \times (0,t)$;
- ▷ f is extendable by setting $f = 0$ on (t_0, ∞);
- ▷ one can define

$$f(\cdot) = 0 = \mathcal{I}(f,\cdot,t) \quad \forall t \in (-\infty, 0).$$

Upon putting

$$e^{t\Delta}(x,0) = (4\pi t)^{-\frac{n}{2}} \exp\left(-\frac{|x|^2}{4t}\right) \quad \& \quad \mathcal{K}(x,t) = \begin{cases} \Delta e^{t\Delta}(x,0) & \forall t > 0 \\ 0 \quad \forall t \le 0, \end{cases}$$

we get

$$\mathcal{I}(f,x,t) = \int_{\mathbb{R}}\int_{\mathbb{R}^n} \mathcal{K}(t-s, x-y) f(y,s)\, \mathrm{d}y\mathrm{d}s,$$

thereby finding that $\mathcal{I}(f,x,t)$ becomes a convolution operator over $\mathbb{R}^{1+n}$. Due to

$$\widehat{\mathcal{K}(\cdot,t)}(\zeta) = \int_{\mathbb{R}^n} \mathcal{K}(x,t)\exp(-2\pi i x\cdot\zeta)\,\mathrm{d}x = -(2\pi)^2|\zeta|^2 \exp(-(2\pi)^2 t|\zeta|^2) \quad \forall t > 0,$$

we have

$$\begin{aligned}\widehat{\mathcal{I}(f,\cdot,t)}(\zeta) &= \int_{\mathbb{R}^{1+n}} f(y,s)\left(\int_{\mathbb{R}^n} \mathcal{K}(u,t-s)\exp(-2\pi i(u+y)\cdot\zeta)du\right)\mathrm{d}y\mathrm{d}s \\ &= -(2\pi)^2 \int_0^t |\zeta|^2 \exp(-(2\pi)^2(t-s)|\zeta|^2)\widehat{f(\cdot,s)}(\zeta)\,\mathrm{d}s.\end{aligned}$$

This last formula, along with the Fubini theorem and the Plancherel formula, derives

$$\begin{aligned}\int_0^\infty \|\mathcal{I}(f,\cdot,t)\|_{L^2}^2 \frac{\mathrm{d}t}{t^\alpha} &\le \int_0^\infty\left(\int_{\mathbb{R}^n}\left(\int_0^t \frac{|\zeta|^2}{\exp((2\pi)^2(t-s)|\zeta|^2)}|\widehat{f(\cdot,s)}(\zeta)|\mathrm{d}s\right)^2 \mathrm{d}\zeta\right)\frac{\mathrm{d}t}{t^\alpha} \\ &= (2\pi)^2 \int_{\mathbb{R}^n}\left(\int_0^\infty\left(\int_0^\infty \frac{\mathbb{1}_{\{0\le s\le t\}}|\zeta|^2|\widehat{f(\cdot,s)}(\zeta)|}{\exp((2\pi)^2(t-s)|\zeta|^2)}\mathrm{d}s\right)^2 \frac{\mathrm{d}t}{t^\alpha}\right)\mathrm{d}\zeta.\end{aligned}$$

This indicates that if

$$\int_0^\infty\left(\int_0^\infty (1_{\{0\le s\le t\}})\frac{|\zeta|^2|\widehat{f(\cdot,s)}(\zeta)|}{\exp((t-s)|\zeta|^2)}\mathrm{d}s\right)^2\frac{\mathrm{d}t}{t^\alpha} \le \int_0^\infty |\widehat{f(\cdot,t)}(\zeta)|^2\frac{\mathrm{d}t}{t^\alpha}$$

is proved then the Plancherel formula can be used once again to produce

$$\int_0^\infty \|\mathcal{I}(f,\cdot,t)\|^2_{L^2(\mathbb{R}^n)} \frac{dt}{t^\alpha} \lesssim \int_0^\infty \|f(\cdot,t)\|^2_{L^2(\mathbb{R}^n)} \frac{dt}{t^\alpha},$$

as required.

In order to prove the required inequality, let us rewrite its left-hand side as

$$\int_0^\infty \left(\int_0^\infty K(s,t)F(\zeta,s)\,ds\right)^2 dt,$$

where

$$\begin{cases} F(\zeta,s) = s^{-\frac{\alpha}{2}}|\widehat{f(\cdot,s)}(\zeta)| \\ K(s,t) = (\mathbb{1}_{\{0\le s\le t\}})\left(\frac{s}{t}\right)^{\frac{\alpha}{2}} |\zeta|^2 \exp(-(t-s)|\zeta|^2). \end{cases}$$

A calculation demonstrates

$$\begin{cases} \int_0^\infty K(s,t)\,ds = |\zeta|^2 \int_0^t \left(\frac{s}{t}\right)^{\frac{\alpha}{2}} \exp(-(t-s)|\zeta|^2)\,ds \le 1 \\ \int_0^\infty K(s,t)\,dt = |\zeta|^2 \int_s^\infty \left(\frac{s}{t}\right)^{\frac{\alpha}{2}} \exp(-(t-s)|\zeta|^2)\,dt \le 1, \end{cases}$$

and then an application of the Schur lemma gives

$$\int_0^\infty \left(\int_0^\infty K(s,t)F(\zeta,s)\,ds\right)^2 dt \lesssim \int_0^\infty (F(t,\zeta))^2 dt,$$

as desired. □

Lemma 6.3. *Let* $(\alpha, n-1) \in (0,1) \times \mathbb{N}$ *and* f *be a function on* $\mathbb{R}^n \times (0,1)$. *Then*

$$\int_0^1 \left\| \sqrt{-\Delta} e^{t\Delta} \int_0^t f(\cdot,s)\,ds \right\|^2_{L^2(\mathbb{R}^n)} \frac{dt}{t^\alpha}$$
$$\lesssim \left(\int_0^1 \|f(\cdot,t)\|_{L^1(\mathbb{R}^n)} \frac{dt}{t^\alpha}\right) \sup_{(x,r)\in\mathbb{R}^n\times(0,1)} r^{2\alpha-n} \int_0^{r^2} \int_{B(x,r)} |f(y,t)| \frac{dy dt}{t^\alpha}.$$

Proof. Recall that $\langle\cdot,\cdot\rangle$ is the inner product in $L^2(\mathbb{R}^n)$. Then

$$\|\cdots\|^2_{L^2(\mathbb{R}^n)} = \int_{\mathbb{R}^n} \left| \sqrt{-\Delta} e^{t\Delta} \int_0^t f(y,s)\,ds \right|^2 dy$$

$$= \left\langle \sqrt{-\Delta} e^{t\Delta} \int_0^t f(y,s)\,\mathrm{d}s, \sqrt{-\Delta} e^{t\Delta} \int_0^t f(y,s)\,\mathrm{d}s \right\rangle$$

$$= \int_0^t \int_0^t \langle \sqrt{-\Delta} e^{t\Delta} f(\cdot,s), \sqrt{-\Delta} e^{t\Delta} f(h,\cdot) \rangle \mathrm{d}s\mathrm{d}h.$$

Also, from [54, p. 163] it follows that

$$\sup_{(z,s)\in\mathbb{R}^n\times(0,1)} \int_0^s e^{2s\Delta} |f(h,z)| \mathrm{d}h \lesssim \sup_{(x,r)\in\mathbb{R}^n\times(0,1)} r^{-n} \int_0^{r^2} \int_{B(x,r)} |f(y,s)| \mathrm{d}y\mathrm{d}s,$$

and so that

$$\sup_{(z,s)\in\mathbb{R}^n\times(0,1)} \int_0^s e^{2s\Delta} |f(h,z)| \mathrm{d}h \lesssim \sup_{(x,r)\in\mathbb{R}^n\times(0,1)} r^{n-2\alpha} \int_0^{r^2} \int_{B(x,r)} |f(y,s)| \frac{\mathrm{d}y\,\mathrm{d}s}{s^\alpha}.$$

Consequently, we obtain

$$\begin{aligned}
\int_0^1 \|\cdots\|^2_{L^2(\mathbb{R}^n)} \frac{\mathrm{d}t}{t^\alpha} &\lesssim \iint_{0<h<s<1} \langle |f(\cdot,s)|, (e^{2\Delta} - e^{2s\Delta}) |f(h,\cdot)| \rangle \mathrm{d}s\mathrm{d}h \\
&\lesssim \int_0^1 \left\langle |f(\cdot,s)|, \int_0^s (e^{2\Delta} - e^{2s\Delta}) |f(h,\cdot)| \mathrm{d}h \right\rangle s^{-\alpha} \mathrm{d}s \\
&\lesssim \left(\int_0^1 \|f(\cdot,s)\|_{L^1(\mathbb{R}^n)} \frac{\mathrm{d}s}{s^\alpha} \right) \sup_{s\in(0,1]} \left\| \int_0^s e^{2s\Delta} |f(h,\cdot)| \mathrm{d}h \right\|_{L^\infty(\mathbb{R}^n)} \\
&\lesssim \left(\int_0^1 \|f(\cdot,s)\|_{L^1(\mathbb{R}^n)} \frac{\mathrm{d}s}{s^\alpha} \right) \sup_{(x,r)\in\mathbb{R}^n\times(0,1)} r^{n-2\alpha} \int_0^{r^2} \int_{B(x,r)} |f(y,s)| \frac{\mathrm{d}y\,\mathrm{d}s}{s^\alpha},
\end{aligned}$$

as desired. □

Below is the existence and uniqueness result for a mild solution to $(\star)$, generalizing the $\mathrm{BMO}^{-1}(\mathbb{R}^n)$-result in [46].

Theorem 6.4. *Let* $(0 \neq \alpha, n-1) \in (-\infty, 1) \times \mathbb{N}$.

(i) $(\star)$ *has a unique small global mild solution*

$$u \in (X_{\alpha,\infty}(\mathbb{R}^{n+1}_+))^n$$

for any initial data a with

$$\nabla \cdot a = 0 \quad \text{and} \quad \|a\|_{Q^{-1}_{\alpha,\infty}(\mathbb{R}^n)} \quad \text{being small.}$$

(ii) *For any* $t_0 \in (0,\infty)$ *there is an* $\epsilon > 0$ *such that* $(\star)$ *has a unique small mild solution*

$$u \text{ in } (X_{\alpha,t_0}(\mathbb{R}^{n+1}_+))^n \text{ on } \mathbb{R}^n \times (0,t_0)$$

when the initial data a satisfies

$$\nabla\cdot a = 0 \quad \& \quad \|a\|_{Q^{-1}_{\alpha,t_0}(\mathbb{R}^n)} \le \epsilon.$$

Consequently, for any

$$a \in \left(\overline{VQ^{-1}_\alpha(\mathbb{R}^n)}\right)^n \text{ with } \nabla\cdot a = 0$$

there exists a unique small local mild solution

$$u \text{ in } (X_{\alpha,t_0}(\mathbb{R}^{n+1}_+))^n \text{ on } \mathbb{R}^n \times (0,t_0).$$

Proof. Since

$$\mathrm{BMO}(\mathbb{R}^n) = Q_{-\infty<\alpha<0}(\mathbb{R}^n),$$

Theorem 6.4 under $\alpha \in (-\infty,0)$ is the main result of [46]. However, the following argument for $\alpha \in (0,1)$ is valid for $\mathrm{BMO}(\mathbb{R}^n)$. This can be seen via taking formally $\alpha = 0$.

In the sequel, $\alpha \in (0,1)$ is assumed. Notice that the following estimate for a distribution f on $\mathbb{R}^n$ (cf. [55, Lemma 16.1])

$$\|e^{t\Delta}f\|_{L^\infty(\mathbb{R}^n)} \lesssim t^{-\frac{1+n}{2}} \sup_{x\in\mathbb{R}^n} \int_0^t \int_{B(x,t)} |e^{s\Delta}f(y)|^2\,dy\,ds \quad \forall t \in (0,\infty)$$

implies

$$\sqrt{t}\|e^{t\Delta}f\|_{L^\infty(\mathbb{R}^n)} \lesssim \|f\|_{Q^{-1}_{0;T}(\mathbb{R}^{n+1}_+)} \lesssim \|f\|_{Q^{-1}_{\alpha,t_0}(\mathbb{R}^{n+1}_+)} \text{ for } 0 < t < t_0 \le \infty.$$

So, according to the Picard contraction principle (see, e. g., [55, p. 145, Theorem 15.1]), we know that verifying Theorem 6.4 via the integral equation induced by $(\star)$ is equivalent to showing that the bilinear operator

$$\mathcal{B}(u,v;t) = \int_0^t e^{(t-s)\Delta}\mathcal{P}\nabla\cdot(u\otimes v)\,ds$$

is bounded

$$\text{from} \quad (X_{\alpha,t_0}(\mathbb{R}^{n+1}_+))^n \times (X_{\alpha,t_0}(\mathbb{R}^{n+1}_+))^n \quad \text{to} \quad (X_{\alpha,t_0}(\mathbb{R}^{n+1}_+))^n \quad \text{where} \quad t_0 \in (0,\infty].$$

Needless to say,

$$u \in (X_{\alpha,t_0}(\mathbb{R}^{n+1}_+))^n \quad \& \quad a \in (Q^{-1}_{\alpha,t_0}(\mathbb{R}^n))^n$$

are respectively equipped with the norms

$$\|u\|_{X_{\alpha,t_0}(\mathbb{R}^{n+1}_+)} = \sum_{j=1}^{n} \|u_j\|_{X_{\alpha,t_0}(\mathbb{R}^{n+1}_+)} \quad \& \quad \|a\|_{Q^{-1}_{\alpha,t_0}(\mathbb{R}^n)} = \sum_{j=1}^{n} \|a_j\|_{Q^{-1}_{\alpha,t_0}(\mathbb{R}^n)}.$$

Step 1. We show L^∞-bound – if $t \in (0, t_0)$ then

$$|\mathcal{B}(u, v; t)| \lesssim t^{-\frac{1}{2}} \|u\|_{(X_{\alpha,t_0}(\mathbb{R}^{n+1}_+))^n} \|v\|_{(X_{\alpha,t_0}(\mathbb{R}^{n+1}_+))^n}.$$

As a matter of fact,

▷ if $\frac{t}{2} \le s < t$, then

$$\begin{aligned} \|e^{(t-s)\Delta}\mathcal{P}\nabla \cdot (u \otimes v)\|_{L^\infty(\mathbb{R}^n)} &\lesssim \|u\|_{L^\infty(\mathbb{R}^n)} \|v\|_{L^\infty(\mathbb{R}^n)} (\sqrt{t-s})^{-1} \\ &\lesssim \|u\|_{(X_{\alpha,t_0}(\mathbb{R}^{n+1}_+))^n} \|v\|_{(X_{\alpha,t_0}(\mathbb{R}^{n+1}_+))^n} (s\sqrt{t-s})^{-1}. \end{aligned}$$

▷ if $0 < s < \frac{t}{2}$, then

$$\begin{aligned} |e^{(t-s)\Delta}\mathcal{P}\nabla \cdot (u \otimes v)| &\lesssim \int_{\mathbb{R}^n} \frac{|u(y, s)||v(y, s)|}{(\sqrt{t} + |x - y|)^{n+1}} dy \\ &\lesssim \sum_{k \in \mathbb{Z}^n} (\sqrt{t} + |k|)^{-(n+1)} \int_{x-y \in \sqrt{t}(k+[0,1]^n)} |u(y, s)||v(y, s)| dy. \end{aligned}$$

Thus, the Cauchy–Schwarz inequality is applied to imply

$$\int_0^t \int_{x-y \in \sqrt{t}(k+[0,1]^n)} |u(y, s)||v(y, s)| dy ds \lesssim t^{\frac{n}{2}} \|u\|_{X_{\alpha,t_0}(\mathbb{R}^{n+1}_+)} \|v\|_{X_{\alpha,t_0}(\mathbb{R}^{n+1}_+)},$$

and consequently,

$$\begin{aligned} |\mathcal{B}(u, v; t)| &\lesssim \int_0^{\frac{t}{2}} |e^{(t-s)\Delta}\mathcal{P}\nabla \cdot (u \otimes v)| ds + \int_{\frac{t}{2}}^{t} |e^{(t-s)\Delta}\mathcal{P}\nabla \cdot (u \otimes v)| ds \\ &\lesssim \left(t^{-\frac{1}{2}} + \int_{\frac{t}{2}}^{t} s^{-1}(t-s)^{-\frac{1}{2}} ds \right) \|u\|_{X_{\alpha,t_0}(\mathbb{R}^{n+1}_+)} \|v\|_{X_{\alpha,t_0}(\mathbb{R}^{n+1}_+)} \\ &\lesssim t^{-\frac{1}{2}} \|u\|_{X_{\alpha,t_0}(\mathbb{R}^{n+1}_+)} \|v\|_{X_{\alpha,t_0}(\mathbb{R}^{n+1}_+)}. \end{aligned}$$

Step 2. We prove L^2-bound – if $x \in \mathbb{R}^n$ and $r^2 \in (0, t_0)$ then

$$r^{2\alpha-n} \int_0^{r^2} \int_{B(x,r)} |\mathcal{B}(u, v; t)|^2 s^{-\alpha} dy ds \lesssim \|u\|^2_{(X_{\alpha,t_0}(\mathbb{R}^{n+1}_+))^n} \|v\|^2_{(X_{\alpha,t_0}(\mathbb{R}^{n+1}_+))^n}.$$

As a matter of fact, let

$$\begin{cases} \mathbb{1}_{x,r} = \mathbb{1}_{B(x,10r)} \\ \mathcal{B}(u,v;t) = \mathcal{B}_1(u,v;t) - \mathcal{B}_2(u,v;t) - \mathcal{B}_3(u,v;t) \\ \mathcal{B}_1(u,v;t) = \int_0^s e^{(s-h)\Delta}\mathcal{P}\nabla\cdot((1-\mathbb{1}_{x,r})u\otimes v)\,dh \\ \mathcal{B}_2(u,v;t) = (-\Delta)^{-\frac{1}{2}}\mathcal{P}\nabla\cdot\int_0^s e^{(s-h)\Delta}\Delta((-\Delta)^{-\frac{1}{2}}((-\Delta)^0 - e^{h\Delta})(\mathbb{1}_{x,r})u\otimes v)\,dh \\ \mathcal{B}_3(u,v;t) = (-\Delta)^{-\frac{1}{2}}\mathcal{P}\nabla\cdot(-\Delta)^{\frac{1}{2}}e^{s\Delta}\Big(\int_0^s (\mathbb{1}_{x,r})u\otimes v\,dh\Big) \\ (-\Delta)^0 = \text{the identity operator.} \end{cases}$$

Then, for

$$0 < s < r^2 \quad \& \quad |y-x| < r$$

we use the Cauchy–Schwarz inequality to achieve

$$\begin{aligned} |\mathcal{B}_1(u,v;t)| &\lesssim \int_0^s \int_{\mathbb{R}^n\setminus B(x,10r)} \frac{|u(z,h)||v(z,h)|}{(\sqrt{s-h}+|y-z|)^{n+1}}\,dzdh \\ &\lesssim \int_0^{r^2} \int_{\mathbb{R}^n\setminus B(x,10r)} \frac{|u(z,h)||v(z,h)|}{|x-z|^{n+1}}\,dzdh \\ &\lesssim \Big(\int_0^{r^2} \int_{\mathbb{R}^n\setminus B(x,10r)} \frac{|u(z,h)|^2}{|x-z|^{n+1}}\,dzdh\Big)^{\frac{1}{2}} \Big(\int_0^{r^2} \int_{\mathbb{R}^n\setminus B(x,10r)} \frac{|v(z,h)|^2}{|x-z|^{n+1}}\,dzdh\Big)^{\frac{1}{2}} \\ &\lesssim r^{-1}\|u\|_{X_{\alpha,t_0}(\mathbb{R}^{n+1}_+)}\|v\|_{X_{\alpha,t_0}(\mathbb{R}^{n+1}_+)}, \end{aligned}$$

thereby obtaining

$$\int_0^{r^2}\int_{B(x,r)} |\mathcal{B}_1(u,v;t)|^2 \frac{dy\,dt}{t^\alpha} \lesssim r^{n-2\alpha}\|u\|^2_{X_{\alpha,t_0}(\mathbb{R}^{n+1}_+)}\|v\|^2_{X_{\alpha,t_0}(\mathbb{R}^{n+1}_+)}.$$

For $\mathcal{B}_2(u,v;t)$ set

$$\mathcal{M}(f)(y,h) = \mathbb{1}_{x,r}(u\otimes v) = \mathbb{1}_{x,r}(y)(u(y,h)\otimes v(y,h)).$$

From the L^2-boundedness of the Riesz transformation and Lemma 6.2 it follows that

$$\int_0^{r^2} \|\mathcal{B}_2(u,v;t)\|^2_{L^2(\mathbb{R}^n)}\frac{dt}{t^\alpha} \lesssim \int_0^{r^2} \|((-\Delta)^{-\frac{1}{2}}((-\Delta)^0 - e^{s\Delta})\mathcal{M}(f)(\cdot,s))\|^2_{L^2(\mathbb{R}^n)}\frac{ds}{s^\alpha}.$$

Note that

$$\sup_{s\in(0,\infty)} s^{-1}(1-\exp(-s^2)) < \infty.$$

So,

$$(-\Delta)^{-\frac{1}{2}}((-\Delta)^0 - e^{s\Delta})$$

is bounded on $\mathrm{L}^2(\mathbb{R}^n)$ with its operator norm $\lesssim \sqrt{s}$. This fact, along with the Cauchy–Schwarz inequality, implies

$$\int_0^{r^2} \|\mathcal{B}_2(u,v;t)\|^2_{\mathrm{L}^2(\mathbb{R}^n)} \frac{dt}{t^\alpha} \lesssim r^{n-2\alpha}\|u\|^2_{\mathrm{X}_{\alpha,t_0}(\mathbb{R}^{n+1}_+)}\|v\|^2_{\mathrm{X}_{\alpha,t_0}(\mathbb{R}^{n+1}_+)}.$$

Similarly, we can establish the following estimate for $\mathcal{B}_3(u,v;t)$:

$$\int_0^{r^2} \|\mathcal{B}_3(u,v;t)\|^2_{\mathrm{L}^2(\mathbb{R}^n)} \frac{dt}{t^\alpha} \lesssim r^{4+n-2\alpha} \int_0^1 \left\|(-\Delta)^{\frac{1}{2}} e^{t\Delta} \int_0^t |\mathcal{M}(f)(r^2\theta, r\cdot)|d\theta\right\|^2_{\mathrm{L}^2(\mathbb{R}^n)} \frac{dt}{t^\alpha}.$$

To see this, note that Lemma 6.3 reveals that if

$$\mathrm{K}(\mathcal{M}(f);\alpha) = \sup_{\rho\in(0,1)} \rho^{-n} \int_0^{\rho^2}\int_{B(x,\rho)} |\mathcal{M}(f)(r^2\theta, rw)|t^{-\alpha} dw\, dt$$

then

$$\int_0^1 \left\|(-\Delta)^{\frac{1}{2}} e^{t\Delta} \int_0^t |\mathcal{M}(f)(r^2\theta, r\cdot)|d\theta\right\|^2_{\mathrm{L}^2(\mathbb{R}^n)} \frac{dt}{t^\alpha}$$
$$\lesssim \mathrm{K}(\mathcal{M}(f);\alpha) \int_0^1 \|\mathcal{M}(f)(r^2\theta, r\cdot)\|_{\mathrm{L}^1(\mathbb{R}^n)} \frac{d\theta}{\theta^\alpha}.$$

So, the easily-verified estimates

$$\begin{cases} \mathrm{K}(\mathcal{M}(f);\alpha) \lesssim r^{-2}\|u\|_{\mathrm{X}_{\alpha,t_0}(\mathbb{R}^{n+1}_+)}\|v\|_{\mathrm{X}_{\alpha,t_0}(\mathbb{R}^{n+1}_+)} \\ \int_0^1 \|\mathcal{M}(f)(r^2\theta, r\cdot)\|_{\mathrm{L}^1(\mathbb{R}^n)} \frac{d\theta}{\theta^\alpha} \lesssim r^{-2}\|u\|_{\mathrm{X}_{\alpha,t_0}(\mathbb{R}^{n+1}_+)}\|v\|_{\mathrm{X}_{\alpha,t_0}(\mathbb{R}^{n+1}_+)} \end{cases}$$

derive

$$\int_0^{r^2} \|\mathcal{B}_3(u,v;t)\|^2_{\mathrm{L}^2(\mathbb{R}^n)} \frac{dt}{t^\alpha} \lesssim r^{n-2\alpha}\|u\|^2_{\mathrm{X}_{\alpha,t_0}(\mathbb{R}^{n+1}_+)}\|v\|^2_{\mathrm{X}_{\alpha,t_0}(\mathbb{R}^{n+1}_+)}.$$

Putting the estimates for $\{\mathcal{B}_j(u,v)\}_{j=1}^3$ together, we find the inequality in *Step 2*.

Finally, the boundedness of

$$\mathcal{B}(\cdot,\cdot;t) : (\mathrm{X}_{\alpha,t_0}(\mathbb{R}^{n+1}_+))^n \times (\mathrm{X}_{\alpha,t_0}(\mathbb{R}^{n+1}_+))^n \mapsto (\mathrm{X}_{\alpha,t_0}(\mathbb{R}^{n+1}_+))^n$$

follows from both steps above. Of course, $t_0 = \infty$ and $t_0 \in (0,\infty)$ assure (i) and (ii), respectively. □

6.2 Evolutionary N-S system in $(\mathrm{M}^{2,2})^n$ or $(\dot{\mathrm{B}}^{-1}_{\infty,\infty})^n$

A careful observation of the analysis carried out in Section 6.1 indicates that we cannot take $\alpha = 1$ in those lemmas and theorem. But, upon observing

$$\begin{cases} (-\Delta)^{-\frac{1}{2}}\mathrm{M}^{2,2}(\mathbb{R}^n) \subseteq \mathrm{BMO}(\mathbb{R}^n) \subseteq \dot{\mathrm{B}}^{1}_{\infty,\infty}(\mathbb{R}^n) \\ \mathrm{M}^{2,2}(\mathbb{R}^n) \subseteq \mathrm{BMO}^{-1}(\mathbb{R}^n) \subseteq \dot{\mathrm{B}}^{-1}_{\infty,\infty}(\mathbb{R}^n) \\ f_\lambda(x) = \lambda f(\lambda x) \quad \forall\,(x,\lambda) \in \mathbb{R}^{n+1}_+ \\ \|f_\lambda\|_{\mathrm{M}^{2,2}(\mathbb{R}^n)} = \|f\|_{\mathrm{M}^{2,2}(\mathbb{R}^n)} \quad \forall\,\lambda \in \mathbb{R}_+ \\ \|f_\lambda\|_{\dot{\mathrm{B}}^{-1}_{\infty,\infty}(\mathbb{R}^n)} = \|f\|_{\dot{\mathrm{B}}^{-1}_{\infty,\infty}(\mathbb{R}^n)} \quad \forall\,\lambda \in \mathbb{R}_+ \\ f \in \dot{\mathrm{B}}^{1}_{\infty,\infty}(\mathbb{R}^n) \Leftrightarrow \|f\|_{\dot{\mathrm{B}}^{1}_{\infty,\infty}(\mathbb{R}^n)} = \sup_{(x,r)\in\mathbb{R}^{n+1}_+} \sqrt{r}|\nabla e^{r\Delta}f(x)| < \infty \\ f \in \dot{\mathrm{B}}^{-1}_{\infty,\infty}(\mathbb{R}^n) \Leftrightarrow \|f\|_{\dot{\mathrm{B}}^{-1}_{\infty,\infty}(\mathbb{R}^n)} = \sup_{(x,r)\in\mathbb{R}^{n+1}_+} \sqrt{r}|e^{r\Delta}f(x)| < \infty, \end{cases}$$

we discover that it is possible to consider a lower or an upper setting of Theorem 6.4 through bringing

$$\text{either} \quad \mathrm{M}^{2,2}(\mathbb{R}^n) \quad \text{or} \quad \dot{\mathrm{B}}^{-1}_{\infty,\infty}(\mathbb{R}^n)$$

into play.

Definition 6.5. Let $(n-1, t_0) \in \mathbb{N} \times (0,\infty]$.

(i) A function $f \in \mathrm{M}^{2,2}(\mathbb{R}^n)$ is said to be in $\mathrm{VM}^{2,2}(\mathbb{R}^n)$ provided that for any $\epsilon > 0$ there is a $\mathrm{C}^\infty_0(\mathbb{R}^n)$ function h such that

$$\|f - h\|_{\mathrm{M}^{2,2}(\mathbb{R}^n)} < \epsilon,$$

namely, $\mathrm{VM}^{2,2}(\mathbb{R}^n)$ is the closure of $\mathrm{C}^\infty_0(\mathbb{R}^n)$ in $\mathrm{M}^{2,2}(\mathbb{R}^n)$.

(ii) A function $g \in \mathrm{L}^2_{\mathrm{loc}}(\mathbb{R}^n \times (0,t_0))$ is said to be in $\mathrm{E}_{t_0}(\mathbb{R}^{n+1}_+)$ provided

$$\|g\|_{\mathrm{E}_{t_0}(\mathbb{R}^{n+1}_+)} = \sup_{t\in(0,t_0)} \sqrt{t}\|g(\cdot,t)\|_{\mathrm{L}^\infty(\mathbb{R}^n)} + \sup_{t\in(0,t_0)} t^{\frac{1}{4}}\|g(\cdot,t)\|_{\mathrm{M}^{4,2}(\mathbb{R}^n)} < \infty.$$

Lemma 6.6. *Let* $(n-1, t_0) \in \mathbb{N} \times (0, \infty]$. *Then:*
(i) $(-\Delta)^{-\frac{1}{2}}\mathsf{M}^{2,2}(\mathbb{R}^n) \subseteq \bigcap_{\alpha\in(0,\frac{1}{2})}\mathsf{Q}_\alpha(\mathbb{R}^n)$;
(ii) $\mathsf{M}^{2,2}(\mathbb{R}^n) \subseteq \bigcap_{\alpha\in[0,\frac{1}{2})}\mathsf{Q}^{-1}_{\alpha,\infty}(\mathbb{R}^n)$;
(iii) $\mathsf{E}_{t_0}(\mathbb{R}^{n+1}_+) \subseteq \bigcap_{\alpha\in[0,\frac{1}{2})}\mathsf{X}^{-1}_{\alpha,1}(\mathbb{R}^n)$.

Proof. (i) For

$$f \in (-\Delta)^{-\frac{1}{2}}\mathsf{M}^{2,2}(\mathbb{R}^n) \subseteq \mathsf{BMO}(\mathbb{R}^n),$$

a function $\chi \in \mathsf{S}(\mathbb{R}^n)$ and an integer j (i. e., $j \in \mathbb{Z}$), let

$$\begin{cases} f = (-\Delta)^{-\frac{1}{2}}g \\ \Psi_j(f)(x) = \chi_j * f(x) \\ \widehat{\Psi'_j(f)}(x) = |2^j x|^\alpha \hat{\chi}(2^{-j}x)\hat{f}(x) \\ \operatorname{supp}(\hat{\chi}) \subseteq \{y \in \mathbb{R}^n : 2^{-1} \le |y| \le 2\} \\ \chi_j(x) = 2^{jn}\chi(2^j x) \\ \sum_j \hat{\chi}_j = 1. \end{cases}$$

A simple computation gives

$$\ell(I)^{2\alpha-n}\int_I\int_I \frac{|f(x)-f(y)|^2}{|x-y|^{n+2\alpha}}\,dx\,dy \lesssim \mathsf{D}_1(I) + \mathsf{D}_2(I) \quad \forall I \in \mathscr{C}(\mathbb{R}^n),$$

where

$$\begin{cases} \mathsf{D}_1(I) = \ell(I)^{2\alpha-n}\int_I\int_I \frac{|\sum_{j<-\log_2\ell(I)}\Psi_j(f)(x) - \sum_{j<-\log_2\ell(I)}\Psi_j(f)(y)|^2}{|x-y|^{n+2\alpha}}\,dx\,dy; \\ \mathsf{D}_2(I) = \ell(I)^{2\alpha-n}\int_I\int_I \frac{|\sum_{j\ge-\log_2\ell(I)}\Psi_j(f)(x) - \sum_{j\ge-\log_2\ell(I)}\Psi_j(f)(y)|^2}{|x-y|^{n+2\alpha}}\,dx\,dy. \end{cases}$$

According to [59, (3.2)], we get

$$\sup_{I\in\mathscr{C}(\mathbb{R}^n)} \mathsf{D}_1(I) \lesssim \|f\|^2_{\mathsf{BMO}(\mathbb{R}^n)} \sup_{I\in\mathscr{C}(\mathbb{R}^n)} \ell(I)^{2\alpha-n-2}\int_I\int_I |x-y|^{2-2\alpha-n}\,dx\,dy \lesssim \|g\|^2_{\mathsf{M}^{2,2}(\mathbb{R}^n)}.$$

Meanwhile, the last estimate for [59, IV] and [4, (22)] yields

$$\sup_{I\in\mathscr{C}(\mathbb{R}^n)} \mathsf{D}_2(I) \lesssim \sup_{I\in\mathscr{C}(\mathbb{R}^n)} \ell(I)^{2\alpha-n} \sum_{j\ge-\log_2\ell(I)} 2^{2\alpha j}\|(-\Delta)^{\frac{1}{2}}\Psi'_j(f)\|^2_{\mathsf{L}^2(I)} \lesssim \|g\|^2_{\mathsf{M}^{2,2}(\mathbb{R}^n)}.$$

Accordingly, $f \in \mathsf{Q}_\alpha(\mathbb{R}^n)$.

(ii) Suppose $f \in \mathrm{M}^{2,2}(\mathbb{R}^n)$. Then an application of the Cauchy–Schwarz inequality implies

$$\|f\|_{\mathrm{M}^{1,1}(\mathbb{R}^n)} = \sup_{(x,r)\in\mathbb{R}^{n+1}_+} r^{1-n} \int_{B(x,r)} |f(y)|\,dy \lesssim \|f\|_{\mathrm{M}^{2,2}(\mathbb{R}^n)} < \infty.$$

This, plus the Minkowski inequality, derives that if $(\alpha,t) \in [0,2^{-1}) \times (0,1)$ then

$$\begin{aligned}\Big\| \int_0^t |e^{s\Delta}f|^2 s^{-\alpha}\,ds \Big\|_{\mathrm{M}^{1,1}(\mathbb{R}^n)} &\lesssim \int_0^t \||e^{s\Delta}f|^2\|_{\mathrm{M}^{1,1}(\mathbb{R}^n)} s^{-\alpha}\,ds \\ &\lesssim \int_0^t \|f\|^2_{\mathrm{M}^{1,1}(\mathbb{R}^n)} s^{-\alpha-\frac{1}{2}}\,ds \\ &\lesssim \|f\|^2_{\mathrm{M}^{1,1}(\mathbb{R}^n)} t^{\frac{1}{2}-\alpha} \\ &\lesssim \|f\|^2_{\mathrm{M}^{2,2}(\mathbb{R}^n)} t^{\frac{1}{2}-\alpha},\end{aligned}$$

where we have used the estimate (cf. [55, Theorems 18.1–18.2])

$$\||e^{s\Delta}f|^2\| \lesssim s^{-\frac{1}{2}} \|f\|_{\mathrm{M}^{1,1}(\mathbb{R}^n)} \|f\|_{\dot{\mathrm{B}}^{-1}_{\infty,\infty}(\mathbb{R}^n)} \lesssim s^{-\frac{1}{2}} \|f\|^2_{\mathrm{M}^{1,1}(\mathbb{R}^n)}$$

and the equivalence

$$\|f\|_{\dot{\mathrm{B}}^{-1}_{\infty,\infty}(\mathbb{R}^n)} < \infty \Leftrightarrow \sup_{s\in(0,\infty)} \sqrt{s}\|e^{s\Delta}f\|_{\mathrm{L}^\infty(\mathbb{R}^n)} < \infty.$$

The above argument yields

$$r^{2\alpha-n}\Big(\int_0^{r^2}\int_{B(x,r)} |e^{s\Delta}f|^2\,dy\Big)\frac{ds}{s^\alpha} \lesssim r^{2\alpha-1}\Big\|\int_0^{r^2} |e^{s\Delta}f|^2 s^{-\alpha}\,ds\Big\|_{\mathrm{M}^{1,1}(\mathbb{R}^n)} \lesssim \|f\|^2_{\mathrm{M}^{2,2}(\mathbb{R}^n)},$$

whence $f \in \mathrm{Q}_{\alpha,\infty}(\mathbb{R}^n)$.

(iii) To do so, we just use the Cauchy–Schwarz inequality and $\alpha \in [0,2^{-1})$ to get that if $g \in \mathrm{E}_{t_0}(\mathbb{R}^{n+1}_+)$ then

$$\int_{B(x,r)} |g(t,y)|^2 dy \lesssim r^{n-1}\|g(\cdot,t)\|^2_{\mathrm{M}^{4,2}(\mathbb{R}^n)} \lesssim t^{-\frac{1}{2}} r^{n-1} \|g\|^2_{\mathrm{E}_{t_0}(\mathbb{R}^{n+1}_+)}$$

and hence $g \in \mathrm{X}_{\alpha,t_0}(\mathbb{R}^{n+1}_+)$. □

As the limiting case $\alpha \to 1$ of Theorem 6.7, we are motivated by Lemma 6.6 to discover the coming-up-next result on the existence and uniqueness for $(\star)$ under

$$\begin{cases} a = (a_1,\dots,a_n) \in \big(\mathrm{M}^{2,2}(\mathbb{R}^n)\big)^n \\ \|a\|_{\mathrm{M}^{2,2}(\mathbb{R}^n)} = \sum_{j=1}^n \|a_j\|_{\mathrm{M}^{2,2}(\mathbb{R}^n)} \\ \|u\|_{\mathrm{E}_{t_0}(\mathbb{R}^{n+1}_+)} = \sum_{j=1}^n \|u_j\|_{\mathrm{E}_{t_0}(\mathbb{R}^{n+1}_+)}. \end{cases}$$

Theorem 6.7. *Let* $(n-1, t_0) \in \mathbb{N} \times (0, \infty]$.

(i) $(\star)$ *has a unique small global mild solution*

$$u \in \left(\mathrm{E}_\infty(\mathbb{R}^{n+1}_+)\right)^n$$

for any initial data

$$a = (a_1, \dots, a_n) \quad \text{with} \quad \nabla \cdot a = 0 \quad \text{and} \quad \|a\|_{(\mathrm{M}^{2,2}(\mathbb{R}^n))^n} \quad \text{being small.}$$

(ii) *For any*

$$a = (a_1, \dots, a_n) \in \left(\mathrm{VM}^{2,2}(\mathbb{R}^n)\right)^n \quad \text{with} \quad \nabla \cdot a = 0$$

there exists a time $t_0 \in (0, \infty)$ *depending on a such that* $(\star)$ *has a unique small local mild solution*

$$u \quad \text{in} \quad \left(\mathrm{E}_{t_0}(\mathbb{R}^{n+1}_+)\right)^n \quad \text{on} \quad \mathbb{R}^n \times (0, t_0).$$

Proof. Note that the following estimate for $f \in \mathrm{M}^{2,2}(\mathbb{R}^n)$ (cf. [55, Theorem 18.1])

$$|e^{t\Delta} f(x)| \lesssim \sum_{k \in \mathbb{Z}^n} \sup_{z \in k+[0,1]^n} \exp\left(-\frac{|z|^2}{4}\right) \int_{k+[0,1]^n} |f(x - \sqrt{t}y)|\, dy \quad \forall\, (x,t) \in \mathbb{R}^{n+1}_+,$$

along with the Cauchy–Schwarz inequality, deduces

$$\sqrt{t}\|e^{t\Delta} f\|_{\mathrm{L}^\infty(\mathbb{R}^n)} \lesssim \|f\|_{\mathrm{M}^{2,2}(\mathbb{R}^n)}.$$

So, this last inequality, plus the uniform boundedness of the map

$$f \mapsto e^{t\Delta} f \quad \text{on } \mathrm{M}^{2,2}(\mathbb{R}^n),$$

namely,

$$\|e^{t\Delta} f\|_{\mathrm{M}^{2,2}(\mathbb{R}^n)} \lesssim \|f\|_{\mathrm{M}^{2,2}(\mathbb{R}^n)} \quad \forall\, t \in (0, \infty),$$

gives

$$\|e^{t\Delta} f\|_{\mathrm{M}^{4,2}(\mathbb{R}^n)} \lesssim \|e^{t\Delta} f\|^{\frac{1}{2}}_{\mathrm{L}^\infty(\mathbb{R}^n)} \|e^{t\Delta} f\|^{\frac{1}{2}}_{\mathrm{M}^{2,2}(\mathbb{R}^n)} \lesssim t^{-\frac{1}{4}} \|f\|_{\mathrm{M}^{2,2}(\mathbb{R}^n)}.$$

Thus

$$\begin{cases} \|e^{t\Delta} f(x)\|_{\mathrm{E}_{t_0}(\mathbb{R}^{n+1}_+)} \lesssim \|f\|_{\mathrm{M}^{2,2}(\mathbb{R}^n)} \\ \lim_{t_0 \to 0} \|e^{t\Delta} f(x)\|_{\mathrm{E}_{t_0}(\mathbb{R}^{n+1}_+)} = 0 \quad \text{as } f \in \mathrm{VM}^{2,2}(\mathbb{R}^n). \end{cases}$$

With the help of the last assertion and the Picard contraction principle, we find that showing Theorem 6.7 via the integral equation induced by ($\star$) amounts to proving the boundedness of the bilinear operator

$$\mathcal{B}(\cdot,\cdot,t):\ (\mathrm{E}_{t_0}(\mathbb{R}^{n+1}_+))^n \times (\mathrm{E}_{t_0}(\mathbb{R}^{n+1}_+))^n \mapsto (\mathrm{E}_{t_0}(\mathbb{R}^{n+1}_+))^n \quad \forall\, t_0 \in (0,\infty].$$

However, this boundedness follows from the following two estimates (cf. [56, (25)–(24)]) for $0 < s < t < t_0$

$$\begin{aligned}&\sqrt{t-s}\|e^{(t-s)\Delta}\mathcal{P}\nabla\cdot(u\otimes v)\|_{\mathrm{L}^\infty(\mathbb{R}^n)}\\ &\quad \lesssim \min\{s^{\frac{3}{2}}(t-s)\|u\|_{\mathrm{M}^{4,2}(\mathbb{R}^n)}\|v\|_{\mathrm{M}^{4,2}(\mathbb{R}^n)}, \|u\|_{\mathrm{L}^\infty(\mathbb{R}^n)}\|v\|_{\mathrm{L}^\infty(\mathbb{R}^n)}\}\end{aligned}$$

and

$$\sqrt{t-s}\|e^{(t-s)\Delta}\mathcal{P}\nabla\cdot(u\otimes v)\|_{\mathrm{M}^{4,2}(\mathbb{R}^n)} \lesssim \|u\|_{\mathrm{M}^{4,2}(\mathbb{R}^n)}\|v\|_{\mathrm{L}^\infty(\mathbb{R}^n)}.$$ □

As an upper situation of Theorem 6.7, we have the following regularity based on the smallness of $\|\cdot\|_{\dot{\mathrm{B}}^{-1}_{\infty,\infty}(\mathbb{R}^n)}$.

Theorem 6.8. *For $n-1 \in \mathbb{N}$, let $u = (u_1,\dots,u_n)$ be a smooth solution to ($\star$) and $t_0 > 0$ be its first possible blow-up time. There is a positive dimensional constant κ_0 such that if*

$$\sup_{t\in(t_0-\epsilon_0,t_0)}\|u(\cdot,t)\|_{\dot{\mathrm{B}}^{-1}_{\infty,\infty}(\mathbb{R}^n)} = \sup_{t\in(t_0-\epsilon_0,t_0)}\sum_{j=1}^n\|u_j(\cdot,t)\|_{\dot{\mathrm{B}}^{-1}_{\infty,\infty}(\mathbb{R}^n)} \le \kappa_0$$

holds for some $\epsilon_0 \in (0,t_0)$ then t_0 is not a blow-up time, and the solution u can be continued beyond t_0.

Proof. First of all, we need the Littlewood–Paley characterization of the Besov space

$$\dot{\mathrm{B}}^{-\infty<\gamma<\infty}_{\infty,\infty}(\mathbb{R}^n).$$

To do so, let $\psi_0 \in \mathrm{C}^\infty_0(\mathbb{R}^n)$ obey

$$\begin{cases} 0 \le \psi_0(x) \le 1 & \text{as } x \in \mathbb{R}^n\\ \psi_0(x) = 1 & \text{as } |x| \le 1\\ \psi_0(x) = 0 & \text{as } |x| \ge \frac{7}{6}\\ \psi(x) = \psi_0(x) - \psi_0(2x) & \text{as } x \in \mathbb{R}^n\\ \widehat{\Omega_{\le j}f}(x) = \psi_0(2^{-j}x)\hat{f}(x) & \text{as } (x,f,j) \in \mathbb{R}^n \times \mathrm{S}(\mathbb{R}^n)\times\mathbb{Z}\\ \widehat{\Omega_j f}(x) = \psi(2^{-j}x)\hat{f}(x) & \text{as } (x,f,j) \in \mathbb{R}^n \times \mathrm{S}(\mathbb{R}^n)\times\mathbb{Z}\\ \Omega_j\Omega_k = 0 & \text{as } |j-k| > 1.\end{cases}$$

For $\gamma \in \mathbb{R}$, the homogeneous Besov space $\dot{B}^{\gamma}_{\infty,\infty}(\mathbb{R}^n)$ can be decided by the Littlewood–Paley-type norm

$$\|f\|_{\dot{B}^{\gamma}_{\infty,\infty}(\mathbb{R}^n),\dagger} = \sup_{j\in\mathbb{Z}} 2^{j\gamma}\|\Omega_j f\|_{L^\infty(\mathbb{R}^n)}.$$

Next, we show that if $\gamma \in (1,\infty)$ then

$$\big\|\Omega_j((u\cdot\nabla)u)\big\|_{L^\infty(\mathbb{R}^n)} \lesssim 2^{j(2-\gamma)}\|u\|_{\dot{B}^{-1}_{\infty,\infty}(\mathbb{R}^n),\dagger}\|u\|_{\dot{B}^{\gamma}_{\infty,\infty}(\mathbb{R}^n),\dagger} \quad \forall j\in\mathbb{Z}.$$

In fact, upon recalling the Bony paraproduct for a pair of functions $\{f,g\}$

$$fg = \sum_{l\in\mathbb{Z}}(\Omega_l f)\tilde{g}_l + \sum_{l\in\mathbb{Z}}(\Omega_l f)(\Omega_{\le l-2}g) + \sum_{l\in\mathbb{Z}}(\Omega_l g)(\Omega_{\le l-2}f)$$

with

$$\tilde{g}_l = \Omega_{l-1}g + \Omega_l g + \Omega_{l+1}g,$$

we have

$$(u\cdot\nabla)u = \big[(u\cdot\nabla)u\big]_1 + \big[(u\cdot\nabla)u\big]_2 + \big[(u\cdot\nabla)u\big]_3,$$

where

$$\begin{cases} [(u\cdot\nabla)u]_1 = \sum_{l\in\mathbb{Z}}((\Omega_{\le l-2}u)\cdot\nabla)\Omega_l u \\ [(u\cdot\nabla)u]_2 = \sum_{l\in\mathbb{Z}}((\Omega_l u)\cdot\nabla)\Omega_{\le l-2}u \\ [(u\cdot\nabla)u]_3 = \sum_{l\in\mathbb{Z}}((\Omega_{\le l-2}u)\cdot\nabla)\Omega_l u\tilde{u}_l \\ \tilde{u}_l = \Omega_{l-1}u + \Omega_l u + \Omega_{l+1}u. \end{cases}$$

Using the frequency localization, we get

$$\begin{aligned} \big\|\Omega_j(\big[(u\cdot\nabla)u\big]_1)\big\|_{L^\infty(\mathbb{R}^n)} &\lesssim \sum_{|l-j|\le 2}\|(\Omega_{\le l-2}u)\cdot\nabla\Omega_l u\|_{L^\infty(\mathbb{R}^n)} \\ &\lesssim 2^{j(2-\gamma)}\|u\|_{\dot{B}^{-1}_{\infty,\infty}(\mathbb{R}^n),\dagger}\|u\|_{\dot{B}^{\gamma}_{\infty,\infty}(\mathbb{R}^n),\dagger} \end{aligned}$$

and similarly,

$$\begin{aligned} \big\|\Omega_j(\big[(u\cdot\nabla)u\big]_2)\big\|_{L^\infty(\mathbb{R}^n)} &\lesssim \sum_{|l-j|\le 2}\|(\Omega_l u)\cdot\nabla\Omega_{\le l-2}u\|_{L^\infty(\mathbb{R}^n)} \\ &\lesssim 2^{j(2-\gamma)}\|u\|_{\dot{B}^{-1}_{\infty,\infty}(\mathbb{R}^n),\dagger}\|u\|_{\dot{B}^{\gamma}_{\infty,\infty}(\mathbb{R}^n),\dagger}. \end{aligned}$$

Moreover, utilizing the divergence-free property of $(\star)$ we obtain

$$(\Omega_l u\cdot\nabla)\tilde{u}_l = \nabla\cdot\big((\Omega_l u)\otimes\tilde{u}_l\big)$$

thereby estimating under $\gamma > 1$

$$\begin{aligned}\|\Omega_j([(u\cdot\nabla)u]_3)\|_{L^\infty(\mathbb{R}^n)} &\lesssim 2^j \sum_{l\ge j-2} \|\Omega_l u\|_{L^\infty(\mathbb{R}^n)}\|\tilde{u}_l\|_{L^\infty(\mathbb{R}^n)}\\ &\lesssim 2^{j(2-\gamma)}\|u\|_{\dot{B}^{-1}_{\infty,\infty}(\mathbb{R}^n),\dagger}\|u\|_{\dot{B}^{\gamma}_{\infty,\infty}(\mathbb{R}^n),\dagger}.\end{aligned}$$

Finally, we prove that if u is a smooth solution of $(\star)$ on some time interval $[0, t_0]$ with smooth initial data $a(\cdot) = u(\cdot, 0)$ and $\gamma = \frac{3}{2}$ then there is dimensional constant $\epsilon_1 > 0$ such that

$$\sup_{0\le t\le t_0}\|u(\cdot,t)\|_{\dot{B}^{-1}_{\infty,\infty}(\mathbb{R}^n),\dagger} \le \epsilon_1 \Rightarrow \sup_{0\le t\le t_0}\|u(\cdot,t)\|_{\dot{B}^{\gamma}_{\infty,\infty}(\mathbb{R}^n),\dagger} \lesssim \|a\|_{\dot{B}^{\gamma}_{\infty,\infty}(\mathbb{R}^n),\dagger}.$$

As a matter of fact, if $\mathcal{P}$ still stands for the Leray projection as seen below

$$\mathcal{P} = \{\mathcal{P}_{jk}\}_{j,k=1,\dots,n} = \{\delta_{jk} + (\partial_j(-\Delta)^{-\frac{1}{2}})(\partial_k(-\Delta)^{-\frac{1}{2}})\}_{j,k=1,\dots,n},$$

then

$$\partial_t(\Omega_j u) - \Delta(\Omega_j u) = -\Omega_j(\mathcal{P}(u\cdot\nabla)u),$$

and hence for $t > 0$, there is a constant $c > 0$ relying only on n such that

$$\|e^{t\Delta}\Omega_j f\|_{L^\infty(\mathbb{R}^n)} \lesssim e^{-c2^{2j}t}\|\Omega_j f\|_{L^\infty(\mathbb{R}^n)} \quad \forall t \in (0,\infty).$$

This in turn implies

$$\begin{aligned}\|\Omega_j u(\cdot,t)\|_{L^\infty(\mathbb{R}^n)} &\lesssim e^{-c2^{2j}t}\|\Omega_j u(\cdot,0)\|_{L^\infty(\mathbb{R}^n)}\\ &\quad + \int_0^t e^{-c2^{2j}(t-s)}2^{j(2-\gamma)}\|u(\cdot,s)\|_{\dot{B}^{-1}_{\infty,\infty}(\mathbb{R}^n),\dagger}\|u(\cdot,s)\|_{\dot{B}^{\gamma}_{\infty,\infty}(\mathbb{R}^n),\dagger}\,ds\\ &\lesssim e^{-c2^{2j}t}\|\Omega_j u(\cdot,0)\|_{L^\infty(\mathbb{R}^n)}\\ &\quad + (1-e^{-c2^{2j}t})2^{-j\gamma}\sup_{0\le s\le t}\|u(\cdot,s)\|_{\dot{B}^{-1}_{\infty,\infty}(\mathbb{R}^n),\dagger}\sup_{0\le s\le t}\|u(\cdot,s)\|_{\dot{B}^{\gamma}_{\infty,\infty}(\mathbb{R}^n),\dagger}.\end{aligned}$$

Accordingly, upon multiplying the last estimation by $2^{j\gamma}$ we achieve

$$\begin{aligned}\sup_{0\le t\le t_0}\|u(\cdot,t)\|_{\dot{B}^{\gamma}_{\infty,\infty}(\mathbb{R}^n),\dagger} &\lesssim \|u(\cdot,0)\|_{\dot{B}^{\gamma}_{\infty,\infty}(\mathbb{R}^n),\dagger}\\ &\quad + \sup_{0\le t\le t_0}\|u(\cdot,t)\|_{\dot{B}^{-1}_{\infty,\infty}(\mathbb{R}^n),\dagger}\sup_{0\le t\le t_0}\|u(\cdot,t)\|_{\dot{B}^{\gamma}_{\infty,\infty}(\mathbb{R}^n),\dagger},\end{aligned}$$

as desired.

Now, upon considering the solution u on the time interval $[t_0-\epsilon, t_0-\eta]$ with $\eta \to 0$ and treating $u(\cdot, t_0-\epsilon)$ as an initial data, we gain a uniform estimate on $\|u(\cdot,t)\|_{\dot{B}^{\gamma}_{\infty,\infty}(\mathbb{R}^n),\dagger}$ independent of η, thereby finding that u must be regular beyond t_0 via a standard regularity argument. □

6.3 Stationary N-S system in $(\mathrm{BMO}^{-1} \cap \mathrm{C}^{2<p<\frac{2n}{n-2},\frac{p(n-2)}{2}})^n$

A deep consideration of the stationary N-S system within $\mathrm{BMO}^{-1}(\mathbb{R}^n)$ reveals the natural nullity of a smooth solution to $(\star\star)$.

Theorem 6.9. *Let $(p, n-1) \in (2, \frac{2n}{n-2}) \times \mathbb{N}$. Then any smooth solution $u = (u_1, \dots, u_n)$ of $(\star\star)$ vanishes under*

$$u_1, \dots, u_n \in \mathrm{BMO}^{-1}(\mathbb{R}^n) \cap \mathrm{C}^{p,\frac{p(n-2)}{2}}(\mathbb{R}^n).$$

Proof. Suppose that $u = (u_1, \dots, u_n)$ is a smooth solution of $(\star\star)$ enjoying the last condition.

Step 1. We prove the Caccioppoli-type estimate for $u \in (\mathrm{BMO}^{-1}(\mathbb{R}^n))^n$

$$\|\nabla u\|_{\mathrm{L}^2(B(x,\frac{r}{2}))}\left(\frac{r}{2}\right)^{\frac{2-n}{2}} \lesssim \|u - u_{B(x,r)}\|_{\mathrm{L}^p(B(x,r))} r^{-\frac{n}{p}} \quad \forall\, (r,x,p) \in (0,\infty) \times \mathbb{R}^n \times (2,\infty).$$

To do so, for

$$2^{-1} r \le t < s \le r$$

we take a nonnegative cut-off function $\varpi \in \mathrm{C}_0^\infty(B(x,r))$ satisfying

$$\begin{cases} \varpi = 1 & \text{in } B(x,t) \\ \varpi = 0 & \text{in } \mathbb{R}^n \setminus B(x,s) \\ |\nabla \varpi| \lesssim (s-t)^{-1} & \text{in } \mathbb{R}^n. \end{cases}$$

For simplicity, upon setting

$$\breve{u} = u - u_{B(x,r)} = \big(u_1 - (u_1)_{B(x,r)}, \dots, u_n - (u_n)_{B(x,r)}\big),$$

we gain that for a given $2 < p < \infty$ there is a function Φ (cf. [75, 2.1]) obeying

$$\begin{cases} \Phi|_{\mathbb{R}^n \setminus B(x,r)} = 0 \\ \operatorname{div} \Phi = \nabla \varpi \cdot \breve{u} = \operatorname{div}(\varpi \breve{u}) \\ \|\nabla \Phi\|_{\mathrm{L}^p(B(x,s))} \lesssim \|\nabla \varpi \cdot \breve{u}\|_{\mathrm{L}^p(B(x,s))} \lesssim (s-t)^{-1} \|\breve{u}\|_{\mathrm{L}^p(B(x,s))}. \end{cases}$$

Now, we utilize $\varpi \breve{u} - \Phi$ to testify $(\star\star)$, integrate by parts over $B(x,s)$, and find

$$\int_{B(x,s)} \varpi(y) |\nabla u(y)|^2 \, dy = \mathrm{E}_1 + \mathrm{E}_2 + \mathrm{E}_3 + \mathrm{E}_4,$$

where

$$\begin{cases} \mathrm{E}_1 = -\int\limits_{B(x,s)} \nabla u(y) : (\nabla \varpi(y) \otimes \check{u}(y))\, dy \\ \mathrm{E}_2 = \int\limits_{B(x,s)} \nabla \Phi(y) : \nabla u(y)\, dy \\ \mathrm{E}_3 = -\int\limits_{B(x,s)} (u(y) \cdot \nabla u(y)) \cdot \varpi(y)\check{u}(y)\, dy \\ \mathrm{E}_4 = \int\limits_{B(x,s)} (u(y) \cdot \nabla u(y)) \cdot \Phi(y)\, dy. \end{cases}$$

It remains to estimate E_1, E_2, E_3, E_4, respectively. From Hölder's inequality it follows that

$$\begin{aligned} |\mathrm{E}_1| + |\mathrm{E}_2| &\lesssim \int\limits_{B(x,s)} |\nabla u(y)||\nabla \varpi(y)||\check{u}(y)| dy \\ &\lesssim (s-t)^{-1} \Big(\int\limits_{B(x,s)} |\nabla u(y)|^2 dy \Big)^{\frac{1}{2}} \Big(\int\limits_{B(x,s)} |\check{u}(y)|^2 dy \Big)^{\frac{1}{2}} \\ &\lesssim \frac{r^{\frac{n(p-2)}{2p}}}{s-t} \Big(\int\limits_{B(x,s)} |\nabla u(y)|^2 dy \Big)^{\frac{1}{2}} \Big(\int\limits_{B(x,r)} |\check{u}(y)|^p dy \Big)^{\frac{1}{p}} \\ &\approx r^{\frac{n(p-2)}{2p}} (s-t)^{-1} \|\nabla u\|_{L^2(B(x,s))} \|\nabla \check{u}\|_{L^p(B(x,r))}. \end{aligned}$$

In order to control E_3 and E_4, we must use the assumption

$$u_1, \dots, u_n \in \mathrm{BMO}^{-1}(\mathbb{R}^n)$$

which induces a skew symmetric matrix

$$\diamond = ((\diamond_{1k})_{k=1}^n, \dots, (\diamond_{nk})_{k=1}^n)$$

such that

$$\begin{cases} u = (u_1, \dots, u_n) = \operatorname{div} \diamond = (\operatorname{div}(\diamond_{1k})_{k=1}^n, \dots, \operatorname{div}(\diamond_{nk})_{k=1}^n) \\ u_j = \nabla \cdot (\diamond_{j1}, \dots, \diamond_{jn}) = \operatorname{div}(\diamond_{jk})_{k=1}^n \\ \diamond_{jk} \in \mathrm{BMO}(\mathbb{R}^n) \\ \| \diamond \|_{\mathrm{BMO}(\mathbb{R}^n)} = \sum_{j,k} \| \diamond_{jk} \|_{\mathrm{BMO}(\mathbb{R}^n)}. \end{cases}$$

Employing integration-by-parts, Hölder's inequality, $p \in (2, \infty)$ and the skew symmetry of $\diamond$, we deduce

$$|\mathrm{E}_3| = \Big| \int\limits_{B(x,s)} u(y) \cdot \nabla \check{u}(y) \cdot \varpi(y)\check{u}(y)\, dy \Big|$$

$$
\begin{aligned}
&= \left| \sum_{i,j,k=1}^{n} \int_{B(x,s)} \check{\diamond}_{jk,k}(y)\check{u}_{i,j}(y)\check{u}_i(y)\varpi(y)\,\mathrm{d}y \right| \\
&= \left| \sum_{i,j,k=1}^{n} \int_{B(x,s)} \check{\diamond}_{jk}(y)\check{u}_{i,j}(y)\check{u}_i(y)\varpi_{,k}(y)\,\mathrm{d}y \right| \\
&\lesssim (s-t)^{-1}\Big(\int_{B(x,s)} |\nabla u(y)|^2\,\mathrm{d}y\Big)^{\frac{1}{2}}\Big(\int_{B(x,s)} |\check{\diamond}(y)|^2|\check{u}(y)|^2\,\mathrm{d}y\Big)^{\frac{1}{2}} \\
&\lesssim (s-t)^{-1}\|\nabla u\|_{L^2(B(x,s))}\|\check{u}\|_{L^p(B(x,s))}\Big(\int_{B(x,s)} |\check{\diamond}(y)|^{\frac{2p}{p-2}}\,\mathrm{d}y\Big)^{\frac{p-2}{2p}} \\
&\lesssim r^{\frac{n(p-2)}{2p}}(s-t)^{-1}\|\nabla u\|_{L^2(B(x,s))}\|\nabla\check{u}\|_{L^p(B(x,r))}\|\diamond\|_{\mathrm{BMO}(\mathbb{R}^n)}.
\end{aligned}
$$

In the above and below, $f_{,k}(y)$ represents $\partial_{y_k}f(y) = \partial_k f(y)$. Similarly, we have

$$
\begin{aligned}
|\mathrm{E}_4| &= \left| \sum_{i,j,k=1}^{n} \int_{B(x,s)} \check{\diamond}_{jk,k}(y)\check{u}_{i,j}(y)\Phi_i(y)\,\mathrm{d}y \right| \\
&= \left| \sum_{i,j,k=1}^{n} \int_{B(x,s)} \check{\diamond}_{jk}(y)\check{u}_{i,j}(y)\Phi_{i,k}(y)\,\mathrm{d}y \right| \\
&\lesssim \Big(\int_{B(x,s)} |\nabla u(y)|^2\,\mathrm{d}y\Big)^{\frac{1}{2}}\Big(\int_{B(x,s)} |\check{\diamond}(y)|^2|\nabla\Phi(y)|^2\,\mathrm{d}y\Big)^{\frac{1}{2}} \\
&\lesssim \Big(\int_{B(x,s)} |\nabla u(y)|^2\,\mathrm{d}y\Big)^{\frac{1}{2}}\Big(\int_{B(x,s)} |\nabla\Phi(y)|^p\,\mathrm{d}y\Big)^{\frac{1}{p}}\Big(\int_{B(x,s)} |\check{\diamond}(y)|^{\frac{2p}{p-2}}\,\mathrm{d}y\Big)^{\frac{p-2}{2p}} \\
&\lesssim (s-t)^{-1}\|\nabla u\|_{L^2(B(x,s))}\|\check{u}(y)\|_{L^p(B(x,r))}\Big(\int_{B(x,s)} |\check{\diamond}(y)|^{\frac{2p}{p-2}}\,\mathrm{d}y\Big)^{\frac{p-2}{2p}} \\
&\approx r^{\frac{n(p-2)}{2p}}(s-t)^{-1}\|\nabla u\|_{L^2(B(x,s))}\|\nabla\check{u}\|_{L^p(B(x,r))}\|\diamond\|_{\mathrm{BMO}(\mathbb{R}^n)}.
\end{aligned}
$$

Accordingly, we obtain

$$
\begin{aligned}
\int_{B(x,t)} |\nabla u(y)|^2\,\mathrm{d}y &\lesssim \int_{B(x,s)} \varpi(y)|\nabla u(y)|^2\,\mathrm{d}y \\
&\lesssim |\mathrm{E}_1| + |\mathrm{E}_2| + |\mathrm{E}_3| + |\mathrm{E}_4| \\
&\lesssim r^{\frac{n(p-2)}{2p}}(s-t)^{-1}\|\nabla u\|_{L^2(B(x,s))}\|\nabla\check{u}\|_{L^p(B(x,r))}(1+\|\diamond\|_{\mathrm{BMO}(\mathbb{R}^n)}).
\end{aligned}
$$

This, along with Young's inequality

$$
\rho\tau \le 2^{-1}(\epsilon^{-1}\rho^2 + \epsilon\tau^2) \quad \forall\,(\rho,\tau,\epsilon) \in (0,\infty)\times(0,\infty)\times(0,\infty),
$$

produces a constant $c_p > 0$ depending only on $p \in (2, \infty)$ such that

$$\|\nabla u\|^2_{\mathrm{L}^2(B(x,t))} \le 2^{-2}\|\nabla u\|^2_{\mathrm{L}^2(B(x,s))} + c_p r^{\frac{n(p-2)}{p}}(s-t)^{-2}\|\check{u}\|^2_{\mathrm{L}^p(B(x,r))}.$$

Now, by iteration we get the desired inequality.

Step 2. We validate $u = 0$ via

$$u_1, \dots, u_n \in \mathrm{C}^{2<p<\frac{2n}{n-2}, \frac{p(n-2)}{2}}(\mathbb{R}^n).$$

On the one hand, if

$$u_1, \dots, u_n \in \mathrm{C}^{2<p<\frac{2n}{n-2}, \frac{p(n-2)}{2}}(\mathbb{R}^n),$$

then the first application of *Step 1* yields

$$\|\nabla u_j\|_{\mathrm{L}^2(\mathbb{R}^n)} \lesssim \|u_j\|_{\mathrm{C}^{p,\frac{p(n-2)}{2}}(\mathbb{R}^n)} < \infty \quad \forall j \in \{1, \dots, n\}.$$

And yet, the second application of *Step 1*, plus the local Gagliardo–Nirenberg–Soblov inequality for $\check{u}_j$:

$$\|\check{u}_j\|_{\mathrm{L}^{\frac{qn}{n-q}}(B(x,r))} \lesssim \|\nabla u_j\|_{\mathrm{L}^q(B(x,r))} \quad \text{under} \quad q \in [1, n),$$

derives the reverse Hölder inequality

$$\|\nabla u_j\|_{\mathrm{L}^2(B(x,\frac{r}{2}))}\left(\frac{r}{2}\right)^{-\frac{n}{2}} \lesssim \left(r^{-n}\int_{B(x,r)} |\nabla u_j(y)|^{\frac{pn}{p+n}}\,dy\right)^{\frac{p+n}{pn}} \quad \text{for} \quad 2 < p < \frac{2n}{n-2}.$$

Recall that $\mathcal{M}(f)$ is the Hardy–Littlewood maximal function of f on $\mathbb{R}^n$. Thus, from

$$\|\nabla u_j\|_{\mathrm{L}^2(\mathbb{R}^n)} < \infty$$

and the last reverse Hölder inequality it follows that if

$$\Pi_j = |\nabla u_j|^{\frac{pn}{p+n}} \quad \text{under} \quad 2 < p < \frac{2n}{n-2} \;\&\; n-1 \in \mathbb{N}$$

then

$$\begin{cases} \Pi_j \in \mathrm{M}^{\frac{2(p+n)}{pn}, n}(\mathbb{R}^n) \\ \mathcal{M}(\Pi_j^{\frac{2(p+n)}{pn}})(x) \lesssim (\mathcal{M}(\Pi_j)(x))^{\frac{2(p+n)}{pn}}. \end{cases}$$

Now, the boundedness of $\mathcal{M}$ on $\mathrm{M}^{\frac{2(p+n)}{pn}, n}(\mathbb{R}^n)$ (see, e. g., [13]) derives

$$\|\Pi_j^{\frac{2(p+n)}{pn}}\|_{\mathrm{M}^{1,n}(\mathbb{R}^n)} \lesssim \|\mathcal{M}(\Pi_j^{\frac{2(p+n)}{pn}})\|_{\mathrm{M}^{1,n}(\mathbb{R}^n)}$$

$$\lesssim \|\mathcal{M}(\Pi_j)\|^{\frac{2(p+n)}{pn}}_{\mathrm{M}^{\frac{2(p+n)}{pn},n}}$$
$$\lesssim \|\Pi_j\|^{\frac{2(p+n)}{pn}}_{\mathrm{M}^{\frac{2(p+n)}{pn},n}}$$
$$\approx \|\Pi_j^{\frac{2(p+n)}{pn}}\|_{\mathrm{M}^{1,n}(\mathbb{R}^n)}.$$

Accordingly,

$$|\nabla u_j|^2 = \Pi_j^{\frac{2(p+n)}{pn}} = 0,$$

otherwise (cf. [32, example 2.1.8] or [30, Remark 3.7])

$$\Pi_j^{\frac{2(p+n)}{pn}} \neq 0 \Rightarrow \mathcal{M}(\Pi_j^{\frac{2(p+n)}{pn}})(x) \approx |x|^{-n} \quad \forall\, |x| \gg 1$$
$$\Rightarrow \|\mathcal{M}(\Pi_j^{\frac{2(p+n)}{pn}})\|_{\mathrm{M}^{1,n}(\mathbb{R}^n)} = \infty,$$

which violates

$$\Pi_j \in \mathrm{M}^{\frac{2(p+n)}{pn},n}(\mathbb{R}^n).$$

Therefore, each u_j is a constant.

With the help of the previous consideration, we use the boundary value condition

$$\lim_{|x|\to\infty} u_j(x) = 0$$

to deduce that the constant u_j is actually zero, and so u vanishes. □

6.4 Notes

Below are some historical remarks on Chapter 6.

Section 6.1. Theorem 6.4 comes from [92, Theorem 1.4] which was naturally generalized in [58, 60] to other related fluid systems over $\mathrm{Q}_\alpha(\mathbb{R}^n)$. Note that

$$\mathrm{Q}_0(\mathbb{R}^n) \subsetneq \mathrm{BMO}(\mathbb{R}^n).$$

So Theorem 6.4 is not valid for $\alpha = 0$, however, according to [96, Theorem 1.1(i)] there is $\epsilon_0 > 0$ such that if the incompressible initial data $a = (a_1, \ldots, a_n)$ satisfies

$$\|a_j\|_{\mathrm{Q}^{-1}_{0,*}(\mathbb{R}^n)} = \sup_{(x,r)\in\mathbb{R}^{n+1}_+} \left(\int_0^{r^2} \int_{B(x,r)} |e^{t\Delta} f(y)|^2 \left(\ln \frac{er}{\sqrt{t}}\right)^2 \frac{dy\,dt}{r^n} \right)^{\frac{1}{2}} < \epsilon_0$$

then $(\star)$ has a unique mild solution $u = (u_1, \dots, u_n)$ with

$$\infty > \|u_j\|_{X_{0,*}(\mathbb{R}^{n+1}_+)} = \sup_{(x,r)\in\mathbb{R}^{n+1}_+} \sqrt{r}|u_j(x,r)| + \sup_{(x,r)\in\mathbb{R}^{n+1}_+} \left(\int_0^{r^2} \int_{B(x,r)} |u_j(y,t)|^2 \left(\ln \frac{er}{\sqrt{t}} \right)^2 \frac{dy\,dt}{r^n} \right)^{\frac{1}{2}}.$$

The idea of treating Theorem 6.4 can be used to handle the heat flow of harmonic maps and the flow of liquid crystals living on $\mathrm{Q}_\alpha(\mathbb{R}^n)$ (cf. [61, Theorems 1.1 and 1.2] and [97, Theorems 1.6 and 1.8] generalizing the main results in [87] under $\alpha \in (-\infty, 0)$). More precisely,

▷ Let

$$(n, k, l, \alpha) \in \mathbb{N} \times \mathbb{N} \times \mathbb{N} \times (0, 1),$$

$\mathfrak{N}$ be a k-dimensional compact Riemannian manifold without boundary, isometrically embedded in some Euclidean space $\mathbb{R}^l$, T_x be the tangent plane at $x \in \mathfrak{N}$ and

$$\mathcal{A}(x) : T_x\mathfrak{N} \times T_x\mathfrak{N} \to (T_x\mathfrak{N})^\perp$$

be the second fundamental form of $\mathfrak{N} \subseteq \mathbb{R}^l$ at $x \in \mathfrak{N}$. Then the system of heat flow of harmonic maps u from $\mathbb{R}^{n+1}_+$ to $\mathfrak{N}$ is

$$(\clubsuit) \quad \begin{cases} u_t - \Delta u = \mathcal{A}(u)(\nabla u, \nabla u) & \text{in } \mathbb{R}^{n+1}_+ \\ u|_{t=0} = u_0 = (u_{0,1}, \dots, u_{0,n}) & \text{in } \mathbb{R}^n. \end{cases}$$

Then there is $\epsilon > 0$ such that provided that $u_0 : \mathbb{R}^n \to \mathfrak{N}$ obeys

$$\|u_0\|_{(\mathrm{Q}_\alpha(\mathbb{R}^n))^l} = \sum_{j=1}^{l} \|u_{0,j}\|_{\mathrm{Q}_\alpha(\mathbb{R}^n)} \le \epsilon,$$

$(\clubsuit)$ has a unique solution

$$u = (u_1, \dots, u_l) \in \mathrm{X}^l_\alpha(\mathbb{R}^{n+1}_+)$$

with

$$|||u|||_{\mathrm{X}^l_\alpha(\mathbb{R}^{n+1}_+)} = \sum_{j=1}^{l} \|u_j\|_{\mathrm{L}^\infty(\mathbb{R}^{n+1}_+)} + \sum_{j=1}^{l} \|u_j\|_{\mathrm{X}_\alpha(\mathbb{R}^{n+1}_+)} \lesssim \epsilon$$

and

$$\|u_j\|_{\mathrm{X}_\alpha(\mathbb{R}^{n+1}_+)} = \sup_{(x,r)\in\mathbb{R}^{n+1}_+} \sqrt{r}|\nabla u_j(x,r)| + \sup_{(x,r)\in\mathbb{R}^{n+1}_+} \left(\int_0^{r^2} \int_{B(x,r)} |\nabla u_j(y,t)|^2 \left(\frac{r^2}{t} \right)^\alpha \frac{dydt}{r^n} \right)^{\frac{1}{2}}.$$

▷ Let $(\alpha, n-1) \in (0,1) \times \mathbb{N}$ and

$$\mathrm{p} : \mathbb{R}^n \to \mathbb{R}$$

be a pressure and

$$\mathrm{d} : \mathbb{R}^n \to \mathbb{S}^2 \text{ – the unit sphere of } \mathbb{R}^3$$

be a unit-vector field representing the macroscopic molecular orientation of the nematic liquid crystal material. Then the incompressible hydrodynamic flow of nematic liquid crystal system is

$$(\spadesuit) \quad \begin{cases} u_t + u\cdot\nabla u - \Delta u + \nabla \mathrm{p} = -\nabla\cdot(\nabla \mathrm{d}\otimes\nabla \mathrm{d}) & \text{in } \mathbb{R}^{n+1}_+ \\ \operatorname{div} u = 0 & \text{in } \mathbb{R}^{n+1}_+ \\ \mathrm{d}_t + u\cdot\nabla \mathrm{d} - \Delta \mathrm{d} = |\nabla \mathrm{d}|^2 \mathrm{d} & \text{in } \mathbb{R}^{n+1}_+ \\ (u|_{t=0}, \mathrm{d}|_{t=0}) = (u_0, \mathrm{d}_0) & \text{in } \mathbb{R}^n. \end{cases}$$

Then there is $\epsilon > 0$ such that if the initial pair

$$(u_0, \mathrm{d}_0) \in (\mathrm{Q}^{-1}(\mathbb{R}^n))^n \times (\mathrm{Q}_\alpha(\mathbb{R}^n))^3$$

satisfies

$$\operatorname{div} u_0 = 0 \quad \& \quad \|u_0\|_{(\mathrm{Q}^{-1}(\mathbb{R}^n))^n} + \|\mathrm{d}_0\|_{(\mathrm{Q}_\alpha(\mathbb{R}^n))^3} \le \epsilon$$

then $(\spadesuit)$ has a unique mild solution

$$(u, \mathrm{d}) \in \mathrm{Z}^n_\alpha(\mathbb{R}^{n+1}_+) \times \mathrm{X}^3_\alpha(\mathbb{R}^{n+1}_+)$$

with

$$\|u\|_{\mathrm{Z}^n_\alpha(\mathbb{R}^{n+1}_+)} + \|\!|\mathrm{d}|\!\|_{\mathrm{X}^3_\alpha(\mathbb{R}^{n+1}_+)} = \sum_{j=1}^{n} \|u_j\|_{\mathrm{Z}_\alpha(\mathbb{R}^{n+1}_+)} + \sum_{j=1}^{3} \|\mathrm{d}_j\|_{\mathrm{X}_\alpha(\mathbb{R}^{n+1}_+)} \lesssim \epsilon$$

and

$$\|u_j\|_{\mathrm{Z}_\alpha(\mathbb{R}^{n+1}_+)} = \sup_{(x,r)\in\mathbb{R}^{n+1}_+} \sqrt{r}|u_j(x,r)| + \sup_{(x,r)\in\mathbb{R}^{n+1}_+} \left(\int_0^{r^2} \int_{B(x,r)} |u_j(y,t)|^2 \left(\frac{r^2}{t}\right)^\alpha \frac{dydt}{r^n} \right)^{\frac{1}{2}}.$$

Section 6.2. Theorem 6.7, as an alternative of [93, Theorem 3.3], is a generalization of [56, Theorem 1 (A)–(B)] from $n = 3$ to $n \ge 2$. In fact, this part is closely related to Kato's work [43]. Also, the idea used in this part has been further explored in [57] handling the global existence and stability for the Keller–Segel equations with small initial data in the critical Morrey space. Moreover, we can use the second inclusion of Lemma 6.6 (taken from [93, Theorem 3.1]) and Theorem 6.4 to obtain that if $\|a\|_{(\mathrm{M}^{2,2}(\mathbb{R}^n))^n}$ is sufficiently small, then there is a unique solution u of $(\star)$ in $(\mathrm{X}_{\alpha,\infty}(\mathbb{R}^{n+1}_+))^n$, which however cannot guarantee

$$u \in \mathrm{E}_\infty(\mathbb{R}^{n+1}_+)$$

due to the second inclusion of Lemma 6.6. In any event, we always have

$$\sup_{t\in(0,\infty)} \sqrt{t}\|u(\cdot,t)\|_{(\mathrm{L}^\infty(\mathbb{R}^n))^n} < \infty$$

and even more general estimate (cf. [56, (49) and Lemma 3])

$$\sup_{t\in(0,\infty)} \sqrt{t}\|u^{(j+1)}(\cdot,t) - u^{(j)}(\cdot,t)\|_{(\mathrm{L}^\infty(\mathbb{R}^n))^n} \lesssim (j+1)^{-2} \quad \forall j+1 \in \mathbb{N},$$

where

$$\begin{cases} u^{(0)}(\cdot,t) = e^{t\Delta}a(\cdot) \\ u^{(j+1)}(\cdot,t) = u^{(0)}(\cdot,t) - \mathcal{B}(u^{(j)}(\cdot,t), u^{(j)}(\cdot,t),). \end{cases}$$

Meanwhile, Theorem 6.8 is a reformulation of [35, Theorem 1.1] which extends the three-dimensional case in [24, Theorem 1.1]; this, together with the three-dimensional ill-posedness result in [9], suggests somewhat that the solution to the open problem of whether a small initial data result is possible in the largest scaling invariant space $\dot{\mathrm{B}}^{-1}_{\infty,\infty}(\mathbb{R}^n)$ might be negative.

Section 6.3. Theorem 6.9 is taken from [96, Theorem 1.1(ii)] which actually extends [75, Theorem 1.1] on

$$\dot{\mathrm{W}}^{1,2}(\mathbb{R}^3) \cap \mathrm{BMO}^{-1}(\mathbb{R}^3)$$

thanks to

$$\dot{\mathrm{W}}^{1,2}(\mathbb{R}^n) \subseteq \mathrm{M}^{\frac{2n}{n-2},n}(\mathbb{R}^n) \subseteq \mathrm{C}^{2<p<\frac{2n}{n-2},\frac{p(n-2)}{2}}(\mathbb{R}^n) \quad \text{as} \quad n \geq 3.$$

7 Left or right composition

In this chapter, we study the left or right composition

$$\mathcal{C}_{l,\phi}f = \phi \circ f \quad \text{or} \quad \mathcal{C}_{r,\phi}f = f \circ \phi$$

operating on $\mathcal{Q}_\alpha(\mathbb{R}^n)$ with a continuous self-map ϕ of $\mathbb{R}$ or a quasiconformal homeomorphism ϕ of $\mathbb{R}^n$ being a symbol, thereby finding a relationship between $\mathcal{Q}_\alpha(\mathbb{R}^n)_*$ and $\mathscr{A}_p(\mathbb{R}^n)$ and solving [22, problem 8.4]. Furthermore, we apply the right composition with $\mathcal{Q}_\alpha(\mathbb{R}^n)$ to an investigation of the well-posedness of the transport equation in $\mathcal{Q}_\alpha(\mathbb{R}^n)$.

7.1 Left composition under Lipschitz condition

Classically, a function $f : \mathbb{R}^n \to \mathbb{R}$ is said to be $(0,1] \ni \gamma$-continuous provided

$$\|f\|_{\mathcal{C}^\gamma(\mathbb{R}^n)} = \sup_{\mathbb{R}^n \ni x \neq y \in \mathbb{R}^n} |x-y|^{-\gamma}|f(x)-f(y)| < \infty.$$

In particular, $\mathcal{C}^1(\mathbb{R}^n)$ is the Lipschitz class. Below is the Campanato and Morrey lemma.

Lemma 7.1. *Let* $\gamma \in (0,1]$. *Then:*
(i) $\mathcal{C}^{2,-2\gamma}(\mathbb{R}^n) = \mathcal{C}^\gamma(\mathbb{R}^n)$;
(ii) $|\nabla f| \in \mathcal{M}^{2,2(1+\gamma)}(\mathbb{R}^n) \Rightarrow f \in \mathcal{C}^\gamma(\mathbb{R}^n)$.

Proof. (i) On the one hand, if $f \in \mathcal{C}^\gamma(\mathbb{R}^n)$, then

$$\fint_{B(x,r)} |f(y) - f_{B(x,r)}|^2\,dy \lesssim r^{-2n} \int_{B(x,r)} \left| \int_{B(x,r)} |f(y)-f(z)|\,dz \right|^2 dy \lesssim r^{2\gamma}\|f\|^2_{\mathcal{C}^\gamma(\mathbb{R}^n)},$$

and hence $f \in \mathcal{C}^{2,-2\gamma}(\mathbb{R}^n)$.

On the other hand, suppose $f \in \mathcal{C}^{2,-2\gamma}(\mathbb{R}^n)$. Then for a given ball $B(x,r) \subseteq \mathbb{R}^n$, we have

$$|f_{B(x,r)} - f_{B(x,2r)}|^2 \lesssim \fint_{B(x,2r)} |f(y) - f_{B(x,2r)}|^2\,dy \lesssim \|f\|^2_{\mathcal{C}^{2,-2\gamma}(\mathbb{R}^n)} r^{2\gamma},$$

thereby getting

$$\begin{aligned}
|f_{B(x,2^{-j}r)} - f_{B(x,r)}| &\leq \sum_{i=0}^{j} |f_{B(x,2^{-i-1}r)} - f_{B(x,2^{-i}r)}| \\
&\lesssim r^\gamma \|f\|_{\mathcal{C}^{2,-2\gamma}(\mathbb{R}^n)} \sum_{i=0}^{j} 2^{-i\gamma} \\
&\lesssim \|f\|_{\mathcal{C}^{2,-2\gamma}(\mathbb{R}^n)} \frac{(1-2^{-(j+1)\gamma})r^\gamma}{1-2^{-\gamma}}
\end{aligned}$$

https://doi.org/10.1515/9783110600285-007

and so

$$|f(x) - f_{B(x,r)}| \lesssim \|f\|_{C^{2,-2\gamma}(\mathbb{R}^n)} r^{\gamma}.$$

Also, if $r = |x - y|$, then an application of the Cauchy–Schwarz inequality and the inclusion

$$B(x,r) \subseteq B(x,2r) \cap B(y,2r)$$

gives

$$\begin{aligned}
|f_{B(x,2r)} - f_{B(y,2r)}| &\leq \fint_{B(x,2r)\cap B(y,2r)} (|f_{B(x,2r)} - f(z)| + |f(z) - f_{B(y,2r)}|)\,dz \\
&\lesssim r^{-n}\Big(\int_{B(x,2r)} |f_{B(x,2r)} - f(z)|\,dz + \int_{B(y,2r)} |f_{B(y,2r)} - f(z)|\,dz \Big) \\
&\lesssim \Big(\fint_{B(x,2r)} |f_{B(x,2r)} - f(z)|^2\,dz \Big)^{2^{-1}} + \Big(\fint_{B(y,2r)} |f_{B(y,2r)} - f(z)|^2\,dz \Big)^{2^{-1}} \\
&\lesssim r^{\gamma} \|f\|_{C^{2,-2\gamma}(\mathbb{R}^n)},
\end{aligned}$$

and hence

$$\begin{aligned}
|f(x) - f(y)| &\leq |f(x) - f_{B(x,2r)}| + |f_{B(x,2r)} - f_{B(y,2r)}| + |f_{B(y,2r)} - f(y)| \\
&\lesssim \|f\|_{C^{2,-2\gamma}(\mathbb{R}^n)} |x - y|^{\gamma}.
\end{aligned}$$

Accordingly, $f \in C^{\gamma}(\mathbb{R}^n)$.

(ii) Since the Poincaré inequality gives

$$\fint_{B(x,r)} |f(y) - f_{B(x,r)}|^2\,dy \lesssim r^2 \fint_{B(x,r)} |\nabla f(y)|^2\,dy \lesssim r^{2\gamma} \||\nabla f|\|^2_{M^{2,2(1+\gamma)}(\mathbb{R}^n)} \quad \forall\,(x,r) \in \mathbb{R}^{n+1}_+,$$

an application of (i) derives $f \in C^{\gamma}(\mathbb{R}^n)$. □

Lemma 7.1 is used to discover the following left composition law for $Q_\alpha(\mathbb{R}^n)$.

Theorem 7.2. *Let $(\alpha, n-1) \in (-\infty, 1) \times \mathbb{N}$ and $\phi : \mathbb{R} \to \mathbb{R}$ be continuous. Then:*

(i) *$C_{l,\phi}$ is bounded on $Q_\alpha(\mathbb{R}^n)$ with*

$$\|C_{l,\phi} f\|_{Q_\alpha(\mathbb{R}^n)} \lesssim \|f\|_{Q_\alpha(\mathbb{R}^n)} \quad \forall f \in Q_\alpha(\mathbb{R}^n)$$

if and only if $\phi \in C^1(\mathbb{R})$;

(ii) *$f \in Q_\alpha(\mathbb{R}^n)$ if and only if there is*

$$g \in Q_\alpha(\mathbb{R}^n) \quad \text{such that} \quad \frac{1}{g} \in Q_\alpha(\mathbb{R}^n) \quad \& \quad f = g - \frac{1}{g}.$$

Proof. (i) On the one hand, if $\phi \in \mathrm{C}^1(\mathbb{R})$, then

$$\|C_{l,\phi}f\|_{\mathrm{Q}_\alpha(\mathbb{R}^n)} \le \|\phi\|_{\mathrm{C}^1(\mathbb{R})}\|f\|_{\mathrm{Q}_\alpha(\mathbb{R}^n)}.$$

On the other hand, suppose that $C_{l,\phi}$ is bounded on $\mathrm{Q}_\alpha(\mathbb{R}^n)$ with

$$\|C_{l,\phi}f\|_{\mathrm{Q}_\alpha(\mathbb{R}^n)} \lesssim \|f\|_{\mathrm{Q}_\alpha(\mathbb{R}^n)} \quad \forall f \in \mathrm{Q}_\alpha(\mathbb{R}^n).$$

Thanks to Theorem 1.2 we may assume

$$\phi(0) = 0 \quad \& \quad I = [-r, r]^n,$$

and then choose

$$f_1(x) = \begin{cases} x_1 & \text{as } x = (x_1, \dots, x_n) \in 2I \\ 0 & \text{as } x = (x_1, \dots, x_n) \in \mathbb{R}^n \setminus 2I. \end{cases}$$

A calculation with Theorem 1.4 shows

$$\|f_1\|_{\mathrm{Q}_\alpha(\mathbb{R}^n)} \lesssim \|f_1\|_{\dot{\mathrm{W}}^{1,n}(\mathbb{R}^n)} \lesssim r$$

and

$$C_{l,\phi}(f_1) = \begin{cases} \phi(x_1) & \text{as } x = (x_1, \dots, x_n) \in 2I \\ 0 & \text{as } x = (x_1, \dots, x_n) \in \mathbb{R}^n \setminus 2I. \end{cases}$$

Meanwhile, an application of Theorem 1.3 derives

$$\begin{aligned} \fint_{[-r,r]} |\phi(x_1) - \phi_{[-r,r]}|^2 \, dx_1 &\le \|C_{l,\phi}f_1\|^2_{\mathrm{BMO}(\mathbb{R}^n)} \\ &\lesssim \|C_{l,\phi}f_1\|^2_{\mathrm{Q}_\alpha(\mathbb{R}^n)} \\ &\lesssim \|f_1\|^2_{\mathrm{Q}_\alpha(\mathbb{R}^n)} \\ &\lesssim r^2. \end{aligned}$$

This, along with Lemma 7.1(i), implies $\phi \in \mathrm{C}^1(\mathbb{R})$.

(ii) Clearly, if f admits the required representation, then $f \in \mathrm{Q}_\alpha(\mathbb{R}^n)$. Conversely, if $f \in \mathrm{Q}_\alpha(\mathbb{R}^n)$ and

$$\begin{cases} \phi(t) = 2^{-1}(\sqrt{t^2 + 2^2} + t) \\ \dfrac{1}{\phi(t)} = 2^{-1}(\sqrt{t^2 + 2^2} - t) \end{cases} \qquad \forall t \in \mathbb{R},$$

then

$$\left(\phi, \frac{1}{\phi}\right) \in \mathrm{C}^1(\mathbb{R}) \times \mathrm{C}^1(\mathbb{R}),$$

and hence

$$\left(\phi \circ f, \frac{1}{\phi} \circ f\right) \in \mathcal{Q}_\alpha(\mathbb{R}^n) \times \mathcal{Q}_\alpha(\mathbb{R}^n)$$

due to (i). A simple calculation gives

$$\begin{aligned} f &= 2^{-1}(f + \sqrt{f^2 + 2^2}) - 2(f + \sqrt{f^2 + 2^2})^{-1} \\ &= C_{l,\phi}(f) - \frac{1}{C_{l,\phi}(f)}, \end{aligned}$$

as desired. □

7.2 Relationship with Muckenhoupt class

Below is a short list of the basic properties of the Muckenhoupt weights $\mathscr{A}_{1\le p\le\infty}(\mathbb{R}^n)$ (see, e. g., [11, 85]).

▷

$$0 \le w \in \mathscr{A}_1(\mathbb{R}^n)$$

amounts to

$$[w]_{\mathscr{A}_1(\mathbb{R}^n)} = \sup_{I\in\mathscr{C}(\mathbb{R}^n)} \frac{⨍_I w(x)\,dx}{\operatorname{essinf}_{x\in I} w(x)} < \infty.$$

▷ In fact,

$$0 \le w \in \mathscr{A}_1(\mathbb{R}^n)$$

is regarded as a limiting case of

$$0 \le w \in \mathscr{A}_{1<p<\infty}(\mathbb{R}^n)$$

which is determined by

$$[w]_{\mathscr{A}_p(\mathbb{R}^n)} = \sup_{I\in\mathscr{C}(\mathbb{R}^n)} \left(⨍_I w(x)\,dx\right)\left(⨍_I (w(x))^{\frac{1}{1-p}}\,dx\right)^{p-1} < \infty$$

and enjoys the monotonicity for $1 < p_1 < p_2 < \infty$

$$\mathscr{A}_1(\mathbb{R}^n) \subset \bigcap_{1<p<\infty} \mathscr{A}_p(\mathbb{R}^n) \subset \mathscr{A}_{p_1}(\mathbb{R}^n) \subset \mathscr{A}_{p_2}(\mathbb{R}^n) \subset \bigcup_{1<p<\infty} \mathscr{A}_p(\mathbb{R}^n) = \mathscr{A}_\infty(\mathbb{R}^n).$$

▷ While,

$$0 \le w \in \mathscr{A}_\infty(\mathbb{R}^n)$$

is equivalent to that there exists a constant τ such that

$$\frac{w(E)}{w(I)} = \frac{\int_E w(x)\,dx}{\int_I w(x)\,dx} \lesssim \left(\frac{\int_E dx}{\int_I dx}\right)^\tau = \left(\frac{m(E)}{m(I)}\right)^\tau \quad \forall \text{ measurable } E \subseteq I \in \mathscr{C}(\mathbb{R}^n)$$

so the least τ can be defined as $[w]_{\mathscr{A}_\infty(\mathbb{R}^n)}$. In other words,

$$\begin{cases} [w]_{\mathscr{A}_\infty(\mathbb{R}^n)} \approx [w]_{\mathscr{A}_\infty(\mathbb{R}^n),\oplus} \approx [w]_{\mathscr{A}_\infty(\mathbb{R}^n),\ominus} < \infty \\ [w]_{\mathscr{A}_\infty(\mathbb{R}^n),\oplus} = \sup\limits_{I\in\mathscr{C}(\mathbb{R}^n)} \dfrac{⨍_I w(x)\,dx}{\exp(⨍_I \ln w(x)\,dx)} \\ [w]_{\mathscr{A}_\infty(\mathbb{R}^n),\ominus} = \sup\limits_{I\in\mathscr{C}(\mathbb{R}^n)} \dfrac{\int_I \mathcal{M}(w\mathbb{1}_I)(x)\,dx}{w(I)}. \end{cases}$$

▷ According to [40, Lemma 2.1], for a nonnegative function w on $\mathbb{R}^n$, one has

$$w \in \mathscr{A}_\infty(\mathbb{R}^n) \quad \& \quad w^\eta \in \mathscr{A}_1(\mathbb{R}^n) \quad \text{for some} \quad \eta > 0 \Rightarrow w \in \mathscr{A}_1(\mathbb{R}^n).$$

Theorem 7.2(ii) leads to a discovery of the following inclusion in connection with the Muckenhoupt weights.

Theorem 7.3. *For $(\alpha, n-1) \in (-\infty, 1) \times \mathbb{N}$, let $\mathrm{Q}_\alpha(\mathbb{R}^n)_*$ be the class of nonnegative functions which and whose reciprocals are in $\mathrm{Q}_\alpha(\mathbb{R}^n)$. Then*

$$\mathrm{Q}_\alpha(\mathbb{R}^n)_* \subseteq \bigcap_{1<p<\infty} \mathscr{A}_p(\mathbb{R}^n).$$

This inclusion is sharp in the sense of

$$\mathrm{BMO}(\mathbb{R}^n)_* = \mathrm{Q}_\alpha(\mathbb{R}^n)_* \not\subseteq \mathscr{A}_1(\mathbb{R}^n) \quad \forall \alpha \in (-\infty, 0).$$

Proof. Note that

$$\mathrm{Q}_\alpha(\mathbb{R}^n)_* \subseteq \mathrm{Q}_{-\infty<\alpha<0}(\mathbb{R}^n)_* = \mathrm{BMO}(\mathbb{R}^n)_*.$$

So, for the desired inclusion it is enough to verify

$$\mathrm{BMO}(\mathbb{R}^n)_* \subseteq \bigcap_{1<p<\infty} \mathscr{A}_p(\mathbb{R}^n).$$

To do so, we first prove that if $1 < p \le 2$ and $w > 0$ then

$$\left(w, \frac{1}{w}\right) \in \mathscr{A}_p(\mathbb{R}^n) \times \mathscr{A}_p(\mathbb{R}^n) \Leftrightarrow \omega_p = \sup_{I\in\mathscr{C}(\mathbb{R}^n)} ⨍_I \left(|w(x) - w_I| \left|\frac{1}{w(x)} - \left(\frac{1}{w}\right)_I\right|\right)^{\frac{1}{p-1}} dx < \infty.$$

On the one hand, suppose

$$\left(w, \frac{1}{w}\right) \in \mathscr{A}_p(\mathbb{R}^n) \times \mathscr{A}_p(\mathbb{R}^n).$$

Using

$$\mathscr{A}_p(\mathbb{R}^n) \subseteq \mathscr{A}_2(\mathbb{R}^n) \quad \forall p \in (1,2],$$

we obtain

$$\begin{aligned}
\omega_p &\le \sup_{I\in\mathscr{C}(\mathbb{R}^n)} \fint_I \left|w(x)^{\frac{1}{p-1}} - w_I^{\frac{1}{p-1}}\right| \left|\left(\frac{1}{w(x)}\right)^{\frac{1}{p-1}} - \left(\left(\frac{1}{w}\right)_I\right)^{\frac{1}{p-1}}\right| dx \\
&\le 1 + \sup_{I\in\mathscr{C}(\mathbb{R}^n)} \left((w^{\frac{1}{p-1}})_I \left(\left(\frac{1}{w}\right)_I\right)^{\frac{1}{p-1}} + (w_I)^{\frac{1}{p-1}} \left(\left(\frac{1}{w}\right)^{\frac{1}{p-1}}\right)_I + (w_I)^{\frac{1}{p-1}} \left(\left(\frac{1}{w}\right)_I\right)^{\frac{1}{p-1}} \right) \\
&\le 1 + \left[\frac{1}{w}\right]_{\mathscr{A}_p(\mathbb{R}^n)} + [w]_{\mathscr{A}_p(\mathbb{R}^n)} + \left([w]_{\mathscr{A}_2(\mathbb{R}^n)}\right)^{\frac{1}{p-1}} \\
&< \infty.
\end{aligned}$$

On the other hand, suppose $\omega_p < \infty$. If $I \in \mathscr{C}(\mathbb{R}^n)$ and $1 < p \le 2$, then

$$\begin{aligned}
\omega_p^{p-1} &\ge \fint_I |w(x) - w_I| \left|\frac{1}{w(x)} - \left(\frac{1}{w}\right)_I\right| dx \\
&\ge \fint_I (w(x) - w_I)\left(\frac{1}{w(x)} - \left(\frac{1}{w}\right)_I\right) dx \\
&= w_I\left(\frac{1}{w}\right)_I - 1,
\end{aligned}$$

and hence $w \in \mathscr{A}_2(\mathbb{R}^n)$. Note that

$$|a^q - b^q| \ge 2^{1-q}a^q - b^q \quad \forall (q,a,b) \in [1,\infty) \times \mathbb{R}_+ \times \mathbb{R}_+.$$

So, for $I \in \mathscr{C}(\mathbb{R}^n)$ one has

$$\begin{aligned}
&\left|(w - w_I)\left(\frac{1}{w} - \left(\frac{1}{w}\right)_I\right)\right|^{\frac{1}{p-1}} \\
&= \left|w - \left(\frac{1}{w}\right)_I + \frac{w_I}{w} - \left(w_I\left(\frac{1}{w}\right)_I + 1\right)\right|^{\frac{1}{p-1}} \\
&\ge 2^{\frac{p-2}{p-1}}\left(w\left(\frac{1}{w}\right)_I + \frac{w_I}{w}\right)^{\frac{1}{p-1}} - \left(1 + w_I\left(\frac{1}{w}\right)_I\right)^{\frac{1}{p-1}} \\
&\ge 2^{\frac{p-2}{p-1}}\left(w^{\frac{1}{p-1}}\left(\left(\frac{1}{w}\right)_I\right)^{\frac{1}{p-1}}\right) + \left(\frac{w_I}{w}\right)^{\frac{1}{p-1}} - \left(w_I\left(\frac{1}{w}\right)_I + 1\right)^{\frac{1}{p-1}}.
\end{aligned}$$

This in turn implies

$$(w^{\frac{1}{p-1}})_I\left(\left(\frac{1}{w}\right)_I\right)^{\frac{1}{p-1}} + \left(\frac{1}{w^{\frac{1}{p-1}}}\right)_I (w_I)^{\frac{1}{p-1}} \le 2^{\frac{2-p}{p-1}}\left(\omega_p + (1 + [w]_{\mathscr{A}_2(\mathbb{R}^n)})^{\frac{1}{p-1}}\right)$$

whence

$$\left(w, \frac{1}{w}\right) \in \mathscr{A}_p(\mathbb{R}^n) \times \mathscr{A}_p(\mathbb{R}^n).$$

Next, if

$$w \in \mathrm{BMO}(\mathbb{R}^n)_*,$$

then an application of the last equivalence and Theorem 2.1 gives

$$\omega_p \leq \|w\|_{\mathrm{BMO}(\mathbb{R}^n), \frac{q}{p-1}}^{\frac{1}{p-1}} \left\|\frac{1}{w}\right\|_{\mathrm{BMO}(\mathbb{R}^n), \frac{q}{(q-1)(p-1)}}^{\frac{1}{p-1}} \lesssim \|w\|_{\mathrm{BMO}(\mathbb{R}^n)}^{\frac{1}{p-1}} \left\|\frac{1}{w}\right\|_{\mathrm{BMO}(\mathbb{R}^n)}^{\frac{1}{p-1}},$$

where

$$\begin{cases} (p,q) \in (1,2] \times (1,\infty) \\ \|f\|_{\mathrm{BMO}(\mathbb{R}^n),q} = \sup_{I \in \mathscr{C}(\mathbb{R}^n)} \left(\fint_I |f(x) - f_I|^q \, \mathrm{d}x \right)^{\frac{1}{q}}, \end{cases}$$

thereby reaching

$$w \in \mathscr{A}_p(\mathbb{R}^n) \quad \forall p \in (1,2].$$

Of course, we always have

$$w \in \mathrm{BMO}(\mathbb{R}^n)_* \Rightarrow w \in \mathscr{A}_2(\mathbb{R}^n) \subset \mathscr{A}_{2<p<\infty}(\mathbb{R}^n).$$

Accordingly,

$$w \in \mathrm{BMO}(\mathbb{R}^n)_* \Rightarrow w \in \mathscr{A}_p(\mathbb{R}^n) \quad \forall p \in (1,\infty).$$

Finally, we show the sharpness of the inclusion. Note that

$$\left(w, \frac{1}{w}\right) \in \mathscr{A}_1(\mathbb{R}^n) \times \mathscr{A}_1(\mathbb{R}^n) \Rightarrow w \approx 1$$

due to

$$\begin{aligned} \frac{\sup_{x\in I} w(x)}{[\frac{1}{w}]_{\mathscr{A}_1(\mathbb{R}^n)}} &= \left(\left[\frac{1}{w}\right]_{\mathscr{A}_1(\mathbb{R}^n)} \inf_{w \in I} \frac{1}{w(x)} \right)^{-1} \\ &\leq \left(\fint_I \frac{\mathrm{d}x}{w(x)} \right)^{-1} \\ &\leq \fint_I w(x) \, \mathrm{d}x \end{aligned}$$

$$\le [w]_{\mathscr{A}_1(\mathbb{R}^n)} \inf_{x\in I} w(x) \quad \forall I \in \mathscr{C}(\mathbb{R}^n).$$

So, if

$$\mathrm{BMO}(\mathbb{R}^n)_* = \mathrm{Q}_\alpha(\mathbb{R}^n)_* \subseteq \mathscr{A}_1(\mathbb{R}^n) \quad \forall \alpha \in (-\infty, 0)$$

then

$$w \in \mathrm{BMO}(\mathbb{R}^n)_* \Rightarrow \left(w, \frac{1}{w}\right) \in \mathscr{A}_1(\mathbb{R}^n) \times \mathscr{A}_1(\mathbb{R}^n) \Rightarrow w \approx 1.$$

However, the following function

$$x \mapsto w(x) = \max\{\ln |x|^{-1}, (\ln(|x|e^2))^{-1}\}$$

is in $\mathrm{BMO}(\mathbb{R}^n)_*$, but is not equivalent to any constant. □

7.3 Local or global self-similar Minkowski dimension

A group of three basic concepts is required below to characterize the right composition acting on $\mathrm{Q}_\alpha(\mathbb{R}^n)$.

Definition 7.4. Let E be a subset of $\mathbb{R}^n$.

(i) Upon denoting by $N_{\mathrm{cov}}(r, E)$ the minimal number of cubes with edge-length r required to cover E, we have that:

▷
$$\overline{\dim}_{\mathrm{lsM}} E = \liminf_{N\to\infty} \limsup_{r\to 0} \sup_{\substack{B(x_B,r_B)\subseteq\mathbb{R}^n \\ Nr\le r_B\le 1}} \frac{\log_2 N_{\mathrm{cov}}(r, E\cap B(x_B, r_B))}{\log_2(\frac{r_B}{r})}$$

is called the local self-similar Minkowski dimension of E;

▷
$$\overline{\dim}_{\mathrm{gsM}} E = \liminf_{N\to\infty} \sup_{r>0} \sup_{\substack{B(x_B,r_B)\subseteq\mathbb{R}^n \\ Nr\le r_B<\infty}} \frac{\log_2 N_{\mathrm{cov}}(r, E\cap B(x_B, r_B))}{\log_2(\frac{r_B}{r})}$$

is called the global self-similar Minkowski dimension of E;

▷
$$\overline{\dim}_M E = \limsup_{r\to 0} \frac{\log_2 N_{\mathrm{cov}}(r, E)}{\log_2(\frac{1}{r})}$$

is called the Minkowski dimension of E.

(ii) $\mathscr{A}_1(\mathbb{R}^n; E)$ is the local Muckenhoupt class of nonnegative functions ω with the following property:

$$\fint_B \omega(z)\,dz \lesssim \inf_{x\in B} \omega(x) \quad \forall B = B(x_B, r_B) \subseteq \mathbb{R}^n \text{ with } 2r_B < d(x_B, E) = \inf_{x\in E} |x - x_B|.$$

Since $\mathscr{A}_1(\mathbb{R}^n; \emptyset)$ is $\mathscr{A}_1(\mathbb{R}^n)$, the set E is said to be the degeneracy set of

$$\omega \in \mathscr{A}_1(\mathbb{R}^n; E).$$

(iii) A homeomorphism $\phi : \mathbb{R}^n \to \mathbb{R}^n$ is said to be quasiconformal provided that

$$\phi \in W^{1,1}_{\mathrm{loc}}(\mathbb{R}^n; \mathbb{R}^n) = \text{the local version of the Sobolev space } W^{1,1}(\mathbb{R}^n; \mathbb{R}^n)$$

and there is a constant $\kappa \geq 1$ obeying

$$|\mathcal{D}\phi(x)|^n = \max_{y \in \mathbb{R}^n \,\&\, |y|=1} |\mathcal{D}\phi(x)y|^n \leq \kappa \mathcal{J}_\phi(x) \quad a.e.\, x \in \mathbb{R}^n,$$

where $\mathcal{D}\phi(x)$ and $\mathcal{J}_\phi(x)$ are the formal derivative and Jacobian determinant of ϕ at x, respectively.

Lemma 7.5. *Let $E \subseteq F \subseteq \mathbb{R}^n$ and $R \in [1, \infty)$. Then:*

(i)

$$\overline{\dim}_{\mathrm{lsM}} E \leq \overline{\dim}_{\mathrm{lsM}} F \quad \& \quad \overline{\dim}_{\mathrm{glM}} E \leq \overline{\dim}_{\mathrm{glM}} F;$$

(ii)

$$\overline{\dim}_{\mathrm{lsM}} E = \liminf_{N \to \infty} \limsup_{r \to 0} \sup_{\substack{B(x_B, r_B) \subseteq \mathbb{R}^n \\ Nr \leq r_B \leq R}} \frac{\log_2 N_{\mathrm{cov}}(r, E \cap B)}{\log_2(\frac{r_B}{r})};$$

(iii)

$$0 \leq \sup_{B(x_B, r_B) \subseteq \mathbb{R}^n} \overline{\dim}_M(E \cap B) \leq \overline{\dim}_{\mathrm{lsM}} E \leq \overline{\dim}_{\mathrm{gsM}} E \leq n.$$

Proof. (i) This is obvious.

(ii) For a given Euclidean ball

$$B = B(x_B, r_B)$$

we always have

$$\overline{\dim}_{\mathrm{lsM}} E \leq \liminf_{N \to \infty} \limsup_{r \to 0} \sup_{\substack{B \subset \mathbb{R}^n \\ Nr \leq r_B \leq R}} \frac{\log_2 N_{\mathrm{cov}}(r, E \cap B)}{\log_2(\frac{r_B}{r})}.$$

Regarding the reverse inequality, there is a dimensional constant c_n such that each Euclidean ball B of radius $r_B \in [1, R]$ can be covered by a family of $c_n R^n$ balls B_i of radius one. So, we have

$$\begin{cases} N_{\mathrm{cov}}(r, E \cap B) \leq c_n R^n \sup\{N_{\mathrm{cov}}(r, E \cap \tilde{B}) : \tilde{B} \subseteq \mathbb{R}^n \text{ with } r_{\tilde{B}} = 1\} \\ \dfrac{\log_2 N_{\mathrm{cov}}(r, E \cap B)}{\log_2 \frac{r_B}{r}} \leq \dfrac{\log_2 c_n R^n}{\log_2 N} + \sup\limits_{\substack{\tilde{B} \subset \mathbb{R}^n \\ r_{\tilde{B}} = 1}} \dfrac{\log_2 N_{\mathrm{cov}}(r, E \cap \tilde{B})}{\log_2(\frac{1}{r})} \end{cases} \forall\, 0 < r < \min\{r_B N^{-1}, 1\}.$$

Since the first term on the right-hand side of this last inequality tends to 0 as $N \to \infty$, the definition of $\overline{\dim}_{\mathrm{lsM}} E$ implies the desired inequality.

(iii) Evidently, the inequality

$$\overline{\dim}_{\mathrm{gsM}}(E) \le n$$

follows from the fact that

$$N_{\mathrm{cov}}(r, E \cap B) \le \left(\frac{2r_B}{r}\right)^n$$

holds for each Euclidean ball B with radius $r_B \ge Nr$. Indeed

$$\overline{\dim}_{\mathrm{gsM}} E \le \liminf_{N\to\infty} \sup_{r>0} \sup_{\substack{B\subset\mathbb{R}^n \\ r_B\ge Nr}} \frac{n\log_2(\frac{r_B}{r}) + n}{n\log_2(\frac{r_B}{r})} = \liminf_{N\to\infty} \frac{n\log_2 N + n}{n\log_2 N} = n.$$

The remaining inequalities follow from Definition 7.4 and (ii) directly. □

If E is a set of finitely many points, then

$$N_{\mathrm{cov}}(r, E \cap B) \lesssim 1$$

holds for each Euclidean ball B with radius $r_B \ge Nr$, and hence for a constant $c > 0$ one has

$$\begin{cases} \overline{\dim}_{\mathrm{gsM}} E \le \liminf_{N\to\infty} \frac{\log_2 c}{\log_2 N} = 0 \\ \overline{\dim}_M E = \overline{\dim}_L E = \overline{\dim}_{\mathrm{gsM}} E = 0. \end{cases}$$

However, for a countable set E and a Euclidean ball B, three dimensions

$$\begin{cases} \overline{\dim}_{\mathrm{lsM}}(E \cap B) \\ \overline{\dim}_{\mathrm{gsM}}(E \cap B) \\ \overline{\dim}_M(E \cap B) \end{cases}$$

may be very different. Write

$$\begin{cases} \mathbb{N}^n = \underbrace{\mathbb{N} \times \cdots \times \mathbb{N}}_{n \text{ copies}} \\ A_{k,\theta} = \{2^k, 2^k + 1, \ldots, 2^k + 2^{[\theta k]}\} \\ 2^{\mathbb{N}_\theta} = \bigcup_{k\in\mathbb{N}\cup\{0\}} A_{k,\theta}; \\ (2^{\mathbb{N}_\theta})^n = \underbrace{2^{\mathbb{N}_\theta} \times \cdots \times 2^{\mathbb{N}_\theta}}_{n \text{ copies}}, \end{cases}$$

where $0 \le \theta \le 1$ and $[\theta k]$ is the largest integer less than or equal to θk.

Note that

$$\begin{cases} 2^{\mathbb{N}_\theta} = \mathbb{N} \cup \{0\} & \text{as } \theta = 1 \\ 2^{\mathbb{N}_\theta} = 2^{\mathbb{N}} \cup \{1\} & \text{as } \theta = 0, \end{cases}$$

and $(2^{\mathbb{N}_\theta})^n \cap B$ only contains finitely many points. So, we get that

$$\overline{\dim}_M((2^{\mathbb{N}_\theta})^n \cap B) = \overline{\dim}_L((2^{\mathbb{N}_\theta})^n \cap B) = \overline{\dim}_{\mathrm{gsM}}((2^{\mathbb{N}_\theta})^n \cap B) = 0$$

holds for any Euclidean ball $B \subseteq \mathbb{R}^n$ and any $\theta \in [0,1]$.

Lemma 7.6. *Let $\theta \in [0,1]$. Then*

$$\begin{cases} \overline{\dim}_{\mathrm{lsM}}(2^{\mathbb{N}_\theta})^n = 0 \\ \overline{\dim}_{\mathrm{gsM}}(2^{\mathbb{N}})^n = 0 \\ \overline{\dim}_{\mathrm{gsM}}\mathbb{N}^n = n \\ \overline{\dim}_{\mathrm{gsM}}(2^{\mathbb{N}_\theta})^n = \theta n. \end{cases}$$

Proof. First of all, we show

$$\overline{\dim}_{\mathrm{lsM}}\left(2^{\mathbb{N}_\theta}\right)^n = 0.$$

Observe that each Euclidean ball

$$B = B(x_B, r_B) \subseteq \mathbb{R}^n \quad \text{with} \quad r_B \le 1$$

contains at most a uniform number of points in $\mathbb{Z}^n$. So for

$$N \ge 1 \quad \& \quad r \in (0, r_B N^{-1}),$$

we can cover $B \cap (2^{\mathbb{N}_\theta})^n$ by a uniform number of balls of radii r, namely

$$N_{\mathrm{cov}}(1, (2^{\mathbb{N}_\theta})^n \cap B) \lesssim 1.$$

This, along with Lemma 7.5, derives the required equation.

Next, we verify

$$\overline{\dim}_{\mathrm{gsM}}\mathbb{N}^n = n \quad \& \quad \overline{\dim}_{\mathrm{gsM}}(2^{\mathbb{N}})^n = 0.$$

To do so, for any ball

$$B = B(x_B, r_B) \subseteq \mathbb{R}^n \quad \text{with} \quad r_B = N,$$

we use Lemma 7.5 to deduce

$$\begin{cases} N_{\mathrm{cov}}(1, \mathbb{N}^n \cap B) = \sharp(\mathbb{N}^n \cap B) \ge (N/\sqrt{n})^n \\ \overline{\dim}_{\mathrm{gsM}}\mathbb{N}^n \ge \liminf\limits_{N\to\infty} \dfrac{\log_2(N/\sqrt{n})^n}{\log_2 N} = n \\ \overline{\dim}_{\mathrm{gsM}}\mathbb{N}^n = n. \end{cases}$$

Meanwhile, for $(N, r) \in \mathbb{N} \times \mathbb{R}_+$ we have that:

▷ if $r \le 1$ and $Nr < r_B$, then

$$N_{\text{cov}}(r, (2^{\mathbb{N}})^n \cap B) \le (\log_2 r_B)^n;$$

▷ if $2^k < r \le 2^{k+1}$ for some $k \ge 0$, then Lemma 7.5 derives

$$\begin{cases} N_{\text{cov}}(r, (2^{\mathbb{N}})^n \cap B) \le \sqrt{n}\left(\log_2\left(\frac{r_B}{r} + 2\right)\right)^n \\ \overline{\dim}_{\text{gsM}}\left(2^{\mathbb{N}}\right)^n \le \liminf_{N\to\infty} \sup_{r\ge 1} \sup_{B, r_B \ge Nr} \frac{n \log_2[\sqrt{n}\log_2(\frac{r_B}{r}+2)]}{\log_2(\frac{r_B}{r})} = 0 \\ \overline{\dim}_{\text{gsM}}(2^{\mathbb{N}})^n = 0. \end{cases}$$

Finally, for

$$\begin{cases} \theta \in (0,1) \\ B = B(0, \sqrt{n}2^{m+1}) \\ m \ge 1 + 2\theta^{-1}, \end{cases}$$

we have

$$N_{\text{cov}}(1, (2^{\mathbb{N}_\theta})^n \cap B) \ge \sharp((2^{\mathbb{N}_\theta})^n \cap B) \ge 2^{n\theta m},$$

whence

$$\overline{\dim}_{\text{gsM}}(2^{\mathbb{N}_\theta})^n \ge \liminf_{N\to\infty} \sup_{2^{m+1}\ge N} \frac{n\theta m}{(m+1)} = \liminf_{N\to\infty} \frac{n\theta(\log_2 N) - n\theta}{\log_2 N} = \theta n.$$

The argument will be concluded by establishing

$$\overline{\dim}_{\text{gsM}}(2^{\mathbb{N}_\theta})^n \le \theta n.$$

Actually, we are led to verify that for every large $N \in \mathbb{N}$, any $r \in \mathbb{R}_+$ and any Euclidean ball B with $r_B \ge Nr$ one has

$$N_{\text{cov}}(r, (2^{\mathbb{N}_\theta})^n \cap B) \lesssim \left(\frac{r_B}{r}\right)^{\theta n},$$

because this last inequality produces a constant $c > 0$ such that

$$\begin{aligned} \overline{\dim}_{\text{gsM}}(2^{\mathbb{N}_\theta})^n &\le \liminf_{N\to\infty} \sup_{r>0} \sup_{B, r_B \ge Nr} \frac{\log_2((\frac{r_B}{r})^{\theta n}) + \log_2 c}{\log_2(\frac{r_B}{r})} \\ &= \liminf_{N\to\infty} \frac{(\theta n)\log_2 N + \theta n \log_2 c}{\log_2 N} \\ &= \theta n. \end{aligned}$$

To do so, we are about to handle two cases under $N \geq 2^5$.

Case $0 < r \leq 1$. If $r_B < 2$, then $(2^{\mathbb{N}_\theta})^n \cap B$ contains no more than a uniform number of points, and hence

$$\sharp((2^{\mathbb{N}_\theta})^n \cap B) \lesssim 1 \lesssim \left(\frac{r_B}{r}\right)^{\theta n}.$$

If $2^m < r_B \leq 2^{m+1}$ for some $m > 1$, then

$$(2^{\mathbb{N}_\theta})^n \cap B \subseteq [0, 2^{m+2}]^n.$$

Notice that the interval $[0, 2^{m+2}]$ contains at most

$$\sum_{k=1}^{m+1} 2^{\theta k} \approx 2^{\theta m}$$

points of $2^{\mathbb{N}_\theta}$. So we have

$$\sharp((2^{\mathbb{N}_\theta})^n \cap B) \lesssim 2^{\theta mn} \lesssim \left(\frac{r_B}{r}\right)^{\theta n},$$

whence

$$N_{\mathrm{cov}}(r, (2^{\mathbb{N}_\theta})^n \cap B) \leq \sharp((2^{\mathbb{N}_\theta})^n \cap B) \lesssim \left(\frac{r_B}{r}\right)^{\theta n}.$$

Case $r > 1$. Assume $2^\ell < r < 2^{l+1}$. Given a Euclidean ball B with $r_B \geq Nr$, suppose that $2^m < r_B \leq 2^{m+1}$ is valid for some $m \geq 5 + \ell$. Then

$$(2^{\mathbb{N}_\theta})^n \cap B \subseteq [0, 2^{m+2}]^n.$$

Observe that $[0, 2^l]$ can be covered by an interval of length r. So, if $l \leq k \leq [l\theta^{-1}]$ then

$$\{2^k, 2^k + 1, \ldots, 2^k + 2^{[\theta k]}\}$$

can be covered by an interval of length r. If $k > [l\theta^{-1}]$, then

$$\{2^k, 2^k + 1, \ldots, 2^k + 2^{[\theta k]}\}$$

can be covered by $2^{[\theta k]-\ell} + 1$ intervals of length r.

▷ If

$$m \leq [l\theta^{-1}] - 2,$$

then

$$2^{\mathbb{N}_\theta} \cap [0, 2^{m+2}]$$

can be covered by $m - l + 2$ intervals of length r.

▷ If

$$m > [l\theta^{-1}] - 2,$$

then

$$2^{\mathbb{N}_\theta} \cap [0, 2^{m+2}]$$

can be covered by $2^{\theta m-\ell}$ intervals of length r.

▷ In both cases,

$$2^{\mathbb{N}_\theta} \cap [0, 2^{m+2}]$$

can be covered by

$$c2^{\theta(m-\ell)} \le c\left(\frac{r_B}{r}\right)^{\theta}$$

intervals of length r for a constant $c > 0$.

Accordingly,

$$N_{\mathrm{cov}}(r, (2^{\mathbb{N}_\theta})^n \cap [0, 2^{m+2}]^n) \lesssim \left(\frac{r_B}{r}\right)^{\theta n}. \qquad \square$$

7.4 Right composition via quasiconformality

Also, a bounded-mean-oscillation variant of $Q_\alpha(\mathbb{R}^n)$ is needed.

Definition 7.7. For a measurable function f on $\mathbb{R}^n$ and $(\alpha, q) \in (0,1) \times (0,\infty)$, set

$$\Theta_{\alpha,q}(f, B(x_0,r)) = \sum_{k=0}^{\infty} 2^{2k\alpha} \fint_{B(x_0,r)} \inf_{c\in\mathbb{R}} \left(\fint_{B(x,2^{-k}r)} |f(z) - c|^q \, dz \right)^{\frac{2}{q}} dx.$$

Define the space $Q_{\alpha,q}(\mathbb{R}^n)$ as the collection of all $f \in L^q_{\mathrm{loc}}(\mathbb{R}^n)$ with

$$\|f\|_{Q_{\alpha,q}(\mathbb{R}^n)} = \sup_{(x_0,r)\in\mathbb{R}^{n+1}_+} (\Theta_{\alpha,q}(f, B(x_0,r)))^{2^{-1}} < \infty.$$

Theorem 7.8. *Let $(\alpha, q, r, x_0) \in (0,1)\times(0,2]\times(0,\infty)\times\mathbb{R}^n$ and f be a measurable function on $\mathbb{R}^n$. If*

$$\Theta_\alpha(f, B(x_0,r)) = r^{2\alpha-n} \int_{B(x_0,r)} \int_{B(x_0,r)} \frac{|f(x) - f(y)|^2}{|x-y|^{n+2\alpha}} \, dx \, dy,$$

then

$$\Theta_{\alpha,q}(f, B(x_0, 2^{-4}r)) \lesssim \Theta_\alpha(f, B(x_0, r)) \lesssim \Theta_{\alpha,q}(f, B(x_0, 2^4 r)).$$

Consequently,

$$\mathrm{Q}_\alpha(\mathbb{R}^n) = \mathrm{Q}_{\alpha,q}(\mathbb{R}^n) \quad \textit{with} \quad \|\cdot\|_{\mathrm{Q}_\alpha(\mathbb{R}^n)} \approx \|\cdot\|_{\mathrm{Q}_{\alpha,q}(\mathbb{R}^n)}.$$

Proof. According to [50], we know that if

$$\sigma \in (0,\infty) \quad \& \quad f \in \mathrm{L}^\sigma_{\mathrm{loc}}(\mathbb{R}^n)$$

then there is a set E with $\mathrm{m}(E) = 0$ such that

$$x, y \in \mathbb{R}^n \setminus E \quad \& \quad |x - y| \in [2^{-k-1}, 2^{-k})$$

implies

$$|f(x) - f(y)| \lesssim \sum_{j\geq k-2} \inf_{c\in\mathbb{R}} \Big(\Big(\fint_{B(x,2^{-j})} |f(w) - c|^\sigma \, dw \Big)^{\frac{1}{\sigma}} + \Big(\fint_{B(y,2^{-j})} |f(w) - c|^\sigma \, dw \Big)^{\frac{1}{\sigma}} \Big).$$

Accordingly,

$$\begin{aligned}\int_{B(x,2r)} \frac{|f(x) - f(y)|^2}{|x - y|^{n+2\alpha}} \, dy &\leq \sum_{j=-1}^{\infty} (2^{-j}r)^{-(n+2\alpha)} \int_{B(x,2^{-j}r)\setminus B(x,2^{-j-1}r)} |f(x) - f(y)|^2 \, dy \\ &\lesssim \mathrm{F}_1(x) + \mathrm{F}_2(x),\end{aligned}$$

where

$$\begin{cases} \mathrm{F}_1(x) = \displaystyle\sum_{j=-1}^{\infty} (2^{-j}r)^{-2\alpha} \Big(\sum_{k\geq j-2} \inf_{c\in\mathbb{R}} \fint_{B(x,2^{-k}r)} |f(w) - c|^q \, dw \Big)^{\frac{2}{q}} \\ \mathrm{F}_2(x) = \displaystyle\sum_{j=-1}^{\infty} \fint_{B(x,2^{-j}r)\setminus B(x,2^{-j-1}r)} \Big(\sum_{k\geq j-2} \inf_{c\in\mathbb{R}} \fint_{B(y,2^{-k}r)} |f(w) - c|^q \, dw \Big)^{\frac{2}{q}} \frac{dy}{(2^{-j}r)^{2\alpha}}. \end{cases}$$

Applying the Hölder inequality and changing the order of summation, we obtain

$$\begin{aligned}\mathrm{F}_1(x) &\lesssim \sum_{j=-1}^{\infty} (2^{-j}r)^{-2\alpha} 2^{-j\alpha} \sum_{k\geq j-2} 2^{k\alpha} \inf_{c\in\mathbb{R}} \Big(\fint_{B(x,2^{-k}r)} |f(w) - c|^q \, dw \Big)^{\frac{2}{q}} \\ &\lesssim \sum_{k\geq -3} 2^{k\alpha} \sum_{j=-1}^{k} (2^{-j}r)^{-2\alpha} 2^{-j\alpha} \inf_{c\in\mathbb{R}} \Big(\fint_{B(x,2^{-k}r)} |f(w) - c|^q \, dw \Big)^{\frac{2}{q}}\end{aligned}$$

$$\lesssim \sum_{k\geq -3} (2^{-k}r)^{-2\alpha} \inf_{c\in\mathbb{R}} \Big(\fint_{B(x,2^{-k}r)} |f(w)-c|^q \,dw \Big)^{\frac{2}{q}},$$

whence

$$\begin{aligned} r^{2\alpha-n} \int_{B(x_0,r)} \mathsf{F}_1(x)\,dx &\lesssim \sum_{k\geq -3} 2^{2k\alpha} \fint_{B(x_0,8r)} \inf_{c\in\mathbb{R}} \Big(\fint_{B(x,2^{-k}r)} |f(w)-c|^q \,dw \Big)^{\frac{2}{q}} dx \\ &\lesssim \Theta_{\alpha,q}(f, B(x_0,8r)). \end{aligned}$$

For $\mathsf{F}_2(x)$, we have

$$\int_{B(x_0,r)} \mathsf{F}_2(x)\,dx \lesssim \int_{B(x_0,4r)} \sum_{j=-1}^{\infty} \Big(\sum_{k\geq j-2} \inf_{c\in\mathbb{R}} \fint_{B(y,2^{-k}r)} |f(w)-c|^q \,dw \Big)^{\frac{2}{q}} \frac{dy}{(2^{-j}r)^{2\alpha}},$$

thereby applying an argument similar to the above estimate for $\mathsf{F}_1(x)$ to derive

$$r^{2\alpha-n} \int_{B(x_0,r)} \mathsf{F}_2(x)\,dx \lesssim \Theta_{\alpha,q}(f, B(x_0,8r)).$$

Combining the estimates on F_1 and F_2, we obtain

$$\Theta_\alpha(f, B(x_0,r)) \lesssim \Theta_{\alpha,q}(f, B(x_0,8r)).$$

On the other hand, noticing that

$$\begin{aligned} &(x,r) \in \mathbb{R}^{n+1}_+ \quad \& \quad k+1 \in \mathbb{N} \\ &\Rightarrow 2^{-k}r \leq |x-w| - |x-z| \leq |z-w| \leq |x-w| + |x-z| \leq 2^{-k+3}r \\ &\Rightarrow z \in B(x,2^{-k}r) \quad \& \quad w \in B(x,2^{-k+2}r) \setminus B(x,2^{-k+1}r), \end{aligned}$$

we utilize $q \in (0,2]$ and the Hölder inequality to achieve

$$\begin{aligned} \inf_{c\in\mathbb{R}} \Big(\fint_{B(x,2^{-k}r)} |f(w)-c|^q \,dw \Big)^{\frac{2}{q}} &\lesssim \fint_{B(x,2^{-k}r)} |f(z) - f_{B(x,2^{-k}r)}|^2 \,dz \\ &\lesssim \fint_{B(x,2^{-k}r)} |f(z) - f_{B(x,2^{-k+2}r)\setminus B(x,2^{-k+1}r)}|^2 \,dz \\ &\lesssim \fint_{B(x,2^{-k}r)} \fint_{B(x,2^{-k+2}r)\setminus B(x,2^{-k+1}r)} |f(z)-f(w)|^2 \,dw\,dz \\ &\lesssim (2^{-k}r)^{2\alpha} \fint_{B(x,2^{-k}r)} \int_{B(x,2^{-k+2}r)\setminus B(x,2^{-k+1}r)} \frac{|f(z)-f(w)|^2}{|z-w|^{n+2\alpha}} \,dw\,dz \\ &\lesssim (2^{-k}r)^{2\alpha} \fint_{B(x,2^{-k}r)} \int_{B(z,2^{-k+3}r)\setminus B(z,2^{-k}r)} \frac{|f(z)-f(w)|^2}{|z-w|^{n+2\alpha}} \,dw\,dz. \end{aligned}$$

Thus, from changing the order of the integrals with respect to dz and dx it follows that

$$\begin{aligned}
\Theta_{\alpha,q}(f,B(x_0,r)) &\lesssim r^{2\alpha}\sum_{k=0}^{\infty}\ \fint_{B(x_0,r)}\ \fint_{B(x,2^{-k}r)}\ \int_{B(z,2^{-k+3}r)\setminus B(z,2^{-k}r)} \frac{|f(z)-f(w)|^2}{|z-w|^{n+2\alpha}}\,\mathrm{d}w\,\mathrm{d}z\,\mathrm{d}x\\
&\lesssim r^{2\alpha}\sum_{k=0}^{\infty}\ \fint_{B(x_0,2r)}\ \fint_{B(z,2^{-k}r)}\ \int_{B(z,2^{-k+3}r)\setminus B(z,2^{-k}r)} \frac{|f(z)-f(w)|^2}{|z-w|^{n+2\alpha}}\,\mathrm{d}w\,\mathrm{d}x\,\mathrm{d}z\\
&\lesssim r^{2\alpha}\sum_{k=0}^{\infty}\ \fint_{B(x_0,2r)}\ \int_{B(z,2^{-k+3}r)\setminus B(z,2^{-k}r)} \frac{|f(z)-f(w)|^2}{|z-w|^{n+2\alpha}}\,\mathrm{d}w\,\mathrm{d}z\\
&\lesssim r^{2\alpha-n}\int_{B(x_0,2r)}\int_{B(z,8r)} \frac{|f(z)-f(w)|^2}{|z-w|^{n+2\alpha}}\,\mathrm{d}w\,\mathrm{d}z\\
&\lesssim \Theta_{\alpha}(f,B(x_0,2^4 r)),
\end{aligned}$$

as required. □

In accordance with [70], if ϕ is a quasiconformal self-map of $\mathbb{R}^n$ with $n-1\in\mathbb{N}$ then the right composition $C_{r,\phi}$ is bounded on

$$\mathrm{BMO}(\mathbb{R}^n)=\mathrm{Q}_{\alpha\in(-\infty,0)}(\mathbb{R}^n).$$

But nevertheless, the next assertion (relying on Theorem 7.8) shows a different $\mathrm{Q}_\alpha(\mathbb{R}^n)$-generalization under $\alpha\in(0,1)$.

Theorem 7.9. *Let $(\alpha,n-1)\in(0,1)\times\mathbb{N}$ and $\phi:\mathbb{R}^n\to\mathbb{R}^n$ be a quasiconformal mapping with its Jacobian*

$$\mathcal{J}_\phi\in\mathscr{A}_1(\mathbb{R}^n;E)\quad\textit{for some closed set}\quad E\subseteq\mathbb{R}^n.$$

If

$$\begin{cases}\overline{\dim}_{\mathrm{lsM}}E\in[0,n) & \textit{as } E \textit{ is bounded}\\ \overline{\dim}_{\mathrm{gsM}}E\in[0,n) & \textit{as } E \textit{ is unbounded,}\end{cases}$$

then C_ϕ is bounded on $\mathrm{Q}_\alpha(\mathbb{R}^n)$ under

$$0<\alpha<\begin{cases}\min\{1,\frac{n-\overline{\dim}_{\mathrm{lsM}}(E)}{2}\} & \textit{as } E \textit{ is bounded}\\ \min\{1,\frac{n-\overline{\dim}_{\mathrm{gsM}}(E)}{2}\} & \textit{as } E \textit{ is unbounded.}\end{cases}$$

Consequently, if

$$\begin{cases}\overline{\dim}_{\mathrm{lsM}}E\in[0,n-2] & \textit{as } E \textit{ is bounded}\\ \overline{\dim}_{\mathrm{gsM}}E\in[0,n-2] & \textit{as } E \textit{ is unbounded,}\end{cases}$$

then C_ϕ is bounded on $\mathrm{Q}_\alpha(\mathbb{R}^n)$ under $0<\alpha<1$.

Proof. Since

$$\begin{cases} \overline{\dim}_{\mathrm{lsM}} E \in [0, n-2] \Rightarrow \min\left\{1, \frac{n - \overline{\dim}_{\mathrm{lsM}}(E)}{2}\right\} = 1 \\ \overline{\dim}_{\mathrm{gsM}} E \in [0, n-2] \Rightarrow \min\left\{1, \frac{n - \overline{\dim}_{\mathrm{gsM}}(E)}{2}\right\} = 1, \end{cases}$$

the second assertion is a straightforward by-product of the first assertion which will be verified under the assumption $\operatorname{diam}(E) < \infty$ while the treatment for $\operatorname{diam}(E) = \infty$ is similar.

Without loss of generality, we may assume

$$\operatorname{diam}(E) = 1 \quad \& \quad E \subseteq B(0,1) = B(o,1).$$

By Theorem 7.8, it suffices to demonstrate

$$\Theta_{\alpha,2}\big(C_{r,\phi}f, B(x_0,r)\big) \lesssim \|f\|^2_{Q_\alpha(\mathbb{R}^n)} \quad \forall\,(x_0,r) \in \mathbb{R}^{n+1}_+.$$

The demonstration is divided into three cases.

The first case: $d(x_0,E) \geq 4r$

This ensures

$$B(x,2r) \cap E = \emptyset \quad \forall\, x \in B(x_0,r).$$

By virtue of

$$\begin{cases} \mathcal{J}_\phi \in \mathscr{A}_1(\mathbb{R}^n, E) \\ \mathcal{J}_\phi(\phi^{-1}(\cdot))\mathcal{J}_{\phi^{-1}}(\cdot) = 1 \quad \text{a. e. on } \mathbb{R}^n, \end{cases}$$

we derive that if

$$(k-1, x) \in \mathbb{N} \times B(x_0,r)$$

then

$$\operatorname{esssup}_{z\in\phi(B(x,2^{-k}r))} \mathcal{J}_{\phi^{-1}}(z) = \left(\operatorname{essinf}_{w\in B(x,2^{-k}r)} \mathcal{J}_\phi(w)\right)^{-1} \lesssim \frac{m(B(x,2^{-k}r))}{m(\phi(B(x,2^{-k}r)))}$$

and hence

$$\begin{aligned} \fint_{B(x,2^{-k}r)} |C_{r,\phi}f(z) - c|^2\,dz &= \frac{m(\phi(B(x,2^{-k}r)))}{m(B(x,2^{-k}r))} \fint_{\phi(B(x,2^{-k}r))} |f(z) - c|^2 \mathcal{J}_{\phi^{-1}}(z)\,dz \\ &\lesssim \fint_{\phi(B(x,2^{-k}r))} |f(z) - c|^2 dz. \end{aligned}$$

Accordingly,

$$\begin{aligned}\Theta_{\alpha,2}(C_{r,\phi}f,B(x_0,r)) &\lesssim \sum_{k=0}^{\infty} 2^{2k\alpha} \fint_{B(x_0,r)} \inf_{c\in\mathbb{R}}\Big(\fint_{\phi(B(x,2^{-k}r))} |f(z)-c|^2\,dz\Big)dx\\ &\lesssim \sum_{k=0}^{\infty}\int_{B(x_0,r)} \frac{m(B(x,r))^{\frac{2\alpha-n}{n}}}{m(\phi(B(x,2^{-k}r)))^{\frac{2\alpha}{n}}} \inf_{c\in\mathbb{R}}\Big(\fint_{\phi(B(x,2^{-k}r))} |f(z)-c|^2\,dz\Big)dx.\end{aligned}$$

Since $\mathcal{J}_\phi \in \mathscr{A}_1(\mathbb{R}^n,E)$ implies that

$$\frac{m(\phi(B(x,2^{-k}r)))}{m(B(x,2^{-k}r))} = \fint_{B(x,2^{-k}r)} \mathcal{J}_\phi(z)\,dz \lesssim \operatorname{essinf}_{z\in B(x,2^{-k}r)} \mathcal{J}_\phi(z) \lesssim \mathcal{J}_\phi(x)$$

holds for almost all $x \in B(x_0,r)$. In other words,

$$\big(m(B(x,2^{-k}r))\big)^{-1} \lesssim \mathcal{J}_\phi(x) m((\phi(B(x,2^{-k}r))))^{-1}$$

holds for almost all $x \in B(x_0,r)$. So, via changing variables we get

$$\begin{aligned}&\Theta_{\alpha,2}(C_{r,\phi}f,B(x_0,r))\\ &\lesssim \sum_{k=0}^{\infty}\int_{B(x_0,r)} \frac{m(B(x,r))^{\frac{2\alpha-n}{n}}}{m(\phi(B(x,2^{-k}r)))^{\frac{2\alpha}{n}}} \inf_{c\in\mathbb{R}}\Big(\fint_{\phi(B(x,2^{-k}r))} |f(z)-c|^2\,dz\Big)(\mathcal{J}_\phi(x))^{\frac{2\alpha}{n}}\,dx\\ &\lesssim \sum_{k=0}^{\infty}\int_{\phi(B(x_0,r))} \frac{m(B(\phi^{-1}(x),r))^{\frac{2\alpha-n}{n}}}{m(\phi(B(\phi^{-1}(x),2^{-k}r)))^{\frac{2\alpha}{n}}}\\ &\quad\times \inf_{c\in\mathbb{R}}\Big(\fint_{\phi(B(\phi^{-1}(x),2^{-k}r))} |f(z)-c|^2\,dz\Big)\mathcal{J}_{\phi^{-1}}(x)(\mathcal{J}_\phi(\phi^{-1}(x)))^{\frac{2\alpha}{n}}\,dx\\ &\lesssim \sum_{k=0}^{\infty}\int_{\phi(B(x_0,r))} \frac{m(B(\phi^{-1}(x),r))^{\frac{2\alpha-n}{n}}}{m(\phi(B(\phi^{-1}(x),2^{-k}r)))^{\frac{2\alpha}{n}}} \inf_{c\in\mathbb{R}}\Big(\fint_{\phi(B(\phi^{-1}(x),2^{-k}r))} |f(z)-c|^2\,dz\Big)\frac{dx}{(\mathcal{J}_{\phi^{-1}}(x))^{\frac{2\alpha-n}{n}}},\end{aligned}$$

thereby finding

$$\operatorname{esssup}_{x\in\phi(B(x_0,r))} \mathcal{J}_{\phi^{-1}}(x) \lesssim \Big(\frac{m(B(x_0,r))}{m(\phi(B(x_0,r)))}\Big)\Big(\frac{m(B(w,r))}{m(\phi(B(w,r)))}\Big) \quad \forall\, w \in B(x_0,r),$$

and so

$$\begin{aligned}&\Theta_{\alpha,2}(C_{r,\phi}f,B(x_0,r))\\ &\lesssim \sum_{k=0}^{\infty}\fint_{\phi(B(x_0,r))} \Big(\frac{m(\phi(B(\phi^{-1}(x),r)))}{m(\phi(B(\phi^{-1}(x),2^{-k}r)))}\Big)^{\frac{2\alpha}{n}} \inf_{c\in\mathbb{R}}\Big(\fint_{\phi(B(\phi^{-1}(x),2^{-k}r))} |f(z)-c|^2\,dz\Big)dx\end{aligned}$$

$$\lesssim \fint_{\phi(B(x_0,r))} \sum_{k=0}^{\infty} \left(\frac{L_\phi(\phi^{-1}(x),r)}{L_\phi(\phi^{-1}(x),2^{-k}r)} \right)^{2\alpha} \inf_{c\in\mathbb{R}} \left(\fint_{\phi(B(\phi^{-1}(x),2^{-k}r))} |f(z)-c|^2\,\mathrm{d}z \right) \mathrm{d}x,$$

where

$$L_\phi(z,r) = \sup\{|f(z)-f(w)| : |z-w| \le r\} \quad \& \quad L_\phi(z,r)^n \approx \mathrm{m}(B(z,r)).$$

Moreover, we use the fact that ϕ is quasiconformal to get

$$\sharp\{k \in \mathbb{Z} : L_\phi(z,2^{-k}r) \in [2^{-j-1}L_\phi(z,r), 2^{-j}L_\phi(z,r))\} \lesssim 1 \quad \forall (j,z) \in \mathbb{Z} \times \mathbb{R}^n.$$

Also, since

$$\phi(B(x_0,r)) \subseteq B(f(x_0), L_\phi(x_0,r)) \quad \& \quad L_\phi(\phi^{-1}(x),r) \le 2^{N_2} L_\phi(x_0,r)$$

holds for some constant $N_2 \ge 1$ (independent of x_0, r; see [48]), we arrive at

$$\begin{aligned}\Theta_{\alpha,2}(C_{r,\phi}f, B(x_0,r)) &\lesssim \sum_{j\ge 0} 2^{2j\alpha} \fint_{B(f(x_0),L_\phi(x_0,r))} \fint_{B(x,2^{-j}2^{N_2}L_\phi(x_0,r))} |f(z)-c|^2\,\mathrm{d}z\,\mathrm{d}x \\ &\lesssim \Theta_{\alpha,2}(f, B(f(x_0), 2^{N_2}L_\phi(x_0,r))),\end{aligned}$$

whence

$$\Theta_{\alpha,2}(C_{r,\phi}f, B(x_0,r)) \lesssim \|f\|^2_{Q_\alpha(\mathbb{R}^n)}$$

follows from Theorem 7.8.

The second case: $d(x_0,E) < 4r \quad \& \quad r \le 1$.

Note that $E^c = \mathbb{R}^n \setminus E$ admits a Whitney decomposition, i. e., there is a family

$$W_{E^c} = \{S_j\}_{j\in\mathbb{N}}$$

of countably many dyadic closed cubes such that:

▷ $$E^c = \bigcup_{j\in\mathbb{N}} S_j \quad \text{and} \quad S_k^\circ \cap S_j^\circ = \emptyset \quad \text{for all } j,k \in \mathbb{N} \text{ with } j \ne k;$$

▷ $$2^7\sqrt{n}\ell(S_j) \le \mathrm{dist}(S_j, \partial E^c) \le 2^9\sqrt{n}\ell(S_j);$$

▷ $$2^{-2}\ell(S_k) \le \ell(S_j) \le 2^2\ell(S_k) \quad \text{as } S_k \cap S_j \ne \emptyset.$$

So, upon assuming

$$2^{-k_0-1} \le 2^4 r < 2^{-k_0} \quad \text{for some} \quad k_0 \in \mathbb{Z},$$

and letting

$$\mathscr{S}_k(2^4B) = \{S_j \in W_{E^c} : S_j \cap 2^4B \ne \emptyset, 2^{-k} \le \ell(S_j) < 2^{-k+1}\} = \{S_{k,i}\}_i \quad \forall\, k \in \mathbb{Z},$$

we can utilize

$$\operatorname{dist}(S_{k,j}, E) \le 2^4 r + \operatorname{dist}(x_0, E) \le 20r,$$

plus the above Whitney decomposition, to get $2^{-k} \lesssim 2^{-k_0}$, namely, there exists an integer N_0 such that

$$k \le k_0 - N_0 \Rightarrow \mathscr{S}_k(2^4B) = \emptyset.$$

Moreover, given

$$\epsilon \in (0, n - \dim_{\mathrm{lsM}}(E) - 2\alpha),$$

we claim

$$\sharp\mathscr{S}_k(2^4B) \lesssim 2^{(k-k_0)(\overline{\dim}_{\mathrm{lsM}}(E)+\epsilon)} \quad \forall\, k \ge k_0 - N_0.$$

To see this, by the definition of $\overline{\dim}_{\mathrm{lsM}}(E)$ there are $N_1 \ge 8$ and $k_1 \in \mathbb{N}$ such that

$$k \ge k_1 + k_0 + N_1 \Rightarrow \begin{cases} \dfrac{\log_2 N_{\mathrm{cov}}(2^{-k}, E \cap 2^5B)}{\log_2(2^5 r/2^{-k})} \le \overline{\dim}_{\mathrm{lsM}}(E) + \epsilon \\ N_{\mathrm{cov}}(2^{-k}, E \cap 2^5B) \lesssim 2^{(k-k_0)(\overline{\dim}_{\mathrm{lsM}}(E)+\epsilon)} \end{cases}.$$

For any $\delta > 0$, denote by $\mathscr{N}_{\mathrm{cov}}(\delta, E \cap 2^5B)$ the collection of cubes of edge-length δ required to cover $E \cap 2^5B$ with

$$\sharp\mathscr{N}_{\mathrm{cov}}(\delta, E \cap 2^5B) = N_{\mathrm{cov}}(\delta, E \cap 2^5B).$$

For

$$k \ge -N_0 \quad \& \quad S_{k,i} \in \mathscr{S}_k(2^4B),$$

we obtain that

$$2^{11}\sqrt{n}S_{k,i} \cap E \ne \emptyset,$$

and $S_{k,i}$ intersects some

$$Q \in \mathscr{N}_{\mathrm{cov}}(2^{-k}, E \cap 2^5B),$$

and so

$$S_{k,i} \subseteq 2^{13} nQ.$$

Notice also that for any

$$Q \in \mathcal{N}_{\text{cov}}(2^{-k}, 2^4 B \cap E)$$

the cube $2^{13}nQ$ can only contain a uniformly bounded number of

$$S_{k,i} \in \mathcal{S}_k(2^4 B).$$

Thus we conclude

$$\begin{cases} \sharp\mathcal{S}_a(2^4B) \lesssim N_{\text{cov}}(2^{-k}, E \cap 2^5 B) & \forall\, k \geq -N_0 \\ \sharp\mathcal{S}_k(2^4B) \lesssim 2^{(k-k_0)(\overline{\dim}_{\text{lsM}}(E)+\epsilon)} & \forall\, k \geq k_1 + k_0 + N_1. \end{cases}$$

Additionally, if

$$k_0 - N_0 \leq k \leq k_1 + k_0 + N_1,$$

then

$$\sharp\mathcal{S}_k(2^4B) \lesssim 2^{n(k-k_0)} \lesssim 2^{(k-k_0)(\overline{\dim}_{\text{lsM}}(E)+\epsilon)},$$

and hence the above claim follows up.

Now, by virtue of Theorem 7.8 we have

$$\begin{aligned} &\Theta_{\alpha,2}(C_{r,\phi}f, B(x_0,r)) \\ &\quad \lesssim \Theta_\alpha(C_{r,\phi}f, B(x_0,2^4r)) \\ &\quad \lesssim r^{2\alpha-n} \sum_{k\geq k_0-N_0} \sum_{i=1}^{\sharp\mathcal{S}_k(2^4B)} \int_{S_{k,i}} \int_{B(x_0,2^4r)} \frac{|C_{r,\phi}f(x) - C_{r,\phi}f(y)|^2}{|x-y|^{n+2\alpha}}\, dx\, dy \\ &\quad \lesssim r^{2\alpha-n} \sum_{k\geq k_0-N_0} \sum_{i=1}^{\sharp\mathcal{S}_k(2^4B)} \int_{S_{k,i}} \int_{2S_{k,i}} \frac{|C_{r,\phi}f(x) - C_{r,\phi}f(y)|^2}{|x-y|^{n+2\alpha}}\, dx\, dy \\ &\qquad + r^{2\alpha-n} \sum_{k\geq k_0-N_0} \sum_{i=1}^{\sharp\mathcal{S}_k(2^4B)} \int_{S_{k,i}} \int_{B(x_0,2^4r)\setminus 2S_{k,i}} \frac{|C_{r,\phi}f(x) - C_{r,\phi}f(y)|^2}{|x-y|^{n+2\alpha}}\, dx\, dy \\ &\quad = \mathsf{G}_1 + \mathsf{G}_2. \end{aligned}$$

For each $S_{k,i}$, let $B_{k,i}$ be the ball with center $x_{k,i}$ and radius $2\sqrt{n}\ell(S_{k,i})$. Then

$$2S_{k,i} \subseteq B_{k,i} \quad \& \quad \operatorname{dist}(x_{k,i}, E) \geq 2^7 \sqrt{n}\ell(S_{k,i}).$$

Upon applying the first case above to

$$E = 2^4 B_{k,i},$$

we have

$$\Theta_\alpha(C_{r,\phi}f, B_{k,i}) \lesssim \Theta_{\alpha,2}(C_{r,\phi}f, 2^4 B_{k,i}) \lesssim \|f\|^2_{Q_\alpha(\mathbb{R}^n)},$$

thereby using

$$n - 2\alpha - \overline{\dim}_{\mathrm{lsM}}(E) - \epsilon > 0$$

to derive

$$\begin{aligned}
\mathsf{G}_1 &\lesssim r^{2\alpha-n} \sum_{k\ge k_0-N_0} \sum_{i=1}^{\sharp\mathscr{S}_\alpha(2^4B)} (\mathrm{m}(B_{k,i}))^{\frac{n-2\alpha}{n}} \Theta_\alpha(C_{r,\phi}f, B_{k,i}) \\
&\lesssim r^{2\alpha-n} \sum_{k\ge k_0-N_0} \sum_{i=1}^{\sharp\mathscr{S}_\alpha(2^4B)} 2^{-(n-2\alpha)k} \|f\|^2_{Q_\alpha(\mathbb{R}^n)} \\
&\lesssim \sum_{k\ge k_0-N_0} 2^{(k-k_0)(\overline{\dim}_{\mathrm{lsM}}(E)+\epsilon)} 2^{(n-2\alpha)(k_0-k)} \|f\|^2_{Q_\alpha(\mathbb{R}^n)} \\
&\lesssim \|f\|^2_{Q_\alpha(\mathbb{R}^n)}.
\end{aligned}$$

In order to estimate G_2, we put

$$\begin{aligned}
&\int_{S_{k,i}} \int_{B(x_0,2^4r)\setminus 2S_{k,i}} \frac{|C_{r,\phi}f(x) - C_{r,\phi}f(y)|^2}{|x-y|^{n+2\alpha}}\,dx\,dy \\
&\lesssim \sum_{l=1}^{k-k_0+5} 2^{(l-k)(-n-2\alpha)} \int_{S_{k,i}} \int_{2^{l+1}S_{k,i}\setminus 2^l S_{k,i}} |C_{r,\phi}f(x) - C_{r,\phi}f(y)|^2\,dx\,dy \\
&\lesssim \sum_{l=1}^{k-k_0+5} 2^{-2\alpha(l-k)} 2^{-kn} \Bigg\{ \fint_{S_{k,i}} |C_{r,\phi}f(x) - (C_{r,\phi}f)_{2^{l+1}S_{k,i}}|^2\,dx \\
&\quad + \fint_{2^{l+1}S_{k,i}} |C_{r,\phi}f(y) - (C_{r,\phi}f)_{2^{l+1}S_{k,i}}|^2\,dy \Bigg\}.
\end{aligned}$$

Upon observing

$$\begin{aligned}
\left(\fint_{S_{k,i}} |C_{r,\phi}f(x) - (C_{r,\phi}f)_{2^{l+1}S_{k,i}}|^2\,dx\right)^{\frac12} &\lesssim \sum_{j=1}^{l+1} \left(\fint_{2^jS_{k,i}} |C_{r,\phi}f(x) - (C_{r,\phi}f)_{2^jS_{k,i}}|^2\,dx\right)^{\frac12} \\
&\lesssim (l+1)\|C_{r,\phi}f\|_{\mathrm{BMO}(\mathbb{R}^n)},
\end{aligned}$$

we find

$$\int_{S_{k,i}} \int_{B(x_0,2^4 r)\setminus 2S_{k,i}} \frac{|C_{r,\phi}f(x) - C_{r,\phi}f(y)|^2}{|x-y|^{n+2\alpha}}\,dx\,dy$$
$$\lesssim \sum_{l=1}^{k-k_0+5} 2^{-2\alpha(l-k)} 2^{-kn} (l+1)^2 \|C_{r,\phi}f\|^2_{\mathrm{BMO}(\mathbb{R}^n)}$$
$$\lesssim 2^{(2\alpha-n)k} \|C_{r,\phi}f\|^2_{\mathrm{BMO}(\mathbb{R}^n)},$$

thereby using

$$n - 2\alpha - \overline{\dim}_{\mathrm{lsM}} E - \epsilon > 0$$

and [70] as well as Theorem 1.2 to get

$$\mathsf{G}_2 \lesssim r^{2\alpha-n} \sum_{k \ge k_0 - N_0} \sum_{i=1}^{\sharp \mathscr{S}_a(2^4 B)} 2^{(2\alpha-n)k} \|C_{r,\phi}f\|^2_{\mathrm{BMO}(\mathbb{R}^n)}$$
$$\lesssim \sum_{k \ge k_0 - N_0} 2^{(k-k_0)(\overline{\dim}_{\mathrm{lsM}}(E)+\epsilon)} 2^{(n-2\alpha)(k_0-k)} \|C_{r,\phi}f\|^2_{\mathrm{BMO}(\mathbb{R}^n)}$$
$$\lesssim \|C_{r,\phi}f\|^2_{\mathrm{BMO}(\mathbb{R}^n)}$$
$$\lesssim \|f\|^2_{\mathrm{BMO}(\mathbb{R}^n)}$$
$$\lesssim \|f\|^2_{\mathcal{Q}_\alpha(\mathbb{R}^n)}.$$

Consequently, the estimates for G_1 and G_2 are put together to derive

$$\Theta_{\alpha,2}\big(C_{r,\phi}f, B(x_0,r)\big) \lesssim \|f\|^2_{\mathcal{Q}_\alpha(\mathbb{R}^n)} \quad \forall\,(x_0,r) \in \mathbb{R}^{n+1}_+.$$

The third case: $d(x_0,E) < 4r \quad \& \quad r > 1.$

Without loss of generality, we may assume $x_0 = 0$ – the origin of $\mathbb{R}^n$. Denote by M the minimum number of Euclidean balls which are centered in $B(0,1) \setminus B(0,2^{-1})$ with radius 2^{-9} and required to cover $B(0,1) \setminus B(0,2^{-1})$. Let $\{B_j\}_{j=1}^M$ be such Euclidean balls centered at $\{x_j\}_{j=1}^M$. Let

$$B_{k,j} = B(2^k x_j, 2^{k-9}) \quad \forall\, k \ge 2 \,\&\, j \in \{1,\dots,M\}.$$

Then an application of

$$2^{k-9} = 2^{k-2}2^{-7} \le 2^{-7} d(2^k x_j, E)$$

implies that if $2^{k_0-1} \le r < 2^{k_0}$ then $k_0 \ge 1$, and hence $B(x_0,2^4 r) \setminus B(x_0,2)$ can be covered by

$$\{B_{k,j} : 2 \le k \le k_0 + 4, 0 \le j \le M\}.$$

Upon setting $B_{1,j} = B(0,2)$, we get

$$\begin{aligned}\Theta_{\alpha,2}(\mathcal{C}_{r,\phi}f, B(x_0,r)) &\lesssim \Theta_\alpha(\mathcal{C}_{r,\phi}f, B(x_0,2^4 r))\\ &\lesssim \sum_{k=1}^{k_0+4}\sum_{j=1}^{M} r^{2\alpha-n}\int_{B_{k,j}}\int_{2^4B}\frac{|\mathcal{C}_{r,\phi}f(x)-\mathcal{C}_{r,\phi}f(y)|^2}{|x-y|^{n+2\alpha}}\,dx\,dy\\ &\lesssim \sum_{k=1}^{k_0+4} r^{2\alpha-n}2^{k(n-2\alpha)}\Theta_\alpha(\mathcal{C}_{r,\phi}f, 2B_{k,j})\\ &\quad+\sum_{k=1}^{k_0+4} r^{2\alpha-n}\int_{B_{k,j}}\int_{2^4B\setminus 2B_{k,j}}\frac{|\mathcal{C}_{r,\phi}f(x)-\mathcal{C}_{r,\phi}f(y)|^2}{|x-y|^{n+2\alpha}}\,dx\,dy\\ &= \mathsf{G}_3+\mathsf{G}_4.\end{aligned}$$

Utilizing Theorem 7.8, the first case above applied to

$$E = 2^5 B_{k,j},$$

and

$$2^{k-2} = 2^{-2}d(2^k x_j, E) \le 2^{-2}d(2^k x_j, E),$$

we get

$$\Theta_\alpha(\mathcal{C}_{r,\phi}f, 2B_{k,j}) \lesssim \Theta_{\alpha,2}(\mathcal{C}_{r,\phi}f, 2^5B_{k,j}) \lesssim \|f\|^2_{\mathcal{Q}_\alpha(\mathbb{R}^n)},$$

whence

$$\mathsf{G}_3 \lesssim \|f\|^2_{\mathcal{Q}_\alpha(\mathbb{R}^n)}\sum_{k=1}^{k_0+4} r^{2\alpha-n}2^{k(n-2\alpha)} \lesssim \|f\|^2_{\mathcal{Q}_\alpha(\mathbb{R}^n)}.$$

Similarly, we have

$$\mathsf{G}_4 \lesssim \|f\|^2_{\mathcal{Q}_\alpha(\mathbb{R}^n)}.$$

Finally, the estimates for both G_3 and G_4 give

$$\Theta_{\alpha,2}(\mathcal{C}_{r,\phi}f, B(x_0,r)) \lesssim \mathsf{G}_3+\mathsf{G}_4 \lesssim \|f\|^2_{\mathcal{Q}_\alpha(\mathbb{R}^n)} \quad \forall\,(x_0,r)\in\mathbb{R}^{n+1}_+,$$

as desired. □

7.5 Reverse right composition

Upon recalling that (cf. [70] or [31, example 2.2]) if

$$\mathcal{C}_{r,\phi} : \mathrm{BMO}(\mathbb{R}^n) \to \mathrm{BMO}(\mathbb{R}^n)$$

is bounded with:

▷ $$\mathrm{m}(E) = 0 \Rightarrow \mathrm{m}(\phi^{-1}(E)) = 0;$$

▷ $\phi : \mathbb{R}^n \to \mathbb{R}^n$ being not only an ACL (absolutely continuous on lines parallel to coordinates of $\mathbb{R}^n$) homeomorphism but also being differentiable almost everywhere,

then ϕ is quasiconformal, we are driven to discover a reverse formulation of Theorem 7.9. In doing so, we need three lemmas.

Lemma 7.10. *Let* $(\alpha, n) \in [0, \min\{1, 2^{-1}n\}) \times \mathbb{N}$. *Then*

$$x \mapsto \ln|x|$$

belongs to $\mathrm{Q}_\alpha(\mathbb{R}^n)$. *Consequently:*

(i) *If* $n = 2$, *then*

$$x = (x_1, x_2) \mapsto \ln\max\{x_1^{-2}, r^2\}$$

belongs to $\mathrm{Q}_\alpha(\mathbb{R}^2)$;

(ii) *If* $n \geq 3$, *then*

$$x = (x_1, x_2, \dots, x_n) \mapsto \ln\max\{(x_1^{-2} + x_2^2)^{-1}, r^2\}$$

belongs to $\mathrm{Q}_\alpha(\mathbb{R}^n)$.

Proof. For $(x_0, r_0) \in \mathbb{R}^{n+1}_+$ and a measurable function f on $\mathbb{R}^n$, let

$$\Phi_\alpha(f, B(x_0, r_0)) = r_0^{2\alpha-n} \int_{B(x_0,r_0)} \int_{B(x_0,r_0)} \frac{|f(x) - f(y)|^2}{|x - y|^{n+2\alpha}}\,dx\,dy.$$

So, in accordance with Theorem 1.2 it suffices to verify

$$f_{\ln}(x) = \ln|x| \Rightarrow \Phi_\alpha(f_{\ln}, B(x_0, r_0)) \lesssim 1.$$

▷ *Case* $|x_0| > 2r_0$. Since there is $t_0 \in (0, 1)$ obeying

$$x, y \in B(x_0, r_0) \Rightarrow r_0 < |x|, |y| \leq 3r_0$$
$$\Rightarrow \big|\ln|x| - \ln|y|\big| = \frac{\big||x| - |y|\big|}{(1 - t_0)|x| + t_0|y|} \leq \frac{|x - y|}{r_0},$$

it follows that

$$\Phi_\alpha(f_{\ln}, B(x_0, r_0)) \leq r_0^{2\alpha-n-2} \int_{B(x_0,r_0)} \int_{B(x,2r_0)} |x - y|^{2-n-2\alpha}\,dy\,dx \lesssim \int_0^{r_0} \frac{t^{1-2\alpha}dt}{r_0^{2-2\alpha}} \lesssim 1,$$

as desired.

▷ *Case* $|x_0| \le 2r_0$. Since

$$B(x_0, r) \subseteq B(0, 3r),$$

it suffices to estimate

$$\Phi_\alpha(f_{\ln}, B(x_0, r_0)).$$

Write

$$\begin{cases} \Phi_\alpha(f_{\ln}, B(x_0, r_0)) = \Phi_{\alpha,1} + \Phi_{\alpha,2} + \Phi_{\alpha,3} \\ \Phi_{\alpha,1} = r_0^{2\alpha-n} \int\limits_{B(x_0,r_0)} \int\limits_{B(x,2^{-1}|x|)} \frac{|\ln|x| - \ln|y||^2}{|x-y|^{n+2\alpha}} \, dydx \\ \Phi_{\alpha,2} = r_0^{2\alpha-n} \int\limits_{B(x_0,r_0)} \int\limits_{B(x_0,r_0)\setminus B(x,2^2|x|)} \frac{|\ln|x| - \ln|y||^2}{|x-y|^{n+2\alpha}} \, dydx \\ \Phi_{\alpha,3} = r_0^{2\alpha-n} \int\limits_{B(x_0,r_0)} \int\limits_{B(x,2^2|x|)\setminus B(x,2^{-1}|x|)} \frac{|\ln|x| - \ln|y||^2}{|x-y|^{n+2\alpha}} \, dydx. \end{cases}$$

Since

$$|x-y| \le 2^{-1}|x| \Rightarrow \big|\ln|x| - \ln|y|\big| \le \frac{2|x-y|}{|x|},$$

it follows that

$$\Phi_{\alpha,1} \lesssim r_0^{2\alpha-n} \int\limits_{B(x_0,r_0)} |x|^{-2} \int\limits_{B(x,2^{-1}|x|)} |x-y|^{2-n-2\alpha} \, dy \, dx \lesssim 1.$$

Also, we have

$$\int\limits_{B(x_0,r_0)\setminus B(x,2^2|x|)} \frac{|\ln|x| - \ln|y||^2}{|x-y|^{n+2\alpha}} \, dy \le \sum_{j\ge 3} \int\limits_{B(x,2^j|x|)\setminus B(x,2^{j-1}|x|)} \frac{|\ln|x| - \ln|y||^2}{|x-y|^{n+2\alpha}} \, dy.$$

Note that if $j - 2 \in \mathbb{N}$ then

$$\begin{aligned} &2^{j-1}|x| \le |x-y| \le 2^j|x| \\ &\quad \Rightarrow 2^{j-2}|x| \le |y| \le 2^{j+1}|x| \\ &\quad \Rightarrow \int\limits_{B(x,\, 2^j|x|)\setminus B(x,\, 2^{j-1}|x|)} \frac{|\ln|x| - \ln|y||^2}{|x-y|^{n+2\alpha}} \, dy \lesssim \frac{2^{j(2-2\alpha)}}{|x|^{2\alpha}}. \end{aligned}$$

So

$$\Phi_{\alpha,2} \lesssim r_0^{2\alpha-n} \int\limits_{B(x_0,r_0)} |x|^{-2\alpha} \sum_{j=3}^{\infty} (\cdots) \, dy \lesssim r_0^{2\alpha-n} \int\limits_{B(x_0,r_0)} |x|^{-2\alpha} \, dx \lesssim 1.$$

Next, note that

$$y \in B(x,\ 2^2|x|) \setminus B(x,\ 2^{-1}|x|) \Rightarrow y \in \overline{B(0,5|x|)}.$$

So

$$\begin{aligned}
\Phi_{\alpha,3} &\lesssim r_0^{2\alpha-n} \int\limits_{B(x_0,r_0)} \int\limits_{B(x,\ 2^2|x|)\setminus B(x,\ 2^{-1}|x|)} \frac{|\ln|x| - \ln|y||^2}{|x-y|^{n+2\alpha}}\,dy\,dx \\
&\lesssim r_0^{2\alpha-n} \int\limits_{B(x_0,r_0)} |x|^{-n-2\alpha} \int\limits_{B(0,\ 5|x|)} \left(\ln\frac{|x|}{|y|}\right)^2 dy\,dx \\
&\lesssim r_0^{2\alpha-n} \int\limits_{B(x_0,r_0)} |x|^{-n-2\alpha} \sum_{i=1}^{\infty} (2^{-i}5|x|)^n i^2\,dx \\
&\lesssim r_0^{2\alpha-n} \int\limits_{B(x_0,r_0)} |x|^{-2\alpha}\,dx \\
&\lesssim 1.
\end{aligned}$$

□

Lemma 7.11. *Let* $(\alpha, n-1) \in (0,1) \times \mathbb{N}$. *If*

$$\begin{cases}
f \in \mathcal{Q}_\alpha(\mathbb{R}^n) \\
\sup\limits_{x_0 \in \mathbb{R}^n\ \&\ r_0 > 1} r_0^{2\alpha} \fint\limits_{B(x_0,r_0)} |f(x)|^2\,dx < \infty \\
g \in C^1(\mathbb{R}) \cap L^\infty(\mathbb{R}),
\end{cases}$$

then

$$\mathbb{R}^n \times \mathbb{R} \ni (x,z) \mapsto f(x)g(z)$$

is in $\mathcal{Q}_\alpha(\mathbb{R}^{n+1})$.

Proof. For convenience, denote by $B_n(x,r)$ and $B_1(z,r)$ the Euclidean balls in $\mathbb{R}^n$ and $\mathbb{R}$, respectively. Set

$$a_{k,r}(x,z) = f_{B_n(x,2^{-k}r)}g(z) \quad \forall k-1 \in \mathbb{N}.$$

It is enough to prove

$$\Psi_\alpha(f, B_n(x,r) \times B_1(z,r)) = \sum_{k=0}^{\infty} 2^{2k\alpha} \fint\limits_{B_n(x,r)\times B_1(z,r)} \fint\limits_{B_n(x,2^{-k}r)\times B_1(z,2^{-k}r)} |f(\tilde{x})g(\tilde{z}) - a_{k,r}(x,z)|^2\,d\tilde{z}\,d\tilde{x}\,dz\,dx$$

$$\lesssim \|f\|^2_{Q_\alpha(\mathbb{R}^n)}.$$

Since g is not only essentially bounded – i. e. – $g \in L^\infty(\mathbb{R})$ but also of the Lipschitz class – i. e. – $g \in C^1(\mathbb{R})$, we can calculate

$$\begin{aligned}
&\fint_{B_n(x,2^{-k}r)\times B_1(z,2^{-k}r)} |f(\tilde{x})g(\tilde{z}) - a_{k,r}(x,z)|^2\, d\tilde{z}\, d\tilde{x}\\
&\quad\lesssim \fint_{B_n(x,2^{-k}r)} |f(\tilde{x}) - f_{B_n(x,2^{-k}r)}|^2\, dx \fint_{b_1(z,r)} |g(\tilde{z})|^2 d\tilde{z}\\
&\qquad + |f_{B_n(x,2^{-k}r)}|^2 \fint_{B_1(z,2^{-k}r)} |g(\tilde{z}) - g(z)|^2\, d\tilde{z}\\
&\quad\lesssim \fint_{B_n(x,2^{-k}r)} |f(\tilde{x}) - f_{B_n(x,2^{-k}r)}|^2\, d\tilde{x} + (\min\{2^{-k}r, 1\})^2 |f_{B_n(x,2^{-k}r)}|^2,
\end{aligned}$$

thereby reaching

$$\begin{aligned}
\Psi_\alpha(f, B_n(x,r)\times B_1(z,r)) &\lesssim \sum_{k=0}^{\infty} 2^{2k\alpha} \fint_{B_n(x,r)} \fint_{B_n(x,2^{-k}r)} |f(\tilde{x}) - f_{B_n(x,2^{-k}r)}|^2\, d\tilde{x}\, dx\\
&\quad + \sum_{k=0}^{\infty} (2^{k\alpha}\min\{2^{-k}r, 1\})^2 \fint_{B_n(x,r)} |f_{B_n(x,2^{-k}r)}|^2\, dx\\
&\lesssim \|f\|^2_{Q_\alpha(\mathbb{R}^n)} + \Sigma(f,\alpha),
\end{aligned}$$

where

$$\Sigma(f,\alpha) = \sum_{k=0}^{\infty} (2^{k\alpha}\min\{2^{-k}r, 1\})^2 \fint_{B_n(x,r)} |f_{B_n(x,2^{-k}r)}|^2\, dx.$$

▷ *Case* $r < 2$. By the hypothesis and the inclusion,

$$Q_\alpha(\mathbb{R}^n) \subseteq \mathrm{BMO}(\mathbb{R}^n),$$

we obtain

$$|f_{B_n(x,2^{-k}r)}| \lesssim |f_{B_n(x,2^{-k}r)} - f_{B_n(x,2r)}| + |f_{B_n(x,2r)}| \lesssim \left(k + \ln\frac{2^2}{r}\right)\|f\|_{Q_\alpha(\mathbb{R}^n)},$$

whence

$$\Sigma(f,\alpha) \lesssim \|f\|^2_{Q_\alpha(\mathbb{R}^n)} \sum_{k=0}^{\infty} 2^{2k(\alpha-1)} r^2 \left(k + \ln\frac{2^2}{r}\right)^2 \lesssim \|f\|^2_{Q_\alpha(\mathbb{R}^n)}.$$

▷ *Case $r \geq 2$.* By the hypothesis and the Hölder inequality, we have

$$\fint_{B_n(x,r)} |f_{B_n(x,2^{-k}r)}|^2\,dx \lesssim \fint_{B_n(x,2r)} |f(y)|^2\,dy \lesssim r^{-2\alpha},$$

thereby achieving

$$\Sigma(f,\alpha) \lesssim (1 + \|f\|^2_{Q_\alpha(\mathbb{R}^n)})\Big(\sum_{k\geq \ln r} 2^{2k(\alpha-1)} r^{2(1-\alpha)} + \sum_{0\leq k\leq \ln r} 2^{2k\alpha} r^{-2\alpha}\Big) \lesssim 1 + \|f\|^2_{Q_\alpha(\mathbb{R}^n)}.$$

A combination of the previous two cases derives

$$\Psi_\alpha(f, B_n(x,r) \times B_1(z,r)) \lesssim \|f\|^2_{Q_\alpha(\mathbb{R}^n)} + \Sigma(f,\alpha) \lesssim 1 + \|f\|^2_{Q_\alpha(\mathbb{R}^n)}.$$ □

Corollary 7.12. *Let $(\alpha, n-1, a_1) \in (0,1) \times \mathbb{N} \times (0,1)$,*

$$g(t) = \begin{cases} 0 & \text{as } t \in (-\infty, -2] \\ 1 - |1+t| & \text{as } t \in [-2, 0] \\ 1 - |1-t| & \text{as } t \in [0, 2] \\ 0 & \text{as } t \in [2, \infty), \end{cases}$$

and $h_2, \ldots, h_n$ be smooth function supported in $(-2, 2)$ and equal to 1 on the open interval $(-1, 1)$.

(i) *If $n = 2$ and*

$$f_\star(x) = (\ln \max\{x_1^{-2}, 1\}) g(x_2),$$

then $f_\star \in Q_\alpha(\mathbb{R}^2)$.

(ii) *If $n \geq 3$ and*

$$f_\star(x) = \begin{cases} (\ln \max\{x_1^{-2}, a_1^2\}) h_2(x_2) \cdots h_{n-1}(x_{n-1}) g(x_n) & \text{as } \alpha \in (0, 2^{-1}) \\ (\ln \max\{(x_1^2 + x_2^2)^{-1}, a_1^2\}) h_2(x_2) \cdots h_{n-1}(x_{n-1}) g(x_n) & \text{as } \alpha \in [2^{-1}, 1), \end{cases}$$

then $f_\star \in Q_\alpha(\mathbb{R}^n)$ with its norm being bounded uniformly on a_1.

Proof. (i) $f_\star \in Q_\alpha(\mathbb{R}^2)$ follows from Lemma 7.10 and the fact that $\ln \max\{x_1{}^{-2}, 1\}$ enjoys the integral condition of Lemma 7.11.

(ii) Notice that

$$\begin{cases} \ln \max\{x_1^{-2}, a_1^2\} = \ln \max\{x_1^{-2}, a_1^2\} - \ln a_1^2 + \ln a_1^2 & \text{as } \alpha \in (0, 2^{-1}) \\ \ln \max\{(x_1^2 + x_2^2)^{-1}, a_1^2\} = \ln \max\{(x_1^2 + x_2^2)^{-1}, a_1^2\} - \ln a_1^2 + \ln a_1^2 & \text{as } \alpha \in [2^{-1}, 1), \end{cases}$$

where

$$\begin{cases} \ln \max\{x_1^{-2}, a_1^2\} - \ln a_1^2 & \text{as } \alpha \in (0, 2^{-1}) \\ \ln \max\{(x_1^2 + x_2^2)^{-1}, a_1^2\} - \ln a_1^2 & \text{as } \alpha \in [2^{-1}, 1), \end{cases}$$

satisfies the intermediate integral condition of Lemma 7.11 and has its norm uniformly bounded on a_1. Applying Lemma 7.10(ii), we obtain that f_* can written as the summation of two $Q_\alpha(\mathbb{R}^n)$-functions with its norm being uniformly bounded on a_1. □

Lemma 7.13. *For $r > 0$ and $0 < a_1 \le \cdots \le a_n = 1$, let*

$$\begin{cases} f_r(x) = f(r^{-1}x) \\ P_{(a_1,\dots,a_n),r} = \{x = (x_1, \dots, x_n) \in \mathbb{R}^n : |x_i| \le a_i r\} \\ f_{(a_1,\dots,a_n),r}(x) = f_r \mathbb{1}_{P_{(a_1,\dots,a_n),r}} ((2r)^n a_1 \cdots a_n)^{-1} \\ c_{(a_1,\dots,a_n)} = \displaystyle\fint_{P_{(a_1,\dots,a_n),1}} |f(x)|\,dx. \end{cases}$$

If $h \in L^1(\mathbb{R}^n)$ and positive numbers $a_1, \dots, a_n$ are rational, then there exists a subsequence $\{r_j\}_j$ converging to 0 such that

$$\begin{cases} f_{(a_1,\dots,a_n),r_j} * h(y) \to 0 \\ |f_{(a_1,\dots,a_n),r_j}| * h(y) \to c_{(a_1,\dots,a_n)} h(y) \end{cases}$$

holds for almost all $y \in \mathbb{R}^n$.

Proof. This follows from [70, Lemma 8 and the corollary]. □

With the help of the above preliminary results, we find out an opposite assertion of Theorem 7.9 as stated below.

Theorem 7.14. *For $(n - 1, \alpha) \in \mathbb{N} \times (0, 1)$, let $\phi : \mathbb{R}^n \to \mathbb{R}^n$ be a homeomorphism. If*

- ▷ *$C_{r,\phi}$ and $C_{r,\phi^{-1}}$ are bijective and bounded on $Q_\alpha(\mathbb{R}^n)$, respectively;*
- ▷ *ϕ is not only ACL but also differentiable almost everywhere,*

then ϕ is quasiconformal.

Proof. Rather than proving that f is quasiconformal, we prove that the inverse ϕ^{-1} of ϕ is quasiconformal. It is enough to show that

$$|\mathcal{D}\phi^{-1}(x)|^n = \max_{y \in \mathbb{R}^n \,\&\, |y|=1} |\mathcal{D}\phi^{-1}(x)\,y|^n \lesssim \mathcal{J}_{\phi^{-1}}(x)$$

holds for almost all $x \in \mathbb{R}^n$ where $\mathcal{D}\phi^{-1}$ and $\mathcal{J}_{\phi^{-1}}$ are the formal derivative and Jacobian determinant of ϕ^{-1}, respectively (cf. [34, p. 250]). Since ϕ^{-1} is absolutely continuous with respect to the n-dimensional Lebesgue measure, one has

$$\mathcal{J}_{\phi^{-1}}(x) = \lim_{r \to 0} \frac{m(\phi^{-1}(B(x,r)))}{m(B(x,r))}$$

almost everywhere and $\mathcal{J}_{\phi^{-1}} \in L^1_{\text{loc}}(\mathbb{R}^n)$. Also, the hypothesis implies that not only ϕ^{-1} is totally differentiable almost everywhere but also $\mathcal{J}_{\phi^{-1}}$ is positive almost everywhere. So, without loss of generality we may assume

$$\mathcal{J}_{\phi^{-1}}(0) > 0 \quad \& \quad h(x) = \mathcal{J}_{\phi^{-1}}(x)\mathbb{1}_{B(0,1)}(x)$$

in Lemma 7.13. Up to some rotation, translation and scaling which preserve the $Q_\alpha(\mathbb{R}^n)$-norm, we may also assume

$$\begin{cases} \phi^{-1}(0) = 0 \\ \mathcal{D}\phi^{-1}(0) = \operatorname{diag}\{\lambda_1, \dots, \lambda_n\} \\ \lambda_1 \geq \cdots \geq \lambda_n = 1, \end{cases}$$

and then verify

$$\lambda_1^n \lesssim \lambda_1 \cdots \lambda_n.$$

For $1 > \epsilon > 0$ and $m \in \mathbb{N}$, we choose rational numbers $a_{m1}, \dots, a_{mn}$ such that

$$|a_{mk}\lambda_k - 1| \leq \epsilon \quad \forall\, k \in \{1, \dots, n\}.$$

Then an application of Lemma 7.13 gives

$$(\dagger) \quad 1 - \ln a_{m1} \lesssim C_{(a_{m1},\dots,a_{mn})} = \fint_{P_{(a_{m1},\dots,a_{mn}),1}} |u(x)|\, dx.$$

In this way, for a sufficiently small $r < \delta_1$ we get that if

$$\begin{cases} R = \{z = (z_1, \dots, z_n) \in \mathbb{R}^n : \ |z_i| \leq r(1 - \epsilon)\ \forall\, i \in \{1, \dots, n\}\} \\ S = \{z = (z_1, \dots, z_n) \in \mathbb{R}^n : \ |z_i| \leq r(1 + \epsilon)\ \forall\, i \in \{1, \dots, n\}\}, \end{cases}$$

then

$$R \subseteq \phi^{-1}(P_{(a_{m1},\dots,a_{mn}),r}) \subseteq S.$$

This implication can be established by the differentials of f and ϕ^{-1} at 0 – the origin of $\mathbb{R}^n$, as well as

$$\mathcal{D}\phi^{-1}(0) = \operatorname{diag}\{\lambda_1, \dots, \lambda_n\} \quad \& \quad \mathcal{D}\phi(0) = \operatorname{diag}\{\lambda_1^{-1}, \dots, \lambda_n^{-1}\}.$$

By virtue of the assumption on ϕ and the function $f_\star$ constructed in Corollary 7.12, we have

$$(\ddagger) \quad \|(f_\star)_r \circ \phi\|_{Q_{-\frac{n}{2}}(\mathbb{R}^n)} \lesssim \|(f_\star)_r \circ \phi\|_{Q_\alpha(\mathbb{R}^n)} \lesssim \|(f_\star)_r\|_{Q_\alpha(\mathbb{R}^n)} \lesssim \|f_\star\|_{Q_\alpha(\mathbb{R}^n)} \lesssim 1.$$

Since

$$\mathrm{Q}_\alpha(\mathbb{R}^n) \subseteq \mathrm{BMO}(\mathbb{R}^n) = \mathrm{Q}_{-\frac{n}{2}}(\mathbb{R}^n)$$

we are required to control

$$\|(f_\star)_r \circ \phi\|_{\mathrm{BMO}(\mathbb{R}^n)} = \|(f_\star)_r \circ \phi\|_{\mathrm{Q}_{-\frac{n}{2}}(\mathbb{R}^n)}$$

from below. Let

$$\begin{cases} P_r = \{z = (z_1, \dots, z_n) \in \mathbb{R}^n : |z_i| \le r \,\forall\, i \in \{1, \dots, n\}\} \\ P_{(a_{m1},\dots,a_{mn}),r} = \{z = (z_1, \dots, z_n) \in \mathbb{R}^n : |z_i| \le a_{mi} r \,\forall\, i \in \{1, \dots, n\}\}. \end{cases}$$

Then

$$\|(f_\star)_r \circ \phi\|_{\mathrm{BMO}(\mathbb{R}^n)} \gtrsim \fint_{\phi^{-1}(P_{(a_{m1},\dots,a_{mn}),r})} \left|(f_\star)_r \circ \phi(x) - \fint_{\phi^{-1}(P_{(a_{m1},\dots,a_{mn}),r})} (f_\star)_r \circ \phi(y)\,dy\right| dx.$$

Notice that the above argument yields

$$\begin{aligned} \fint_{\phi^{-1}(P_{(a_{m1},\dots,a_{mn}),r})} (f_\star)_r \circ \phi(x)\,dx &= \frac{\mathrm{m}(P_{(a_{m1},\dots,a_{mn}),r})}{\mathrm{m}(\phi^{-1}(P_{(a_{m1},\dots,a_{mn}),r}))} \fint_{P_{(a_{m1},\dots,a_{mn}),r}} (f_\star)_r(z) \mathcal{J}_{\phi^{-1}}(z)\,dz \\ &= \frac{\mathrm{m}(P_{(a_{m1},\dots,a_{mn}),r})}{\mathrm{m}(\phi^{-1}(P_{(a_{m1},\dots,a_{mn}),r}))} (f_\star)_{(a_{m1},\dots,a_{mn}),r} * h(0). \end{aligned}$$

So, according to Lemma 7.13 for each $j \in \mathbb{N}$ there are $\delta_2 \in (0, \delta_1)$ and $r_j < \delta_2$ such that

$$\left| \fint_{\phi^{-1}(P_{(a_{m1},\dots,a_{mn}),r})} (f_\star)_{r_j} \circ \phi(x)\,dx \right| \le \epsilon \quad \forall\, m \in \mathbb{N}.$$

Accordingly,

$$\|(f_\star)_r \circ \phi\|_{\mathrm{BMO}(\mathbb{R}^n)} \ge \fint_{\phi^{-1}(P_{(a_{m1},\dots,a_{mn}),r})} |(f_\star)_r \circ \phi(x)|\,dx - \epsilon \quad \forall\, r \in (0, \infty).$$

Similarly,

$$\begin{aligned} \fint_{\phi^{-1}(P_{(a_{m1},\dots,a_{mn}),r})} |(f_\star)_r \circ \phi(x)|\,dx &= \frac{\mathrm{m}(P_{(a_{m1},\dots,a_{mn}),r})}{\mathrm{m}(\phi^{-1}(P_{(a_{m1},\dots,a_{mn}),r}))} \fint_{P_{(a_{m1},\dots,a_{mn}),r}} |(f_\star)_r(z)| \mathcal{J}_{\phi^{-1}}(z)\,dz \\ &= \frac{\mathrm{m}(P_{(a_{m1},\dots,a_{mn}),r})}{\mathrm{m}(\phi^{-1}(P_{(a_{m1},\dots,a_{mn}),r}))} |(f_\star)_{(a_{m1},\dots,a_{mn}),r}| * h(0). \end{aligned}$$

By Lemma 7.13, we have

$$\liminf_{r_j\to 0} \fint_{\phi^{-1}(P_{(a_1,\dots,a_n),r_j})} |(f_\star)_{r_j}\circ\phi(x)|\,dx = \left(\frac{c_{(a_1,\dots,a_n)}}{(h(0))^{-1}}\right)\left(\liminf_{r_j\to 0}\frac{m(P_{(a_1,\dots,a_n),r_j})}{m(\phi^{-1}(P_{(a_1,\dots,a_n),r_j}))}\right).$$

For $r_j < \delta_1$, we utilize

$$1-\epsilon \le a_{mk}\lambda_k \le 1+\epsilon$$

to derive

$$\left(\frac{m(P_{(a_{m1},\dots,a_{mn}),r_j})}{m(\phi^{-1}(P_{(a_{m1},\dots,a_{mn}),r_j}))}\right)h(0) \ge \frac{(a_{m1}\lambda_1)\cdots(a_{mn}\lambda_n)}{(1+\epsilon)^n} \ge \left(\frac{1-\epsilon}{1+\epsilon}\right)^n.$$

Accordingly, we have

$$\liminf_{r_j\to 0} \fint_{\phi^{-1}(P_{(a_{m1},\dots,a_{mn}),r})} |(f_\star)_{r_j}\circ\phi(x)|\,dx \ge \left(\frac{1-\epsilon}{1+\epsilon}\right)^n c_{(a_1,\dots,a_n)}$$

thereby reaching

$$\|(f_\star)_r\circ\phi\|_{\mathrm{BMO}(\mathbb{R}^n)} \ge \left(\frac{1-\epsilon}{1+\epsilon}\right)^n c_{(a_{m1},\dots,a_{mn})} - \epsilon.$$

Upon combining this last estimation under $\epsilon\to 0$ with (†) and (‡), we achieve

$$\begin{cases} 1-\ln a_{m1} \lesssim 1; \\ a_{m1} \gtrsim 1; \\ \lambda_1 \lesssim 1; \\ \lambda_1^n \lesssim \lambda_1\cdots\lambda_n. \end{cases}$$ □

7.6 Transport equation in $Q_\alpha \supset \dot{B}^{\alpha}_{\frac{n}{\alpha},1\le q\le 2}$

This section utilizes Theorem 7.9 and Theorem 1.4(i) to establish the existence and uniqueness of a weak solution to the Cauchy problem for the transport equation on $\mathbb{R}^n\times[0,T]$ which assumes a given data in

$$Q_\alpha(\mathbb{R}^n) \supset \dot{B}^{\alpha}_{\frac{n}{\alpha},1\le q\le 2}(\mathbb{R}^n).$$

Theorem 7.15. *Given*

$$\begin{cases} (\alpha, n-1, q, T) \in (0,1)\times\mathbb{N}\times[1,2]\times(0,\infty) \\ b:\mathbb{R}^n\times[0,T]\to\mathbb{R}^n, \end{cases}$$

let

$$\begin{cases} b \in \mathrm{L}^1(\mathrm{W}^{1,1}_{\mathrm{loc}}(\mathbb{R}^n);[0,T]) \Leftrightarrow \int_0^T \int_K (|b(x,t)| + |\nabla_x b(x,t)|)\mathrm{d}x\mathrm{d}t < \infty \quad \forall \text{ compact } K \subset \mathbb{R}^n \\ \|b\|_{\mathrm{L}^\infty(\mathrm{Lip}(\mathbb{R}^n);[0,T])} = \left\| \sup_{\mathbb{R}^n \ni x \neq y \in \mathbb{R}^n} |x-y|^{-1}|b(x,\cdot) - b(y,\cdot)| \right\|_{\mathrm{L}^\infty([0,T])} < \infty. \end{cases}$$

If

$$\begin{cases} u_0 \in \mathrm{Q}_\alpha(\mathbb{R}^n) \\ \text{or} \\ u_0 \in \dot{\mathrm{B}}^{\alpha}_{\frac{n}{\alpha},q}(\mathbb{R}^n), \end{cases}$$

then the transport equation

$$\begin{cases} \partial_t u(x,t) - b(x,t)\cdot \nabla_x u(x,t) = 0 & \forall\, (x,t) \in \mathbb{R}^n \times [0,T] \\ u(x,0) = u_0(x) & \forall\, x \in \mathbb{R}^n \end{cases}$$

has a unique weak solution

$$u : \mathbb{R}^n \times [0,T] \to \mathbb{R}$$

with

$$\begin{cases} \|u\|_{\mathrm{L}^\infty(\mathrm{Q}_\alpha(\mathbb{R}^n);[0,T])} = \operatorname{esssup}_{t\in[0,T]} \|u(\cdot,t)\|_{\mathrm{Q}_\alpha(\mathbb{R}^n)} < \infty \\ \text{or} \\ \|u\|_{\mathrm{L}^\infty(\dot{\mathrm{B}}^{\alpha}_{\frac{n}{\alpha},q}(\mathbb{R}^n);[0,T])} = \operatorname{esssup}_{t\in[0,T]} \|u(\cdot,t)\|_{\dot{\mathrm{B}}^{\alpha}_{\frac{n}{\alpha},q}(\mathbb{R}^n)} < \infty \end{cases}$$

in the sense that

$$\int_{\mathbb{R}^n} \int_0^T u(x,t)\partial_t \zeta(x,t)\,\mathrm{d}t\mathrm{d}x + \int_{\mathbb{R}^n} u_0(x)\zeta(x,0)\,\mathrm{d}x$$
$$= \int_{\mathbb{R}^n} \int_0^T u(x,t)\,\mathrm{div}(b(x,t)\zeta(x,t))\,\mathrm{d}t\mathrm{d}x$$

holds for any

$$\zeta \in \mathrm{C}^\infty(\mathbb{R}^n \times [0,T]) \quad \text{with} \quad \mathrm{supp}(\zeta) \subseteq \mathbb{R}^n \times [0,T).$$

Proof. The argument consists of two steps.

Existence. Since

$$\|b\|_{\mathrm{L}^\infty(\mathrm{Lip}(\mathbb{R}^n);[0,T])} < \infty,$$

according to the classical result (cf. [71, p. 247]),

$$\begin{cases} \partial_t \chi(x,t) = b(\chi(x,t),t) \\ \chi(x,0) = x, \end{cases}$$

has a unique solution

$$(x,t) \mapsto \chi(x,t).$$

This solution is a bi-Lipschitzian, namely

$$|\chi(x,t) - \chi(y,t)| \approx |x-y| \quad \forall\,(x,y) \in \mathbb{R}^n \times \mathbb{R}^n,$$

and so quasiconformal self-homeomorphism of $\mathbb{R}^n$ for any given $t \in [0,T]$, depends differentiably on the initial value for fixed t and a. e. x, and obeys

$$\begin{cases} \chi(x,t) = x + \int\limits_0^t b(\chi(x,s),s)\,\mathrm{d}s \\ |\chi(x,t)| \le |x| + \int\limits_0^t \left\| \frac{b(\cdot,s)}{1+|\cdot|\ln^+|\cdot|} \right\|_{L^\infty(\mathbb{R}^n)} (1 + |\chi(x,s)|)\big|\ln^+|\chi(x,s)|\big|\,\mathrm{d}s \\ \ln^+ r = \max\{0, \ln r\} \quad \forall r > 0. \end{cases}$$

This inequality, along with the fact (cf. [71, Theorem 3])

$$\int\limits_0^T \left\| \frac{b(\cdot,s)}{1+|\cdot|\ln^+|\cdot|} \right\|_{L^\infty(\mathbb{R}^n)} \mathrm{d}s \lesssim T\big(1 + \|b\|_{L^\infty(\mathrm{Lip}(\mathbb{R}^n);[0,T])}\big),$$

and the Gronwall-type inequality on [21, p. 3], derives

$$|\chi(x,t)| \lesssim 1 \quad \forall\,(x,t) \in B(0,R) \times [0,T]$$

where the constant relies on R and the last integral over $[0,T]$, and consequently, $\chi(\cdot,t)$ maps bounded sets to bounded sets in $\mathbb{R}^n$ for any finite time $t \in [0,T]$.

On the one hand, if

$$u_0 \in \mathrm{Q}_\alpha(\mathbb{R}^n) \quad \text{or} \quad u_0 \in \dot{\mathrm{B}}^{\alpha}_{\frac{n}{\alpha},q}(\mathbb{R}^n)$$

then

$$\mathcal{J}_{\chi(x,t)} \approx 1 \quad \forall x \in \mathbb{R}^n$$

and hence Theorem 7.9 or [45, Theorem 1.1] can be employed to imply that the right composition

$$x \mapsto u(x,t) = u_0(\chi(x,t))$$

belongs to

$$Q_\alpha(\mathbb{R}^n) \quad \text{or} \quad \dot{B}^{\alpha}_{\frac{n}{\alpha},q}(\mathbb{R}^n)$$

with

$$\begin{aligned}&\operatorname{esssup}_{t\in[0,T]} \|u(\cdot,t)\|_{Q_\alpha(\mathbb{R}^n)} \lesssim \|u_0\|_{Q_\alpha(\mathbb{R}^n)} \quad \text{or}\\ &\operatorname{esssup}_{t\in[0,T]} \|u(\cdot,t)\|_{\dot{B}^{\alpha}_{\frac{n}{\alpha},q}(\mathbb{R}^n)} \lesssim \|u_0\|_{\dot{B}^{\alpha}_{\frac{n}{\alpha},q}(\mathbb{R}^n)}.\end{aligned}$$

On the other hand, for any

$$(\phi,\psi) \in C_c^\infty(\mathbb{R}^n) \times C_c^\infty([0,T))$$

we have not only

$$\begin{aligned}&\int_{\mathbb{R}^n}\int_0^T u(x,t)\partial_t(\phi(x)\psi(t))\,dtdx\\
&= \int_{\mathbb{R}^n}\int_0^T u_0(\chi(x,t))\phi(x)\partial_t\psi(t)\,dt\,dx\\
&= \int_{\mathbb{R}^n}\int_0^T u_0(x)(\partial_t\psi(t))\phi(\chi(x,-t))\mathcal{J}_{\chi(x,-t)}\,dxdt\\
&= -\int_{\mathbb{R}^n} u_0(x)\psi(0)\phi(x)\,dx - \int_0^T\int_{\mathbb{R}^n} u_0(x)\psi(t)\partial_t\big(\phi(\chi(x,-t))\mathcal{J}_{\chi(x,-t)}\big)\,dxdt,\end{aligned}$$

but also

$$\begin{aligned}&\int_{\mathbb{R}^n}\int_0^T u(x,t)\operatorname{div}(b(x,t)\phi(x)\psi(t))\,dxdt\\
&= \int_{\mathbb{R}^n}\int_0^T u_0(\chi(x,t))\psi(t)\operatorname{div}(b(x,t)\phi(x))\,dxdt\\
&= \int_{\mathbb{R}^n}\int_0^T u_0(x)\psi(t)\operatorname{div}(b(\chi(x,-t),t)\phi(\chi(x,-t)))\mathcal{J}_{\chi(x,-t)}\,dxdt.\end{aligned}$$

Upon recalling (cf. [2])

$$\partial_t\mathcal{J}_{\chi(x,-t)} = -\mathcal{J}_{\chi(x,-t)}\operatorname{div}\big(b(\chi(x,-t),t)\big)$$

we calculate

$$\begin{aligned}&\partial_t\phi(\chi(x,-t))\mathcal{J}_{\chi(x,-t)}\\&\quad=-\nabla_x\phi(\chi(x,-t))\cdot b(\chi(x,-t),t)\mathcal{J}_{\chi(x,-t)}-\phi(\chi(x,-t))\operatorname{div}b(\chi(x,-t),t)\mathcal{J}_{\chi(x,-t)}\\&\quad=-\operatorname{div}\big(b(\chi(x,-t),t)\big)\phi(\chi(x,-t))\mathcal{J}_{\chi(x,-t)},\end{aligned}$$

thereby getting

$$\begin{aligned}&\int_{\mathbb{R}^n}\int_0^T u(x,t)\partial_t\big(\phi(x)\psi(t)\big)\,dtdx+\int_{\mathbb{R}^n}u_0(x)\psi(0)\phi(x)\,dx\\&\quad=\int_{\mathbb{R}^n}\int_0^T u(x,t)\operatorname{div}\big(b(x,t)\phi(x)\psi(t)\big)\,dtdx,\end{aligned}$$

and so that

$$u(x,t)=u_0(\chi(x,t))$$

is a weak solution of the transport equation with

$$u\in L^\infty(Q_\alpha(\mathbb{R}^n);[0,T])\quad\text{or}\quad u\in L^\infty(\dot{B}^{\alpha}_{\frac{n}{\alpha},q}(\mathbb{R}^n);[0,T]).$$

Uniqueness. Suppose that

$$u\in L^\infty(\dot{B}^{\alpha}_{\frac{n}{\alpha},q}(\mathbb{R}^n);[0,T])\quad\text{or}\quad u\in L^\infty(Q_\alpha(\mathbb{R}^n);[0,T])$$

is a weak solution to the transport equation with the initial data $u_0=0$. Utilizing

$$\|b\|_{L^\infty(\mathrm{Lip}(\mathbb{R}^n);[0,T])}<\infty$$

we achieve that if

$$\begin{cases}0\le\eta\in C_c^\infty(\mathbb{R}^n)\\\eta_\epsilon(x)=\epsilon^{-n}\eta(\epsilon^{-1}x)\quad\forall\,(\epsilon,x)\in(0,\infty)\times\mathbb{R}^n\\u_\epsilon(x,t)=u(\cdot,t)*\eta_\epsilon(x),\end{cases}$$

then according to [19, Theorem 2.1 and Lemma 2.1]

$$\begin{cases}\partial_t u_\epsilon(x,t)+b(x,t)\cdot\nabla_x u_\epsilon(x,t)=r_\epsilon\\u_\epsilon\to0\quad\&\quad r_\epsilon\to0\quad\text{in}\quad L^1(L^1_{\mathrm{loc}}(\mathbb{R}^n);[0,T])\quad\text{as}\quad\epsilon\to0,\end{cases}$$

and hence

$$\partial_t\Phi(u(x,t))+b(x,t)\cdot\nabla_x\Phi(u(x,t))=0$$

for

$$\begin{cases} \Phi \in C^1(\mathbb{R}) \\ \Phi(0) = 0 \\ \Phi' \in L^\infty(\mathbb{R}), \end{cases}$$

owing to the renormalization property of the transport equation and Theorem 7.2(i). For $j \in \mathbb{N}$ set

$$\Phi_j(t) = \min\{|t|, j\}.$$

An approximation yields that $\Phi_j(u)$ is a weak solution to the transport equation with the initial data $\Phi_j(u_0) = 0$ and so that $\Phi_j(u) = 0$ due to the well-posedness of the transport equation in $L^\infty(\mathbb{R}^n)$ (cf. [14, Theorem 2.2]). So, $u = 0$ follows from

$$\lim_{j\to\infty} \Phi_j(u) = 0. \qquad \square$$

7.7 Notes

Below are some historical remarks on Chapter 7.

Section 7.1. Lemma 7.1 can be found in [29, pp. 72–73] (cf. [78]). However, Theorem 7.2(i) is taken from [94, Proposition 2.2]; see also [40, 39, 6] for $BMO(\mathbb{R}^n)$-setting.

Section 7.2. Theorem 7.3 is produced by using [39, Theorem 3] and [40, Lemma 2.2].

Section 7.3. This part comes basically from [49, Section 2].

Section 7.4. Theorem 7.8 is the same as [49, Proposition 3.1], however, it can be compared with Theorem 1.5. Theorem 7.9 is taken from [49, Theorem 1.3]. This assertion not only shows that if

$$\phi(x) = |x|^{\beta-1}x \quad \text{with} \quad \beta \neq 0$$

(existing as a singular quasiconformal self-map of $\mathbb{R}^n$) then $C_{r,\phi}$ is bounded on $Q_\alpha(\mathbb{R}^n)$ – in particular – $Q_\alpha(\mathbb{R}^n)$ is conformally invariant in the sense that $f \in Q_\alpha(\mathbb{R}^n)$ if and only if

$$x \mapsto f(x|x|^{-2})$$

belongs to $Q_\alpha(\mathbb{R}^n)$; see [49, Corollary 1.4] which naturally complements Theorem 1.2, but also is optimal in the sense (cf. [49, Theorems 1.6 and 1.7]) that for a given pair $(\alpha_0, n-1) \in (0,1) \times \mathbb{N}$:

▷ $\exists$ a bounded set

$$E_{\alpha_0} \quad \text{with} \quad \overline{\dim}_{\mathrm{lsM}} E_{\alpha_0} = n - 2\alpha_0$$

and a quasiconformal mapping

$$\phi : \mathbb{R}^n \to \mathbb{R}^n \quad \text{with} \quad \mathcal{J}_\phi \in \mathscr{A}_1(\mathbb{R}^n; E_{\alpha_0})$$

under which C_ϕ is unbounded on $\mathsf{Q}_\alpha(\mathbb{R}^n)$ for any $\alpha \in (\alpha_0, 1)$;

▷ $\exists$ a unbounded set

$$E_{\alpha_0} \subseteq \mathbb{Z}^n \quad \text{with} \quad \overline{\dim}_{\mathrm{gsM}} E_{\alpha_0} = n - 2\alpha_0 \quad \text{while} \quad \overline{\dim}_{\mathrm{lsM}} E_{\alpha_0} = 0$$

and a quasiconformal mapping

$$f : \mathbb{R}^n \to \mathbb{R}^n \quad \text{with} \quad \mathcal{J}_\phi \in \mathscr{A}_1(\mathbb{R}^n; E_{\alpha_0})$$

under which C_ϕ is unbounded on $\mathsf{Q}_\alpha(\mathbb{R}^n)$ for any $\alpha \in (\alpha_0, 1)$.

Section 7.5. Theorem 7.14 is taken from [98, Theorem 1.2]. Furthermore, [98, Theorem 1.5] shows that if $\phi : \mathbb{R}^2 \to \mathbb{R}^2$ is a homeomorphism and C_ϕ is bounded on $\mathsf{Q}_{2^{-1}<\alpha<1}(\mathbb{R}^2)$ then ϕ is quasiconformal.

Section 7.6. Theorem 7.15 comes from a suitable modification of [94, Theorem 1.1].

On the one hand, regarding $\mathsf{BMO}(\mathbb{R}^n)$ [15, Theorem 1] reveals that if $\mathcal{D}b$ and $(\mathcal{D}b)^{\mathrm{t}}$ are the spatial gradient matrix of b and its transpose, and

$$\begin{cases} b \in \mathsf{L}^1(\mathsf{W}^{1,1}_{\mathrm{loc}}(\mathbb{R}^n); [0,T]) \\ \displaystyle\int_0^T \left\| \frac{b(\cdot,t)}{1+|\cdot|\ln^+|\cdot|} \right\|_{\mathsf{L}^\infty(\mathbb{R}^n)} dt < \infty \\ \mathcal{S}b = 2^{-1}(\mathcal{D}b + (\mathcal{D}b)^{\mathrm{t}}) - n^{-1}(\operatorname{div} b) I_{n\times n} \text{ obeys } \displaystyle\int_0^T \|\mathcal{S}b(\cdot,t)\|_{\mathsf{L}^\infty(\mathbb{R}^n)}\, dt < \infty, \end{cases}$$

then for any

$$u_0 \in \mathsf{BMO}(\mathbb{R}^n)$$

the transport equation has a unique weak solution

$$u \in \mathsf{L}^\infty(\mathsf{BMO}(\mathbb{R}^n); [0,T]) \text{ – i. e. – } \sup_{t\in[0,T]} \|u(\cdot,t)\|_{\mathsf{BMO}(\mathbb{R}^n)} < \infty.$$

Interestingly, this last result is comparable with [66, Theorem A] yielding that if

$$\begin{cases} b \in \mathsf{L}^1(\mathsf{W}^{1,1}_{\mathrm{loc}}(\mathbb{R}^n); [0,T]) \\ \int_0^T \left\| \frac{b(\cdot,t)}{1+|\cdot|} \right\|_{\mathbb{L}^1(\mathbb{R}^n)} dt < \infty \\ \mathrm{div}(b) \in \mathsf{L}^1(\mathsf{BMO}(\mathbb{R}^n); [0,T]) \\ \mathrm{supp}\,(\,\mathrm{div}\, b(\cdot,t)) \subseteq B(0,R) \quad \text{for a fixed} \quad R > 0, \end{cases}$$

then for each

$$u_0 \in \mathsf{L}^\infty(\mathbb{R}^n)$$

the transport equation has a unique weak solution

$$u \in \mathsf{L}^\infty(\mathsf{L}^\infty(\mathbb{R}^n); [0,T]).$$

On the other hand, concerning $\mathsf{L}^\infty(\mathbb{R}^n)$ it has been proved in [19] that if

$$\begin{cases} b \in \mathsf{L}^1(\mathsf{W}^{1,1}_{\mathrm{loc}}(\mathbb{R}^n); [0,T]) \\ \int_0^T \left\| \frac{b(x,\cdot)}{1+|\cdot|} \right\|_{\mathsf{L}^1(\mathbb{R}^n)+\mathsf{L}^\infty(\mathbb{R}^n)} < \infty \\ \mathrm{div}(b) \in \mathsf{L}^\infty(\mathsf{L}^\infty(\mathbb{R}^n); [0,T]), \end{cases}$$

then for each

$$u_0 \in \mathsf{L}^{1\le p\le\infty}(\mathbb{R}^n)$$

the transport equation has a unique renormalized solution

$$u \in \mathsf{L}^\infty(\mathsf{L}^{1\le p\le\infty}(\mathbb{R}^n); [0,T])$$

according to the partial differential left composition law that

$$\partial_t \Phi\big(u(x,t)\big) + b(x,t)\cdot\nabla_x\Phi\big(u(x,t)\big) = 0 \quad \forall\,(x,t) \in \mathbb{R}^n \times [0,T]$$

holds in the distributional sense for any bounded $\mathsf{C}^1(\mathbb{R}^n)$-function $\Phi(\cdot)$ with bounded derivatives. The special case $p = \infty$ of this last result has been extended to the case when $\mathsf{W}^{1,1}_{\mathrm{loc}}(\mathbb{R}^n)$ is replaced by $\mathsf{BV}(\mathbb{R}^n)$ in [2].

8 Extension or restriction principle

In this chapter, we study the extension or restriction principle for $Q_\alpha(\mathbb{R}^n)$ via the boundedness of the Hardy–Littlewood maximal operator on $Q_\alpha(\mathbb{R}^n)$ and the concept of $Q_\alpha(D)$ which is a Q space on a domain $D \subseteq \mathbb{R}^n$ by considering only coordinate cubes $I \subseteq D$ used in determining $\|\cdot\|_{Q_\alpha(\mathbb{R}^n)}$ in Definition 1.1, thereby resolving [22, problem 8.5] – a generalization of the Jones theorem in [41] saying that for a Jordan domain $D \subseteq \mathbb{R}^2$,

$$\mathrm{BMO}(D) = \mathrm{BMO}(\mathbb{R}^2)|_D$$

if and only if ∂D is a quasi-circle.

8.1 Hardy–Littlewood maximal operator

Since the Hardy–Littlewood maximal operator $\mathcal{M}$ plays an important role in the forthcoming investigation of the extension or restriction problem for $Q_\alpha(\mathbb{R}^n)$. It is better to start with understanding its action on $Q_\alpha(\mathbb{R}^n)$.

Theorem 8.1. *Let* $(\alpha, n-1) \in (-\infty, 1) \times \mathbb{N}$. *Then the Hardy–Littlewood maximal operator* $\mathcal{M}$ *is bounded on* $Q_\alpha(\mathbb{R}^n)$.

Proof. If

$$\begin{cases} f \in Q_\alpha(\mathbb{R}^n) \\ f_y(x) = f(x+y) \quad \forall x, y \in \mathbb{R}^n \\ I \in \mathscr{C}(\mathbb{R}^n), \end{cases}$$

then

$$|\mathcal{M}f(x+y) - \mathcal{M}f(x)| \leq \mathcal{M}(f_y - f)(x) \quad \text{a. e. } x, y \in \mathbb{R}^n,$$

and hence

$$\int_{|y|<\ell(I)} \int_I \frac{|\mathcal{M}f(x+y) - \mathcal{M}f(x)|^2}{|y|^{n+2\alpha}}\, dx\, dy \leq \int_{|y|<\ell(I)} \int_I \frac{|\mathcal{M}(f_y - f)(x)|^2}{|y|^{n+2\alpha}}\, dx\, dy.$$

Upon writing

$$f_y - f = (f_y - f)\mathbb{1}_{3I} + (f_y - f)\mathbb{1}_{(3I)^c} = F_{1,y} + F_{2,y}$$

and using the L^2-boundedness of $\mathcal{M}$ and Theorem 1.2, we obtain

$$\int_{|y|<\ell(I)} \int_I |\mathcal{M}F_{1,y}(x)|^2 \frac{dx\, dy}{|y|^{n+2\alpha}}$$

https://doi.org/10.1515/9783110600285-008

$$
\begin{aligned}
&= \int\limits_{|y|<\ell(I)} \int\limits_{I} \left|(\mathcal{M}(f_y - f)\mathbb{1}_{3I})(x)\right|^2 \frac{dx\,dy}{|y|^{n+2\alpha}} \\
&\lesssim \int\limits_{|y|<\ell(I)} \int\limits_{3I} |f(x+y) - f(x)|^2 \frac{dx\,dy}{|y|^{n+2\alpha}} \\
&\lesssim \ell(I)^{n-2\alpha} \|f\|^2_{Q_\alpha(\mathbb{R}^n)}.
\end{aligned}
$$

Meanwhile, upon noticing (cf. [47] or [102, (3.4)])

$$
\left(\mathcal{M}F_{2,y}(x)\right)^2 \lesssim \sup_{I \subseteq 3J \in \mathscr{C}(\mathbb{R}^n)} \frac{1}{m(J)} \int\limits_{J} |F_{2,y}(z)|^2\,dz \quad \forall x \in I
$$

and using the Fubini theorem and Theorem 1.2, we get

$$
\begin{aligned}
&\int\limits_{|y|<\ell(I)} \int\limits_{I} |\mathcal{M}F_{2,y}(x)|^2 \frac{dx\,dy}{|y|^{n+2\alpha}} \\
&\quad\lesssim \int\limits_{|y|<\ell(I)} \int\limits_{I} \left(\sup_{I \subseteq 3J \in \mathscr{C}(\mathbb{R}^n)} \frac{1}{m(J)} \int\limits_{J} |F_{2,y}(z)|^2\,dz \right) \frac{dx\,dy}{|y|^{n+2\alpha}} \\
&\quad\lesssim \int\limits_{I} \left(\sup_{I \subseteq 3J \in \mathscr{C}(\mathbb{R}^n)} \frac{1}{m(J)} \int\limits_{|y|<\ell(I)} \int\limits_{J} |F_{2,y}(z)|^2 \frac{dz\,dy}{|y|^{n+2\alpha}} \right) dx \\
&\quad\lesssim \int\limits_{I} \left(\sup_{I \subseteq 3J \in \mathscr{C}(\mathbb{R}^n)} \frac{1}{m(J)} \int\limits_{|y|<\ell(I)} \int\limits_{J} |f(y+z) - f(z)|^2 \mathbb{1}_{(3I)^c}(z) \frac{dz\,dy}{|y|^{n+2\alpha}} \right) dx \\
&\quad\lesssim \|f\|^2_{Q_\alpha(\mathbb{R}^n)} \int\limits_{I} \sup_{I \subseteq 3J \in \mathscr{C}(\mathbb{R}^n)} \frac{\ell(J)^{n-2\alpha}}{m(J)}\,dx \\
&\quad\lesssim \ell(I)^{n-2\alpha} \|f\|^2_{Q_\alpha(\mathbb{R}^n)}.
\end{aligned}
$$

Putting all together gives

$$
\int\limits_{|y|<\ell(I)} \int\limits_{I} |\mathcal{M}f(x+y) - \mathcal{M}f(x)|^2 \frac{dx\,dy}{|y|^{n+2\alpha}} \lesssim \ell(I)^{n-2\alpha} \|f\|^2_{Q_\alpha(\mathbb{R}^n)},
$$

namely,

$$
\mathcal{M}f \in Q_\alpha(\mathbb{R}^n) \quad \& \quad \|\mathcal{M}f\|_{Q_\alpha(\mathbb{R}^n)} \lesssim \|f\|_{Q_\alpha(\mathbb{R}^n)}. \qquad \square
$$

8.2 Extension principle

Such an extension principle relies on introducing the $\mathfrak{Q}$-spaces on a subdomain of $\mathbb{R}^n$.

Definition 8.2. For $(\alpha, n-1) \in \mathbb{R} \times \mathbb{N}$ and a domain $D \subseteq \mathbb{R}^n$ let $\mathfrak{Q}_\alpha(D)$ consist of all measurable functions f on D with

$$\|f\|_{\mathfrak{Q}_\alpha(D)} = \sup_{I \in \mathscr{C}(\mathbb{R}^n)} \left(\mathrm{m}(I)^{\frac{2\alpha-n}{n}} \int_{I\cap D}\int_{I\cap D} \frac{|f(x)-f(y)|^2}{|x-y|^{n+2\alpha}}\,dx\,dy \right)^{\frac{1}{2}} < \infty.$$

If the above $I \cap D$ is replaced by $I \subseteq D$, then the corresponding space and its seminorm are written as $\mathrm{Q}_\alpha(D)$ and $\|\cdot\|_{\mathrm{Q}_\alpha(D)}$, respectively.

It is easy to check:

▷

$$\mathrm{Q}_\alpha(\mathbb{R}^n) = \mathfrak{Q}_\alpha(\mathbb{R}^n) \subseteq \mathfrak{Q}_\alpha(D) \subseteq \mathrm{Q}_\alpha(D);$$

▷

$$\mathrm{Q}_\alpha(D) = \mathrm{Q}_\alpha(\mathbb{R}^n)|_D \Leftrightarrow \mathrm{Q}_\alpha(D) = \mathfrak{Q}_\alpha(D) \quad \& \quad \mathfrak{Q}_\alpha(D) = \mathrm{Q}_\alpha(\mathbb{R}^n)|_D.$$

Below is an extension principle for $\mathrm{Q}_\alpha(\mathbb{R}^n)$.

Theorem 8.3. *Let $(\alpha, n-1) \in [0,1) \times \mathbb{N}$ and $D \subseteq \mathbb{R}^n$ be a domain. Then the following three statements are equivalent:*

(i) *D is a $\mathfrak{Q}_\alpha$-extension domain – i. e. –*

$$\mathrm{Q}_\alpha(\mathbb{R}^n)|_D = \{f = g|_D : g \in \mathrm{Q}_\alpha(\mathbb{R}^n)\} = \mathfrak{Q}_\alpha(D)$$

with

$$\|f\|_{\mathrm{Q}_\alpha(\mathbb{R}^n)|_D} = \inf\{\|g\|_{\mathrm{Q}_\alpha(\mathbb{R}^n)} : g|_D = f \,\&\, g \in \mathrm{Q}_\alpha(\mathbb{R}^n)\} \approx \|f\|_{\mathfrak{Q}_\alpha(D)};$$

(ii) *D is a $\mathfrak{Q}_\alpha$-imbedding domain – i. e. – there are two positive constants C_1 and C_2 such that*

$$0 < \|f\|_{\mathfrak{Q}_\alpha(D)} < \infty \Rightarrow \inf_{c\in\mathbb{R}} \sup_{I\in\mathscr{C}(\mathbb{R}^n)} \int_{I\cap D} \exp\left(\frac{C_1|f(x)-c|}{\|f\|_{\mathfrak{Q}_\alpha(D)}}\right)\frac{dx}{\mathrm{m}(I)} \le C_2;$$

(iii) *D is an Ahlfors n-regular domain – i. e. – there is a constant $c > 0$ such that for any coordinate cube I centered in D with its edge-length*

$$\ell(I) < 2\,\mathrm{diam}(D)$$

one has

$$\mathrm{m}(I \cap D) \ge c\mathrm{m}(I).$$

Proof. The argument consists of three implications.

(i)⇒(ii) Suppose that (i) is valid, that is, D is a $\mathfrak{Q}_\alpha$-extension domain. Then for any function $f \in \mathfrak{Q}_\alpha(D)$ we can find a function

$$\tilde{f} = \mathcal{E}f \in \mathfrak{Q}_\alpha(\mathbb{R}^n)$$

and two constants $C_3, C_4 > 0$ such that

$$\begin{cases} \mathcal{E}f(x) = f(x) \quad \forall x \in D \\ \|\tilde{f}\|_{\mathrm{BMO}(\mathbb{R}^n)} \le C_3 \|\tilde{f}\|_{\mathfrak{Q}_\alpha(\mathbb{R}^n)} \\ \|\mathcal{E}f\|_{\mathfrak{Q}_\alpha(\mathbb{R}^n)} \le C_4 \|f\|_{\mathfrak{Q}_\alpha(D)}. \end{cases}$$

Using Theorem 2.1, we can take a positive constant $C_1 < 1/(2^n e)$ to reach

$$\begin{aligned} \inf_{c\in\mathbb{R}} \int_{I\cap D} \exp\left(\frac{C_1|f(x)-c|}{C_3C_4\|f\|_{\mathfrak{Q}_\alpha(\mathbb{R}^n)}}\right) dx &\le \inf_{c\in\mathbb{R}} \int_{I\cap D} \exp\left(\frac{C_1|f(x)-c|}{\|\tilde{f}\|_{\mathrm{BMO}(\mathbb{R}^n)}}\right) dx \\ &\le \int_I \exp\left(\frac{C_1|\tilde{f}(x)-\tilde{f}_I|}{\|\tilde{f}\|_{\mathrm{BMO}(\mathbb{R}^n)}}\right) dx \\ &\le \int_0^\infty \mathrm{m}\left(\left\{x \in I : \frac{C_1|\tilde{f}(x)-\tilde{f}_I|}{\|\tilde{f}\|_{\mathrm{BMO}(\mathbb{R}^n)}} > t\right\}\right) de^t \\ &\le \mathrm{m}(I) \int_0^\infty \exp(t(1-(2^n eC_1)^{-1}))\, dt \\ &\lesssim \mathrm{m}(I). \end{aligned}$$

(ii)⇒(iii) In order to proceed, we need the following space embedding – if $0 \le \alpha < \beta < 1$ and $\dot{W}^{\beta,\frac{n}{\beta}}(D)$ is the space of all measurable functions f on D with

$$\|f\|_{\dot{W}^{\beta,\frac{n}{\beta}}(D)} = \left(\int_D \int_D \frac{|f(x)-f(y)|^{\frac{n}{\beta}}}{|x-y|^{2n}}\, dxdy\right)^{\frac{\beta}{n}} < \infty,$$

then

$$\|f\|_{\mathfrak{Q}_\alpha(D)} \lesssim \|f\|_{\dot{W}^{\beta,\frac{n}{\beta}}(D)} \quad \forall f \in \dot{W}^{\beta,\frac{n}{\beta}}(D).$$

This result follows easily from

$$\begin{aligned} &\int_{I\cap D}\int_{I\cap D} \frac{|f(x)-f(y)|^2}{|x-y|^{n+2\alpha}}\, dy\, dx \\ &\le \left(\int_{I\cap D}\int_{I\cap D} \frac{|f(x)-f(y)|^{\frac{n}{\beta}}}{|x-y|^{2n}} dy\, dx\right)^{\frac{2\beta}{n}} \left(\int_{I\cap D}\int_{I\cap D} |x-y|^{-(n+2\alpha-4\beta)\cdot\frac{n}{n-2\beta}}\, dy\, dx\right)^{\frac{n-2\beta}{n}} \end{aligned}$$

$$\lesssim \left(\int_{I\cap D}\int_{I\cap D} \frac{|f(x)-f(y)|^{\frac{n}{\beta}}}{|x-y|^{2n}}\,dy\,dx\right)^{\frac{2\beta}{n}} \left(\int_{I\cap D} \ell(I)^{\frac{2\beta-2\alpha}{n-2\beta}\cdot n}\,dx\right)^{\frac{n-2\beta}{n}}$$

$$\lesssim \left(\int_{I\cap D}\int_{I\cap D} \frac{|f(x)-f(y)|^{\frac{n}{\beta}}}{|x-y|^{2n}}\,dy\,dx\right)^{\frac{2\beta}{n}} \ell(I)^{n-2\alpha}$$

$$\lesssim \left(\int_{D}\int_{D} \frac{|f(x)-f(y)|^{\frac{n}{\beta}}}{|x-y|^{2n}}\,dy\,dx\right)^{\frac{\beta}{n}} \ell(I)^{n-2\alpha}.$$

Now, assume that (ii) holds. Then the previous embedding assertion gives two constants $C_5, C_6 > 0$ such that

$$\inf_{c\in\mathbb{R}} \sup_{I\in\mathscr{C}(\mathbb{R}^n)} \mathrm{m}(I)^{-1} \int_{I\cap D} \exp\left(\frac{C_5|f(x)-c|}{\|f\|_{\dot{W}^{\beta,\frac{n}{\beta}}(D)}}\right) dx \le C_6 \quad \forall\, 0 < \|f\|_{\dot{W}^{\beta,\frac{n}{\beta}}(D)} < \infty$$

holds for $\beta \in (\alpha, 1)$. Also, given

$$z \in D \quad \& \quad 0 < t < r \le 1$$

let

$$f_{t<r}(x) = \begin{cases} 1 & \forall\, x \in B(z,t)\cap D \\ \frac{r-|x-z|}{r-t} & \forall\, x \in (B(z,r)\setminus B(z,t))\cap D \\ 0 & \forall\, x \in D\setminus B(z,t). \end{cases}$$

Then, according to [104, Lemma 2.4] we have

$$0 < \|f_{t<r}\|_{\dot{W}^{\beta,\frac{n}{\beta}}(D)} \lesssim (\mathrm{m}(B(z,r)\cap D))^{\frac{\beta}{n}}(r-t)^{-\beta},$$

thereby producing a constant $C_7 > 0$ such that

$$\inf_{c\in\mathbb{R}} \int_{B(z,r)\cap D} \exp\left(\frac{C_7|f_{t<r}(x)-c|(r-t)^{\beta}}{(\mathrm{m}(B(z,r)\cap D))^{\frac{\beta}{n}}}\right) dx \lesssim r^n.$$

Upon fixing $r_0 \in \mathbb{R}_+$ and choosing

$$c_0 = 1 \quad \& \quad c_j \in (0,1)$$

such that

$$\mathrm{m}(B(z,c_j r_0)\cap D) = 2^{-1}\mathrm{m}(B(z,c_{j-1}r_0)\cap D) = 2^{-j}\mathrm{m}(B(z,r_0)\cap D),$$

we get

$$\begin{cases} 1 \ge r = c_1 r_0 > t = c_2 r_0 > 0 \\ \inf_{c\in\mathbb{R}} \int_{B(z,r)\cap D} \exp\left(\frac{C_7|f_{c_2r_0<c_1r_0}(x)-c|(r_0(c_1-c_2))^{\beta}}{(\mathrm{m}(B(x,c_1r_0)\cap D))^{\frac{\beta}{n}}}\right) dx \lesssim (c_1 r_0)^n. \end{cases}$$

Since

$$x \in ((B(z,r_0) \setminus B(z,c_1r_0)) \cap D) \cup (B(z,c_2r_0) \cap D) \Rightarrow |f_{c_2r_0<c_1r_0}(x) - c| \geq 2^{-1},$$

we achieve

$$\mathrm{m}(B(z,c_1r_0) \cap D) \exp\left(\frac{(r_0(c_1 - c_2))^\beta}{(\mathrm{m}(B(x,c_1r_0) \cap D))^{\frac{\beta}{n}}} \right) \lesssim r_0^n,$$

and its counterpart for $c_jr_0 - c_{j+1}r_0$. Accordingly, there is a constant $C_8 > 0$ such that

$$c_1r_0 = \sum_{j=1}^{\infty}(c_j - c_{j+1})r_0 \lesssim \mathrm{m}(B(x,r_0) \cap D)^{\frac{1}{n}} \left(\ln\left(\frac{C_8r_0^n}{\mathrm{m}(B(z,r_0) \cap D)} \right) \right)^{\frac{1}{\beta}}.$$

This in turn implies

$$\left(\frac{(c_1r_0)^n}{\mathrm{m}(B(x,r_0) \cap D)} \right)^\beta \lesssim \ln\left(\frac{C_8r_0^n}{\mathrm{m}(B(z,r_0) \cap D)} \right),$$

and so

$$r_0^n \lesssim \mathrm{m}(B(z,r_0) \cap D),$$

namely, (iii) holds.

(iii)⇒(i) Suppose that (iii) is valid, namely, D is Ahlfors n-regular. Then [33, 104] ensures

$$\mathrm{m}(\overline{D} \setminus D) = 0.$$

Consequently, the open set

$$\overline{D}^c = \mathbb{R}^n \setminus \overline{D}$$

allows the Whitney decomposition – i. e. – a collection

$$W = \{I_i\}_{i\in\mathbb{N}}$$

of countably many closed cubes with

$$x_i,\ \ell_i = \ell(I_i) \quad \text{and} \quad I_i^\circ$$

being the center, edge-length and interior of I_i respectively, such that

$$\begin{cases} \overline{D}^c = \bigcup_{i\in\mathbb{N}} I_i \quad \& \quad I_i^\circ \cap I_j^\circ = \emptyset \quad \forall i \neq j \\ 2^2\sqrt{n}\ell_i \leq \operatorname{dist}(I_i, D) = \operatorname{dist}(I_i, D) \leq 2^4\sqrt{n}\ell_i \quad \forall i \in \mathbb{N} \\ 2^{-2}\ell I_i \leq \ell(I_j) \leq 4\ell_i \quad \forall I_i \cap I_j \neq \emptyset \\ \forall i \in \mathbb{N} \quad \exists x_i^* \in D \quad \text{such that} \quad \operatorname{dist}(x_i^*, I_i) \leq 17\sqrt{n}\ell_i \\ \#\{j : I_j \cap I_i \neq \emptyset\} \leq 12^n \quad \forall I_i. \end{cases}$$

Meanwhile, for the above decomposition there exists a family of smooth functions $\{\varphi_i\}$ such that

$$\begin{cases} \operatorname{supp}(\varphi_i) \subseteq 3I(x_i, \ell_i) = 3I_i \quad \forall\, i \in \mathbb{N} \\ |\nabla \varphi_i| \lesssim \ell_i^{-1} \quad \forall\, i \in \mathbb{N} \\ \sum_{i\in\mathbb{N}} \varphi_i = \mathbb{1}_{\overline{D}^c}. \end{cases}$$

Here and henceforth, $I(x_i, \ell_i)$ stands for the cube with center x_i and edge-length ℓ_i. Moreover, if

$$\mathbb{N}_x = \{i \in \mathbb{N} : x \in 3I_i\} \quad \forall\, x \in \overline{D}^c,$$

then

$$\sharp\mathbb{N}_x \lesssim 1 \quad \forall\, x \in \overline{D}^c.$$

Now, for each $f \in \mathfrak{Q}_\alpha(D)$ and the above pair $\{x_i^*, \ell_i\}$ let

$$\mathcal{E}f(x) = \begin{cases} f(x) & \forall\, x \in D \\ 0 & \forall\, x \in \overline{D} \setminus D \\ \sum_{i\in\mathbb{N}_x} \varphi_i(x) \fint_{I(x_i^*, \ell_i)\cap D} f(z)\,\mathrm{d}z & \forall\, x \in \overline{D}^c. \end{cases}$$

Obviously,

$$\mathcal{E}f = f \quad \text{on } D.$$

We are about to show

$$\mathcal{E}f \in \mathfrak{Q}_\alpha(\mathbb{R}^n) \quad \& \quad \|\mathcal{E}f\|_{\mathfrak{Q}_\alpha(\mathbb{R}^n)} \lesssim \|f\|_{\mathfrak{Q}_\alpha(D)}.$$

By the definition of $\|\mathcal{E}f\|_{\mathfrak{Q}_\alpha(\mathbb{R}^n)}$, this last action is reduced to checking

$$\sup_{I\in\mathscr{C}(\mathbb{R}^n)} \ell(I)^{2\alpha-n} \int_I \int_I \frac{|\mathcal{E}f(x) - \mathcal{E}f(y)|^2}{|x-y|^{n+2\alpha}}\,\mathrm{d}x\mathrm{d}y \lesssim \|f\|^2_{\mathfrak{Q}_\alpha(D)}.$$

So, we are required to handle the forthcoming unbounded and bounded cases.

The unbounded case: $\operatorname{diam} D = \infty$.

We consider three subcases

$$\begin{cases} I \cap D \neq \emptyset \\ I \cap D = \emptyset \quad \text{but} \quad \operatorname{dist}(I, D) \leq 2^2 5^2 \sqrt{n}\,\ell(I) \\ I \cap D = \emptyset \quad \text{but} \quad \operatorname{dist}(I, D) > 2^2 5^2 \sqrt{n}\,\ell(I). \end{cases}$$

Subcase $I \cap D \neq \emptyset$. Write

$$\int_I \int_I \frac{|\mathcal{E}f(x) - \mathcal{E}f(y)|^2}{|x-y|^{n+2\alpha}}\,dx\,dy = \mathsf{H}_1 + 2\mathsf{H}_2 + \mathsf{H}_3,$$

where

$$\begin{cases} \mathsf{H}_1 = \displaystyle\int_{I\cap D}\int_{I\cap D} \frac{|f(x)-f(y)|^2}{|x-y|^{n+2\alpha}}\,dx\,dy \\ \mathsf{H}_2 = \displaystyle\int_{I\setminus D}\int_{I\cap D} \frac{|\mathcal{E}f(x)-f(y)|^2}{|x-y|^{n+2\alpha}}\,dx\,dy \\ \mathsf{H}_3 = \displaystyle\int_{I\setminus D}\int_{I\setminus D} \frac{|\mathcal{E}f(x)-\mathcal{E}f(y)|^2}{|x-y|^{n+2\alpha}}\,dx\,dy. \end{cases}$$

It suffices to prove

$$\mathsf{H}_i \lesssim \mathrm{m}(I)^{\frac{n-2\alpha}{n}} \|f\|^2_{\mathcal{Q}_\alpha(D)} \quad \forall i = 1,2,3.$$

- ▷ Obviously, H_1 satisfies the above inequality.
- ▷ To estimate H_2, for

$$\begin{cases} x \in I \setminus \overline{D} \\ y \in I \cap D \\ \displaystyle\sum_{i\in\mathbb{N}_x} \varphi_i(x) = 1, \end{cases}$$

we have

$$\begin{aligned} \mathcal{E}f(x) - f(y) &\le \Big(\sum_{i\in\mathbb{N}_x} \varphi_i(x) \fint_{I(x_i^*,\ell_i)\cap D} f(z)\,dz\Big) - f(y) \\ &= \sum_{i\in\mathbb{N}_x} \varphi_i(x)\Big(\fint_{I(x_i^*,\ell_i)\cap D} f(z)\,dz - f(y)\Big) \\ &= \sum_{i\in\mathbb{N}_x} \varphi_i(x) \fint_{I(x_i^*,\ell_i)\cap D} (f(z)-f(y))\,dz, \end{aligned}$$

thereby getting

$$|\mathcal{E}f(x) - f(y)| \le \sum_{i\in\mathbb{N}_x} \varphi_i(x) \fint_{I(x_i^*,\ell_i)\cap D} |f(z)-f(y)|\,dz.$$

Moreover, we claim

$$|\mathcal{E}f(x) - f(y)| \lesssim \fint_{I(x,2^2 5\sqrt{n}\,\mathrm{dist}\,(x,D))\cap D} |f(z)-f(y)|\,dz.$$

To do so, note that

$$x \in I \cap \overline{D}^c \Rightarrow \operatorname{dist}(x,D) \le \sqrt{n}\ell(I) \Rightarrow I(x, 2^2 5\sqrt{n}\operatorname{dist}(x,D)) \subseteq I(x_I, 21n\ell(I)).$$

So, for each $i \in \mathbb{N}_x$ we use Lemma 6.2 to obtain

$$\begin{cases} \operatorname{dist}(I_i,D) + 2\sqrt{n}\ell_i \le 18\sqrt{n}\ell_i \ge \operatorname{dist}(x,D) \ge \operatorname{dist}(I_i,D) - 2\sqrt{n}\ell_i \ge 2\sqrt{n}\ell_i \\ \ell_i \approx \operatorname{dist}(I_i,D) \approx \operatorname{dist}(x, D) \\ |x_i^* - x| \le \operatorname{dist}(x_i^*, I_i) + 2\sqrt{n}\ell_i \le 19\sqrt{n}\ell_i \\ I(x_i^*, \ell_i) \subseteq I(x, 2^2 5\sqrt{n}\operatorname{dist}(x,D)). \end{cases}$$

Meanwhile, for

$$\begin{cases} x \in D^c \\ I(x, \lambda \operatorname{dist}(x,D)) \\ \lambda \ge 2^3, \end{cases}$$

we utilize the Ahlfors n-regular condition to derive that

$$\mathrm{m}(I(x,\lambda \operatorname{dist}(x,D)) \cap D) \ge \mathrm{m}(I(\check{x}, (2^{-1}\lambda - 2)\operatorname{dist}(x,D))) \gtrsim \mathrm{m}(I(x,\lambda \operatorname{dist}(x,D)))$$

is valid for

$$\check{x} \in D \quad \& \quad \operatorname{dist}(x,\check{x}) = 2\operatorname{dist}(x,D),$$

and so that the desired claim follows.

For

$$y \in I \cap D \quad \& \quad z \in I(x, 2^2 5\sqrt{n}\operatorname{dist}(x,D)) \cap D,$$

we have

$$|x-y| \ge \operatorname{dist}(x,D) \quad \& \quad |z-y| \le |z-x| + |x-y| \le 2^2 5\sqrt{n}\operatorname{dist}(x,D) + |x-y| \lesssim |x-y|.$$

This, along with $0 \le \alpha < 1$, yields

$$\begin{aligned} \frac{|\mathcal{E}f(x) - f(y)|}{|x-y|^{\frac{n+2\alpha}{2}}} &\lesssim \fint_{I(x,2^2 5\sqrt{n}\operatorname{dist}(x,D))\cap D} \frac{|f(z) - f(y)|}{|z-y|^{\frac{n+2\alpha}{2}}}\,\mathrm{d}z \\ &= \fint_{I(x,2^2 5\sqrt{n}\operatorname{dist}(x,D))\cap D} \frac{|f(z) - f(y)|}{|z-y|^{\frac{n+2\alpha}{2}}} \mathbb{1}_{I(x_I,21n\ell(I))\cap D}(z)\,\mathrm{d}z \\ &\lesssim \fint_{B(x,2^2 5n\operatorname{dist}(x,D))} \frac{|f(z) - f(y)|}{|z-y|^{\frac{n+2\alpha}{2}}} \mathbb{1}_{I(x_I,21n\ell(I))\cap D}(z)\,\mathrm{d}z \end{aligned}$$

$$\lesssim \mathcal{M}\left(\frac{|f(\cdot)-f(y)|}{|\cdot-y|^{\frac{n+2\alpha}{2}}}\mathbb{1}_{I(x_I,21n\ell(I))\cap D}(\cdot)\right)(x).$$

Upon applying the L^2-boundedness of $\mathcal{M}$, we achieve

$$\begin{aligned}
\mathsf{H}_2 &\lesssim \int_{I\setminus D}\int_{I\cap D}\left(\mathcal{M}\left(\frac{|f(\cdot)-f(y)|}{|\cdot-y|^{\frac{n+2\alpha}{2}}}\mathbb{1}_{I(x_I,21n\ell(I))\cap D}(\cdot)\right)(x)\right)^2 dy\,dx\\
&\lesssim \int_{I\cap D}\int_{\mathbb{R}^n}\left(\mathcal{M}\left(\frac{|f(\cdot)-f(y)|}{|\cdot-y|^{\frac{n+2\alpha}{2}}}\mathbb{1}_{I(x_I,21n\ell(I))\cap D}(\cdot)\right)(x)\right)^2 dx\,dy\\
&\lesssim \int_{I\cap D}\int_{\mathbb{R}^n}\frac{|f(x)-f(y)|^2}{|x-y|^{n+2\alpha}}\mathbb{1}_{I(x_I,21n\ell(I))\cap D}(x)\,dx\,dy\\
&\lesssim \int_{I\cap D}\int_{I(x_I,21n\ell(I))\cap D}\frac{|f(x)-f(y)|^2}{|x-y|^{n+2\alpha}}\,dx\,dy\\
&\lesssim \int_{I(x_I,21n\ell(I))\cap D}\int_{I(x_I,21n\ell(I))\cap D}\frac{|f(x)-f(y)|^2}{|x-y|^{n+2\alpha}}\,dx\,dy,
\end{aligned}$$

whence

$$\mathsf{H}_2 \lesssim \mathrm{m}(I)^{\frac{n-2\alpha}{n}}\|f\|^2_{\mathfrak{Q}_\alpha(D)}.$$

▷ Regarding H_3, for each $x\in I\setminus D$ we divide $I\setminus D$ into two parts

$$\begin{cases}X_1(x)=\{y\in I\setminus D: |x-y|\ge 2^{-1}\max\{\operatorname{dist}(x,D),\operatorname{dist}(y,D)\}\}\\ X_2(x)=\{y\in I\setminus D: |x-y|< 2^{-1}\max\{\operatorname{dist}(x,D),\operatorname{dist}(y,D)\}\},\end{cases}$$

and write

$$\begin{cases}\mathsf{H}_3=\mathsf{H}_{3,1}+\mathsf{H}_{3,2}\\ \mathsf{H}_{3,1}=\int_{I\setminus D}\int_{X_1(x)}\frac{|\mathcal{E}f(x)-\mathcal{E}f(y)|^2}{|x-y|^{n+2\alpha}}\,dy\,dx\\ \mathsf{H}_{3,2}=\int_{I\setminus D}\int_{X_2(x)}\frac{|\mathcal{E}f(x)-\mathcal{E}f(y)|^2}{|x-y|^{n+2\alpha}}\,dy\,dx.\end{cases}$$

For

$$x\in I\setminus D \quad \& \quad y\in X_1(x),$$

we have

$$\sum_{i\in\mathbb{N}_x}\varphi_i(x)=1=\sum_{i\in\mathbb{N}_y}\varphi_i(y),$$

whence

$$
\begin{aligned}
|\mathcal{E}f(x)-\mathcal{E}f(y)| &= \left|\sum_{i\in\mathbb{N}_x}\sum_{j\in\mathbb{N}_y}\varphi_i(x)\varphi_j(y)\left(\fint_{I(x_i^*,\ell_i)\cap D} f(z)\,\mathrm{d}z - \fint_{I(x_j^*,\ell_j)\cap D} f(w)\,\mathrm{d}w\right)\right| \\
&\le \sum_{i\in\mathbb{N}_x}\sum_{j\in\mathbb{N}_y}\varphi_i(x)\varphi_j(y) \fint_{I(x_i^*,\ell_i)\cap D}\fint_{I(x_j^*,\ell_j)\cap D} |f(z)-f(w)|\,\mathrm{d}w\,\mathrm{d}z.
\end{aligned}
$$

As a consequence, we have

$$
\begin{aligned}
&|\mathcal{E}f(x)-\mathcal{E}f(y)| \\
&\quad \lesssim \sum_{i\in\mathbb{N}_x}\sum_{j\in\mathbb{N}_y}\varphi_i(x)\varphi_j(y) \fint_{I(x,2^2 5\sqrt{n}\,\mathrm{dist}\,(x,D))\cap D}\fint_{I(y,2^2 5\sqrt{n}\,\mathrm{dist}\,(y,D))\cap D} |f(z)-f(w)|\,\mathrm{d}w\,\mathrm{d}z \\
&\quad \lesssim \fint_{I(x,2^2 5\sqrt{n}\,\mathrm{dist}\,(x,D))\cap D}\fint_{I(y,2^2 5\sqrt{n}\,\mathrm{dist}\,(y,D))\cap D} |f(z)-f(w)|\,\mathrm{d}w\,\mathrm{d}z.
\end{aligned}
$$

Note that

$$
x\in I\setminus D \quad \& \quad y\in X_1(x) \Rightarrow |x-y|\ge 2^{-1}\max\{\mathrm{dist}\,(x,D),\ \mathrm{dist}\,(x,D)\}.
$$

So

$$
\begin{cases} z\in I(x,2^2 5\sqrt{n}\,\mathrm{dist}\,(x,D))\cap D \\ w\in I(y,2^2 5\sqrt{n}\,\mathrm{dist}\,(y,D))\cap D \end{cases} \Rightarrow |z-w|\lesssim |x-y|.
$$

This, plus $0\le\alpha<1$, leads to

$$
\begin{aligned}
\frac{|\mathcal{E}f(x)-\mathcal{E}f(y)|}{|x-y|^{\frac{n+2\alpha}{2}}} &\lesssim \fint_{I(x,2^2 5\sqrt{n}\,\mathrm{dist}\,(x,D))\cap D}\fint_{I(y,2^2 5\sqrt{n}\,\mathrm{dist}\,(y,D))\cap D} \frac{|f(z)-f(w)|}{|z-w|^{\frac{n+2\alpha}{2}}}\,\mathrm{d}w\,\mathrm{d}z \\
&\lesssim \fint_{I(x,2^2 5\sqrt{n}\,\mathrm{dist}\,(x,D))}\fint_{I(x,2^2 5\sqrt{n}\,\mathrm{dist}\,(x,D))} \frac{|f(z)-f(w)|}{|z-w|^{\frac{n+2\alpha}{2}}}\mathbb{1}_D(z)\mathbb{1}_D(w)\,\mathrm{d}w\,\mathrm{d}z.
\end{aligned}
$$

Upon setting

$$
G(w,z)=\frac{|f(z)-f(w)|}{|w-z|^{\frac{n+2\alpha}{2}}}\mathbb{1}_{I(x_I,21\sqrt{n}\ell(I))\cap D}(w)\mathbb{1}_{I(x_I,21\sqrt{n}\ell(I))\cap D}(z),
$$

we gain

$$
\begin{aligned}
\frac{|\mathcal{E}f(x)-\mathcal{E}f(y)|}{|x-y|^{\frac{n+2\alpha}{2}}} &\lesssim \fint_{I(x,2^2 5\sqrt{n}\,\mathrm{dist}\,(x,D))}\fint_{I(x,2^2 5\sqrt{n}\,\mathrm{dist}\,(x,D))} G(z,w)\,\mathrm{d}w\,\mathrm{d}z \\
&\lesssim (\mathcal{M}\times\mathcal{M})(G)(x,y),
\end{aligned}
$$

where

$$(\mathcal{M} \times \mathcal{M})(G)(x, y)$$

denotes the iterated Hardy–Littlewood maximal function of G. By the L^2-boundedness of $\mathcal{M}$, we obtain

$$\begin{aligned}
\mathsf{H}_{3,1} &\lesssim \int_{\mathbb{R}^n}\int_{\mathbb{R}^n} \big((\mathcal{M} \times \mathcal{M})(G)(x, y)\big)^2 \,\mathrm{d}x\,\mathrm{d}y \\
&\lesssim \int_{\mathbb{R}^n}\int_{\mathbb{R}^n} G(x, y)^2 \,\mathrm{d}x\,\mathrm{d}y \\
&\lesssim \int_{I(x_I,21n\ell(I))\cap D}\ \int_{I(x_I,21n\ell(I))\cap D} \frac{|f(x) - f(y)|^2}{|x - y|^{n+2\alpha}} \,\mathrm{d}x\,\mathrm{d}y \\
&\lesssim \mathrm{m}(I)^{\frac{n-2\alpha}{n}} \|f\|^2_{\mathfrak{Q}_\alpha(D)}.
\end{aligned}$$

In order to estimate $\mathsf{H}_{3,2}$, notice that

$$x \in I \setminus D \quad \& \quad y \in X_2(x) \Rightarrow \sum_{i \in \mathbb{N}_x \cup \mathbb{N}_y} (\varphi_i(x) - \varphi_i(y)) = \sum_{i \in \mathbb{N}_x} \varphi_i(x) - \sum_{j \in \mathbb{N}_y} \varphi_j(y) = 0.$$

So, Lemma 6.2 implies

$$\begin{aligned}
&|\mathcal{E}f(x) - \mathcal{E}f(y)| \\
&= \Big| \sum_{i \in \mathbb{N}_x \cup \mathbb{N}_y} (\varphi_i(x) - \varphi_i(y)) \fint_{I(x_i^*, \ell_i)\cap D} f(z)\,\mathrm{d}z \Big| \\
&= \Big| \sum_{i \in \mathbb{N}_x \cup \mathbb{N}_y} (\varphi_i(x) - \varphi_i(y)) \Big(\fint_{I(x_i^*, \ell_i)\cap D} f(z)\,\mathrm{d}z - \fint_{I(x, 2^2 5\sqrt{n}\,\mathrm{dist}\,(x,D))\cap D} f(w)\,\mathrm{d}w \Big) \Big| \\
&= \Big| \sum_{i \in \mathbb{N}_x \cup \mathbb{N}_y} (\varphi_i(x) - \varphi_i(y)) \fint_{I(x_i^*, \ell_i)\cap D} \fint_{I(x, 2^2 5\sqrt{n}\,\mathrm{dist}\,(x,D))\cap D} (f(z) - f(w))\,\mathrm{d}w\,\mathrm{d}z \Big| \\
&\lesssim \sum_{i \in \mathbb{N}_x \cup \mathbb{N}_y} \frac{|x - y|}{\ell_i} \fint_{I(x_i^*, \ell_i)\cap D} \fint_{I(x, 2^2 5\sqrt{n}\,\mathrm{dist}\,(x,D))\cap D} |f(z) - f(w)|\,\mathrm{d}w\,\mathrm{d}z.
\end{aligned}$$

Choosing

$$\bar{y} \in \overline{D} \quad \& \quad |y - \bar{y}| = \mathrm{dist}\,(y, D)$$

and using $y \in X_2(x)$, we achieve

$$|x - y| \le 2^{-1} \max\{\mathrm{dist}\,(x, D), \mathrm{dist}\,(y, D)\},$$

whence

$$\begin{aligned}\operatorname{dist}(x,D) &\le |x-\bar{y}| + |y-\bar{y}| \\ &\le 2^{-1}\operatorname{dist}(x,D) + 2^{-1}\operatorname{dist}(y,D) + \operatorname{dist}(y,D) \\ &= 2^{-1}\operatorname{dist}(x,D) + 2^{-1}3\operatorname{dist}(y,D).\end{aligned}$$

Similarly, we get

$$3^{-1}\operatorname{dist}(y,D) \le \operatorname{dist}(x,D) \le 3\operatorname{dist}(y,D).$$

This in turns derives

$$\begin{cases} I(y, 2^2 5\sqrt{n}\operatorname{dist}(y,D)) \subseteq I(x, 80\sqrt{n}\operatorname{dist}(x,D)) \\ \mathrm{m}(I(y, 2^2 5\sqrt{n}\operatorname{dist}(y,D)) \cap D) \approx \mathrm{m}(I(x, 80\sqrt{n}\operatorname{dist}(x,D)) \cap D). \end{cases}$$

For

$$y \in X_2(x) \quad \& \quad i \in \mathbb{N}_x \cup \mathbb{N}_y,$$

we have

$$\ell_i \approx \operatorname{dist}(x,D)$$

and

$$\mathrm{m}(I(x_i^*, \ell_i) \cap D) \approx \mathrm{m}(I(y, 2^2 5\sqrt{n}\operatorname{dist}(y,D)) \cap D) \approx \mathrm{m}(I(x, 80\sqrt{n}\operatorname{dist}(x,D)) \cap D).$$

Next, we utilize

$$\sharp\mathbb{N}_x + \sharp\mathbb{N}_y \lesssim 1$$

and the Hölder inequality to calculate

$$\begin{aligned}&\frac{|\mathcal{E}f(x) - \mathcal{E}f(y)|}{\frac{|x-y|}{\operatorname{dist}(x,D)}} \\ &\lesssim \fint_{I(x,80\sqrt{n}\operatorname{dist}(x,D))\cap D} \fint_{I(x,2^2 5\sqrt{n}\operatorname{dist}(x,D))\cap D} |f(z) - f(w)|\,\mathrm{d}w\,\mathrm{d}z \\ &\lesssim \fint_{I(x,80\sqrt{n}\operatorname{dist}(x,D))\cap D} \left(\fint_{I(x,2^2 5\sqrt{n}\operatorname{dist}(x,D))\cap D} |f(z) - f(w)|^2\,\mathrm{d}w\right)^{\frac{1}{2}} \mathrm{d}z \\ &\lesssim \operatorname{dist}(x,D)^{\alpha} \fint_{I(x,80\sqrt{n}\operatorname{dist}(x,D))\cap D} \left(\int_{I(x,2^2 5\sqrt{n}\operatorname{dist}(x,D))\cap D} \frac{|f(z) - f(w)|^2}{|z-w|^{n+2\alpha}}\,\mathrm{d}w\right)^{\frac{1}{2}} \mathrm{d}z.\end{aligned}$$

Due to

$$I(x, 80\sqrt{n}\,\mathrm{dist}\,(x, D)) \subseteq I(x_I, 81n\ell(I)),$$

we estimate

$$\begin{aligned}
&\frac{|\mathcal{E}f(x) - \mathcal{E}f(y)|}{\frac{|x-y|}{\mathrm{dist}\,(x,D)^{1-\alpha}}} \\
&\lesssim \fint_{I(x,80\sqrt{n}\,\mathrm{dist}\,(x,D))} \mathbb{1}_{I(x_I,81n\ell(I))\cap D}(z) \left(\int_{I(x_I,81n\ell(I))\cap D} \frac{|f(z) - f(w)|^2}{|z - w|^{n+2\alpha}}\,\mathrm{d}w \right)^{\frac{1}{2}} \mathrm{d}z \\
&\lesssim \left(\left(\int_{I(x_I,81n\ell(I))\cap D} \frac{|f(\cdot) - f(w)|^2}{|\cdot - w|^{n+2\alpha}}\,\mathrm{d}w \right)^{\frac{1}{2}} \mathbb{1}_{I(x_I,81n\ell(I))\cap D}(\cdot) \right)(x),
\end{aligned}$$

thereby getting

$$\begin{aligned}
\mathsf{H}_{3,2} &\lesssim \int_{I\setminus D} \left(\int_{X_2(x)} \frac{|x - y|^{2-n-2\alpha}}{\mathrm{dist}\,(x, D)^{2-2\alpha}}\,\mathrm{d}y \right) \\
&\quad \times \left(\mathcal{M}\left(\left(\int_{I(x_I,81n\ell(I))\cap D} \frac{|f(\cdot) - f(w)|^2}{|\cdot - w|^{n+2\alpha}}\,\mathrm{d}w \right)^{\frac{1}{2}} \mathbb{1}_{I(x_I,81n\ell(I))\cap D}(\cdot) \right)(x) \right)^2 \mathrm{d}x.
\end{aligned}$$

Thanks to

$$X_2(x) \subseteq B(x, 15\,\mathrm{dist}\,(x, D)),$$

we have

$$\int_{X_2(x)} \frac{|x - y|^{2-n-2\alpha}}{\mathrm{dist}\,(x, D)^{2-2\alpha}}\,\mathrm{d}y \lesssim \int_{B(x,15\,\mathrm{dist}\,(x,D))} \frac{|x - y|^{2-n-2\alpha}}{\mathrm{dist}\,(x, D)^{2-2\alpha}}\,\mathrm{d}y \lesssim 1.$$

This, along with the L^2-boundedness of $\mathcal{M}$, yields

$$\begin{aligned}
\mathsf{H}_{3,2} &\lesssim \int_{\mathbb{R}^n} \left(\mathcal{M}\left(\left(\int_{I(x_I,81n\ell(I))\cap D} \frac{|f(\cdot) - f(w)|^2}{|\cdot - w|^{n+2\alpha}}\,\mathrm{d}w \right)^{\frac{1}{2}} \mathbb{1}_{I(x_I,81n\ell(I))\cap D}(\cdot) \right)(x) \right)^2 \mathrm{d}x \\
&\lesssim \int_{I(x_I,81n\ell(I))\cap D} \int_{I(x_I,81n\ell(I))\cap D} \frac{|f(x) - f(w)|^2}{|x - w|^{n+2\alpha}}\,\mathrm{d}w\,\mathrm{d}x \\
&\lesssim \mathrm{m}(I)^{\frac{n-2\alpha}{n}} \|f\|^2_{\mathfrak{Q}_\alpha(D)}.
\end{aligned}$$

Combining the estimates for $\mathsf{H}_{3,1}$ and $\mathsf{H}_{3,2}$ yields

$$\mathsf{H}_3 = \mathsf{H}_{3,1} + \mathsf{H}_{3,2} \lesssim \mathrm{m}(I)^{\frac{n-2\alpha}{n}} \|f\|^2_{\mathfrak{Q}_\alpha(D)}.$$

Subcase $I \cap D = \emptyset$ *but* $\operatorname{dist}(I, D) \le 2^2 5^2 \sqrt{n}\ell(I)$.

This subcase reduces to the last *Subcase* $I \cap D \neq \emptyset$ if we consider a large cube $\tilde{I} = 2^3 5^2 I$ which enjoys

$$\begin{cases} \tilde{I} \cap D \neq \emptyset; \\ \ell(I)^{2\alpha-n} \int_I \int_I \frac{|\mathcal{E}f(x) - \mathcal{E}f(y)|^2}{|x-y|^{n+2\alpha}}\,dx\,dy \lesssim \ell(\tilde{I})^{2\alpha-n} \int_{\tilde{I}} \int_{\tilde{I}} \frac{|\mathcal{E}f(x) - \mathcal{E}f(y)|^2}{|x-y|^{n+2\alpha}}\,dx\,dy. \end{cases}$$

Subcase $I \cap D = \emptyset$ *but* $\operatorname{dist}(I, D) \ge 2^2 5^2 \sqrt{n}\ell(I)$.

For $(x, y) \in I \times I$, we utilize

$$\sum_{i \in \mathbb{N}_x \cup \mathbb{N}_y} (\varphi_i(x) - \varphi_i(y)) = 0$$

and Lemma 6.2 to obtain

$$|\mathcal{E}f(x) - \mathcal{E}f(y)| \lesssim \sum_{i \in \mathbb{N}_x \cup \mathbb{N}_y} \frac{|x-y|}{\ell_i} \fint_{I(x_i^*, \ell_i) \cap D} \fint_{I(x, 2^2 5 \sqrt{n} \operatorname{dist}(x, D)) \cap D} |f(z) - f(w)|\,dw\,dz.$$

Observe that

$$\operatorname{dist}(I, D) \le \operatorname{dist}(x, D), \operatorname{dist}(y, D) \le \operatorname{dist}(I, D) + \sqrt{n}\ell(I) \le \frac{11\sqrt{n}}{10} \operatorname{dist}(I, D)$$

ensures

$$\begin{cases} \operatorname{dist}(x, D) \approx \operatorname{dist}(y, D) \approx \operatorname{dist}(I, D) \\ I(x, 2^2 5 \sqrt{n} \operatorname{dist}(x, D)) \subseteq I(x_I, 30n \operatorname{dist}(I, D)) \\ I(y, 2^2 5 \sqrt{n} \operatorname{dist}(y, D)) \subseteq I(x_I, 30n \operatorname{dist}(I, D)). \end{cases}$$

So, for $i \in \mathbb{N}_x$ we have

$$x \in I(x_i, 2\ell_i) \quad \& \quad \operatorname{dist}(x, D) \approx \operatorname{dist}(I_i, D).$$

An application of Lemma 6.2 gives

$$\begin{cases} \ell_i \approx \operatorname{dist}(x, D) \approx \operatorname{dist}(I, D); \\ m(I(x_i^*, \ell_i) \cap D) \approx m(I(x, 2^2 5 \sqrt{n} \operatorname{dist}(x, D)) \cap D) \approx m(I(x_I, 30n \operatorname{dist}(I, D)) \cap D) \end{cases}$$

and

$$\begin{aligned} &|\mathcal{E}f(x) - \mathcal{E}f(y)| \\ &\lesssim \frac{|x-y|}{\operatorname{dist}(I, D)} \fint_{I(x_I, 30n \operatorname{dist}(I,D)) \cap D} \fint_{I(x_I, 30n \operatorname{dist}(I,D)) \cap D} |f(z) - f(w)|\,dw\,dz \end{aligned}$$

$$\lesssim \frac{|x-y|}{\operatorname{dist}(I,D)}\Big(\fint_{I(x_I,30n\operatorname{dist}(I,D))\cap D}\fint_{I(x_I,30n\operatorname{dist}(I,D))\cap D}|f(z)-f(w)|^2\,dw\,dz\Big)^{\frac{1}{2}}$$

$$\lesssim \frac{|x-y|}{\operatorname{dist}(I,D)^{1-\alpha+n/2}}\Big(\int_{I(x_I,30n\operatorname{dist}(I,D))\cap D}\int_{I(x_I,30n\operatorname{dist}(I,D))\cap D}\frac{|f(z)-f(w)|^2}{|z-w|^{n+2\alpha}}\,dw\,dz\Big)^{\frac{1}{2}}$$

$$\lesssim \frac{|x-y|}{\operatorname{dist}(I,D)}\|f\|_{\mathcal{Q}_\alpha(D)}.$$

Accordingly, we read off

$$\ell(I)^{2\alpha-n}\int_I\int_I\frac{|\mathcal{E}f(x)-\mathcal{E}f(y)|^2}{|x-y|^{n+2\alpha}}\,dx\,dy\lesssim\Big(\frac{\|f\|_{\mathcal{Q}_\alpha(D)}\ell(I)}{\operatorname{dist}(I,D)}\Big)^2\lesssim\|f\|^2_{\mathcal{Q}_\alpha(D)}.$$

The bounded case: $\operatorname{diam} D<\infty$.

Without loss of generality, we may assume

$$\begin{cases}\operatorname{diam} D=1\\ D\subseteq I(0,1)\\ \mathrm{m}(D)\approx 1,\end{cases}$$

and then consider two situations as seen below.

▷ The first is $\ell(I)\le 2^3 5^3\sqrt{n}$.
  - If $I\cap D\neq\emptyset$, then the same argument for verifying the subcase $I\cap D\neq\emptyset$ of *Case* $\operatorname{diam} D=\infty$ induces the desired estimation

$$\ell(I)^{2\alpha-n}\int_I\int_I\frac{|\mathcal{E}f(x)-\mathcal{E}f(y)|^2}{|x-y|^{n+2\alpha}}dxdy\lesssim\|f\|^2_{\mathcal{Q}_\alpha(D)}.$$

  - However, if $I\cap D=\emptyset$, then two more subcases are taken into account: when

$$\operatorname{dist}(I,D)\le 2^2 5^2\sqrt{n}\ell(I),$$

considering the cube $2^3 5^2 I$ and using

$$2^3 5^2 I\cap D\neq\emptyset,$$

we have

$$\ell(I)^{2\alpha-n}\int_I\int_I\frac{|\mathcal{E}f(x)-\mathcal{E}f(y)|^2}{|x-y|^{n+2\alpha}}dxdy\lesssim\|f\|^2_{\mathcal{Q}_\alpha(D)};$$

when

$$\operatorname{dist}(I, D) \geq 2^2 5^2 \sqrt{n}\ell(I),$$

a similar argument for the subcase of

$$\operatorname{dist}(I, D) \geq 2^2 5^2 \sqrt{n}\ell(I)$$

of *Case* $\operatorname{diam} D = \infty$ derives

$$\ell(I)^{2\alpha-n} \int_I \int_I \frac{|\mathcal{E}f(x) - \mathcal{E}f(y)|^2}{|x-y|^{n+2\alpha}} \mathrm{d}x\mathrm{d}y \lesssim \|f\|^2_{\mathfrak{Q}_\alpha(D)}.$$

▷ The second is $\ell(I) \geq 2^3 5^3 \sqrt{n}$.

– If

$$\ell(I) \geq 2^3 5^3 \sqrt{n} \quad \& \quad \operatorname{dist}(I, D) \geq 2^2 5^2 \sqrt{n},$$

then for $x \in I$ and $i \in \mathbb{N}_x$ (ensuring $x \in 3I_i$), we claim

$$\mathcal{E}f(x) = \fint_D f(z)\,\mathrm{d}z \quad \forall x \in I.$$

Clearly, the claim deduces the desired result. In order to validate the claim, for every $x \in I$ and $i \in \Lambda_x$ we employ Lemma 6.2 to estimate

$$\begin{cases} \operatorname{dist}(3I_i, D) \leq \operatorname{dist}(I_i, D) + 2\sqrt{n}\ell_i \leq 23^2\sqrt{n}\ell_i; \\ \operatorname{dist}(I, D) \leq \operatorname{dist}(x, D) \leq \operatorname{dist}(3I_i, D) + 2\sqrt{n}\ell_i \leq 2^2 5\sqrt{n}\ell_i, \end{cases}$$

thereby finding

$$2^2 5^2 \sqrt{n} \leq 2^2 5 \sqrt{n}\ell_i \text{ – i.e. – } \ell_i \geq 5.$$

By

$$x_i^* \in D \quad \& \quad \operatorname{diam} D = 1,$$

we get

$$I(x_i{}^*, \ell_i) \cap D = D.$$

So, for $i \in \mathbb{N}_x$ we achieve

$$\mathcal{E}f(x) = \sum_{i\in\mathbb{N}_x} \varphi_i(x) \fint_{I(x_i{}^*,\ell_i)\cap D} f(z)\,\mathrm{d}z = \sum_{i\in\mathbb{N}_x} \varphi_i(x) \fint_D f(z)\,\mathrm{d}z = \fint_D f(z)\,\mathrm{d}z,$$

thereby validating the above claim.

– If

$$\ell(I) \geq 2^3 5^3 \sqrt{n} \quad \& \quad I \cap D \neq \emptyset,$$

then

$$\int_I \int_I \frac{|\mathcal{E}f(x) - \mathcal{E}f(y)|^2}{|x-y|^{n+2\alpha}}\, dx\, dy = \mathsf{H}_{4,1} + \mathsf{H}_{4,2} + \mathsf{H}_{4,3},$$

where

$$\begin{cases} \mathsf{H}_{4,1} = \displaystyle\int_{I \setminus I(0,2^2 5^2\sqrt{n})} \int_{I \setminus I(0,2^2 5^2 \sqrt{n})} \frac{|\mathcal{E}f(x) - \mathcal{E}f(y)|^2}{|x-y|^{n+2\alpha}}\, dx\, dy \\ \mathsf{H}_{4,2} = 2 \displaystyle\int_{I \setminus I(0,2^3 5^2\sqrt{n})} \int_{I(0,2^2 5^2 \sqrt{n})} \frac{|\mathcal{E}f(x) - \mathcal{E}f(y)|^2}{|x-y|^{n+2\alpha}}\, dx\, dy \\ \mathsf{H}_{4,3} = \displaystyle\int_{I(0,2^3 5^2\sqrt{n})} \int_{I(0,2^3 5^2 \sqrt{n})} \frac{|\mathcal{E}f(x) - \mathcal{E}f(y)|^2}{|x-y|^{n+2\alpha}}\, dx\, dy. \end{cases}$$

From

$$\mathrm{m}(I(0, 2^3 5^2 \sqrt{n})) \approx 1$$

and the treatment of

$$\ell(I) \leq 2^3 5^3 \sqrt{n} \quad \& \quad I \cap D \neq \emptyset$$

it follows that

$$\mathsf{H}_{4,3} \lesssim \|f\|^2_{\mathfrak{Q}_\alpha(D)} \lesssim \mathrm{m}(I)^{\frac{n-2\alpha}{n}} \|f\|^2_{\mathfrak{Q}_\alpha(D)}.$$

Now, an application of the argument for

$$\ell(I) \geq 2^3 5^3 \sqrt{n} \quad \& \quad \mathrm{dist}\,(I, D) \geq 2^2 5^2 \sqrt{n}$$

derives

$$\mathcal{E}f(x) = \fint_D f(z)\, dz \quad \forall\, x \in I \setminus I(0, 2^3 5^2 \sqrt{n}),$$

and so

$$\mathsf{H}_{4,1} \lesssim \mathrm{m}^{\frac{n-2\alpha}{n}} \|f\|^2_{\mathfrak{Q}_\alpha(D)}.$$

Also, if

$$x \in I \setminus I(0, 2^3 5^2 \sqrt{n}) \quad \& \quad y \in I(0, 2^2 5^2 \sqrt{n}),$$

then

$$\begin{cases} |x-y| \geq 2^{-1}|x| \\ \mathcal{E}f(x) = \fint_D f(z)\,\mathrm{d}z \\ |\mathcal{E}f(x) - \mathcal{E}f(y)|^2 \leq \fint_D |f(z) - \mathcal{E}f(y)|^2\,\mathrm{d}z, \end{cases}$$

and hence

$$\begin{aligned} &\int_{I(0,2^2 5^2\sqrt{n})} \frac{|\mathcal{E}f(x) - \mathcal{E}f(y)|^2}{|x-y|^{n+2\alpha}}\,\mathrm{d}y \\ &\lesssim |x|^{-n-2\alpha} \int_{I(0,2^2 5^2\sqrt{n})} \int_D |f(z) - \mathcal{E}f(y)|^2\,\mathrm{d}z\,\mathrm{d}y \\ &\lesssim |x|^{-n-2\alpha} \int_{I(0,2^2 5^2\sqrt{n})} \int_{I(0,2^2 5^2\sqrt{n})} \frac{|\mathcal{E}f(z) - \mathcal{E}f(y)|^2}{|y-z|^{n+2\alpha}}\,\mathrm{d}z\,\mathrm{d}y \\ &\lesssim |x|^{-n-2\alpha} \|f\|^2_{\mathfrak{Q}_\alpha(D)}, \end{aligned}$$

whose last inequality follows from the result for

$$\ell(I) \leq 2^3 5^3 \sqrt{n} \quad \& \quad I \cap D \neq \emptyset.$$

Thus, we obtain

$$\mathsf{H}_{4,2} \lesssim \|f\|^2_{\mathfrak{Q}_\alpha(D)} \int_{I \setminus I(0,2^3 5^2\sqrt{n})} |x|^{-n-2\alpha}\,\mathrm{d}x \lesssim \|f\|^2_{\mathfrak{Q}_\alpha(D)} \lesssim \mathrm{m}(I)^{\frac{n-2\alpha}{n}} \|f\|^2_{\mathfrak{Q}_\alpha(D)}.$$

Putting the estimates for $\mathsf{H}_{4,1}$, $\mathsf{H}_{4,2}$ and $\mathsf{H}_{4,3}$ together derives

$$\ell(I)^{2\alpha-n} \int_I \int_I \frac{|\mathcal{E}f(x) - \mathcal{E}f(y)|^2}{|x-y|^{n+2\alpha}}\,\mathrm{d}x\,\mathrm{d}y \lesssim \|f\|^2_{\mathfrak{Q}_\alpha(D)}.$$

- If

$$\ell(I) \geq 2^3 5^3 \sqrt{n} \quad \& \quad \mathrm{dist}\,(I, D) \leq 2^2 5^2 \sqrt{n},$$

then a consideration of the cube

$$2\sqrt{n}I \quad \text{with} \quad 2\sqrt{n}I \cap D \neq \emptyset,$$

also yields the last integral inequality.

A combination of the above cases validates (i). □

8.3 Restriction principle

We start with a basic lemma.

Lemma 8.4. *Let* $(n-1, k, I) \in \mathbb{N} \times \mathbb{Z} \times \mathscr{C}(\mathbb{R}^n)$.

(i) *If* $D \subseteq \mathbb{R}^n$ *is a domain, then the following two statements hold:*

(i)-a *there exists a collection* $W_D = \{S_j\}_{j\in\mathbb{N}}$ *of countably many dyadic closed cubes such that*

$$\begin{cases} D = \bigcup_{j\in\mathbb{N}} S_j \\ S_i^\circ \cap S_j^\circ = \emptyset \; \forall i \neq j \\ 2^7 \sqrt{n}\ell(S_j) \le \operatorname{dist}(S_j, \partial D) \le 2^9 \sqrt{n}\ell(S_j) \\ 2^{-2}\ell(S_i) \le \ell(S_j) \le 2^2\ell(S_i) \quad \forall\, S_i \cap S_j \neq \emptyset; \end{cases}$$

(i)-b *for disjoint Whitney cubes* S_i *and* S_j, *a family of Whitney cubes* $\{S_{i_s}\}_{s=1}^t$ *is called a joining* S_i *and* S_j *provided that*

$$S_{i_1} = S_i \quad \& \quad S_{i_t} = S_j,$$

and $S_{i_s} \cap S_{i_{s+1}}$ *is an* $(n-1)$*-dimensional closed cube for each* $s \in \{1, \dots, t-1\}$, *and the constant* t *is called the length of the chain. There exists a Whitney chain*

$$F(S_i, S_j) = \{S_{i_s}\}_{s=1}^t$$

joining S_i *and* S_j *with its length*

$$t \lesssim \left|\log_2 \frac{\ell(S_i)}{\ell(S_j)}\right| + \log_2\left(\frac{\operatorname{dist}(S_i, S_j)}{\ell(S_i) + \ell(S_j)} + 2\right).$$

(ii) *If* $D \subseteq \mathbb{R}^n$ *is a uniform domain,* $0 < \epsilon$ *is such a small number that*

$$n - 2\alpha - \epsilon > \begin{cases} \overline{\dim}_{\mathrm{lsM}} \partial\partial D & \text{as } \partial D \text{ is bounded} \\ \overline{\dim}_{\mathrm{gsM}} \partial D & \text{as } \partial D \text{ is unbounded,} \end{cases}$$

and

$$\mathscr{S}_k(I) = \{S_j \in W_D : \; S_j \cap I \neq \emptyset, \ell(S_j) = 2^{-k}\} = \{S_{k,i}\}_i,$$

then the following two statements hold:

(ii)-a *under* $\operatorname{diam} D < \infty$, *for any cube* I *with*

$$\begin{cases} x_I \in D \\ \ell(I) \le 2 \\ I \cap D^c \neq \emptyset \end{cases}$$

one has

$$\begin{cases} \sharp\mathscr{S}_k(I) = 0 & \forall\, k < -\log_2 \ell(I) + 7 \\ \sharp\mathscr{S}_k(I) \lesssim (2^k \ell(I))^{\overline{\dim}_{\mathrm{lsM}}\partial D+\epsilon} & \forall\, k \geq -\log_2 \ell(I) + 7 \end{cases}$$

while under $\operatorname{diam} D = \infty$, *for any cube* I *with*

$$\begin{cases} x_I \in D \\ I \cap D^c \neq \emptyset \end{cases}$$

one has

$$\sharp\mathscr{S}_k(I) \lesssim (2^k \ell(I))^{\overline{\dim}_{\mathrm{gsM}}\partial D+\epsilon};$$

(ii)-b *under* $\operatorname{diam} D < \infty$, *for any Whitney cube* S_j *with*

$$\begin{cases} \ell(S_j) \leq 1 \\ m \geq -\log_2 S_j \\ 1 \leq \delta \leq -\log_2 S_j + 2^2 \end{cases}$$

one has

$$\sharp\mathscr{S}_m(2^\delta S_j) \lesssim (2^{m+\delta} \ell(S_j))^{\overline{\dim}_{\mathrm{lsM}}\partial D+\epsilon}$$

while under $\operatorname{diam} D = \infty$, *for any Whitney cube* S_j *with*

$$\begin{cases} m \geq -\log_2 S_j \\ \delta \geq 1 \end{cases}$$

one has

$$\sharp\mathscr{S}_m(2^\delta S_j) \lesssim (2^{m+\delta} \ell(S_j))^{\overline{\dim}_{\mathrm{gsM}}\partial D+\epsilon}.$$

Proof. It is enough to verify (ii)-a.

Up to dividing I into 2^n subcubes, we may assume $\ell(I) \leq 1$. If

$$S_j \cap I \neq \emptyset \quad \& \quad I \cap D^c \neq \emptyset,$$

then

$$\operatorname{dist}(S_j, D^c) \leq \operatorname{diam} I \leq \sqrt{n}\ell(I),$$

which, plus Lemma 8.4 (i)-a, yields

$$\ell(S_j) \leq 2^{-7}\ell(I).$$

Consequently,

$$k < -\log_2 \ell(I) + 7 \Rightarrow \mathscr{S}_k(I) = \emptyset.$$

When

$$k \geq -\log_2 \ell(I) + 7,$$

for $\delta > 0$ denote by $\mathscr{N}_{\rm cov}(\delta, \partial D \cap J)$ the collection of cubes of edge-length δ required to cover $\partial D \cap J$. Then

$$\sharp \mathscr{N}_{\rm cov}(\delta, \partial D \cap J) = N_{\rm cov}(\delta, \partial D \cap J).$$

For $S_{k,i} \in \mathscr{S}_k(I)$, we have

$$2^{11}\sqrt{n} S_{k,i} \cap \partial D \neq \emptyset,$$

thereby finding that $2^{11}\sqrt{n} S_{k,i}$ intersects some cube

$$R \in \mathscr{N}_{\rm cov}(2^{-k}, \partial D \cap J),$$

which implies

$$S_{k,i} \subseteq 2^{13} n R.$$

Note also that for each cube

$$R \in \mathscr{N}_{\rm cov}(2^{-k}, \partial D \cap J),$$

the cube $2^{13} nR$ can only contain a uniformly bounded number of

$$S_{k,i} \in \mathscr{S}_k(J).$$

So, we reach

$$k \geq -\log_2 \ell(I) + 7 \Rightarrow \sharp \mathscr{S}_k(J) \lesssim N_{\rm cov}(2^{-k}, \partial D \cap J).$$

Let ϵ be a sufficiently small number satisfying

$$0 < \epsilon < n - 2\alpha - \overline{\dim}_{\rm lsM} \partial D.$$

It suffices to show

$$\begin{cases} I \in \mathscr{C}(\mathbb{R}^n) \\ \ell(J) \leq 1 \\ k > -\log_2 \ell(J) \end{cases} \Rightarrow \quad N_{\rm cov}(2^{-k}, \partial D \cap J) \lesssim [2^k \ell(J)]^{\overline{\dim}_{\rm lsM} \partial D + \epsilon}.$$

In fact, from

$$\overline{\dim}_{\mathrm{lsM}}\partial D \quad \& \quad k \geq -\log_2 \ell(J),$$

it follows that for sufficiently small ϵ there are constants

$$N_1 \geq 2^3 \quad \& \quad k_1 \in \mathbb{N}$$

enjoying the implication

$$\begin{cases} k \geq k_1 \\ J \in \mathscr{C}(\mathbb{R}^n) \\ \ell(J) \in [2^{-k+N_1}, 1] \end{cases} \Rightarrow \begin{cases} \frac{\log_2 N_{\mathrm{cov}}(2^{-k}, \partial D \cap J)}{\log_2(\ell(J)2^k)} \leq \overline{\dim}_{\mathrm{lsM}}\partial D + \epsilon \\ N_{\mathrm{cov}}(2^{-k}, \partial D \cap J) \leq (2^k \ell(J))^{\overline{\dim}_{\mathrm{lsM}}\partial D + \epsilon}. \end{cases}$$

Consequently,

$$\text{under} \quad I \in \mathscr{C}(\mathbb{R}^n) \quad \text{with} \quad \ell(J) \leq 1 \ \& \ k > -\log_2 \ell(J),$$

we have

$$k \geq k_1 \quad \& \quad k \geq -\log_2 \ell(J) + N_1 \Rightarrow N_{\mathrm{cov}}(2^{-k}, \partial D \cap J) \leq (2^k \ell(J))^{\overline{\dim}_{\mathrm{lsM}}\partial D + \epsilon}.$$

If

$$-\log_2 \ell(J) \leq k < k_1 \quad \text{or} \quad -\log_2 \ell(J) \leq k < -\log_2 \ell(J) + N_1,$$

then

$$\begin{cases} 2^k \ell(J) \lesssim 1 \\ N_{\mathrm{cov}}(2^{-k}, \partial D \cap I) \leq (2^k \ell(I))^n \lesssim (2^k \ell(I))^{\overline{\dim}_{\mathrm{lsM}}\partial D + \epsilon}. \end{cases}$$

□

We then utilize Lemma 8.4 to resolve [22, Problem 8.5] as recalled in the beginning of this chapter.

Theorem 8.5. *Let $(\alpha, n-1) \in [0,1) \times \mathbb{N}$. If $D \subseteq \mathbb{R}^n$ is a unifrom domain – i. e. – there exists a constant $\kappa > 0$ such that for every pair of points $x, y \in D$ one can find a curve $\zeta \subset D$ joining x, y and obeying*

$$l(\zeta) \leq \kappa |x-y| \quad \& \quad \mathrm{dist}\,(z, D^c) \geq \kappa^{-1}|x-z||y-z||x-y|^{-1} \quad \forall z \in \zeta$$

with $l(\zeta)$ being the Euclidean length of ζ, and if

$$n > \begin{cases} 2\alpha + \overline{\dim}_{\mathrm{lsM}}\partial D & \text{as } \partial D \text{ is bounded} \\ 2\alpha + \overline{\dim}_{\mathrm{gsM}}\partial D & \text{as } \partial D \text{ is unbounded,} \end{cases}$$

then

$$\mathcal{Q}_\alpha(\mathbb{R}^n)|_D = \mathcal{Q}_\alpha(D).$$

Proof. Since

$$\mathfrak{Q}_\alpha(D) \subseteq \mathcal{Q}_\alpha(D),$$

it suffices to prove

$$\mathcal{Q}_\alpha(D) \subseteq \mathfrak{Q}_\alpha(D).$$

We are suggested to consider three situations as seen below.

The first case: $\operatorname{diam} \partial D < \infty$ & $\operatorname{diam} D < \infty$.

Without loss of generality, we may assume that $I(0,1)$ is the origin-centered cube with the unit edge-length and D obeys

$$\operatorname{diam} \partial D = 1 \quad \& \quad D \subseteq I(0,1).$$

We need to show

$$\Psi_\alpha(f, I \cap D) = \int_{I\cap D}\int_{I\cap D} \frac{|f(x)-f(y)|^2}{|x-y|^{n+2\alpha}}\,dy\,dx \lesssim \ell(I)^{n-2\alpha}\|f\|^2_{\mathcal{Q}_\alpha(D)} \quad \forall I \in \mathscr{C}(\mathbb{R}^n).$$

If

$$I \subseteq D \quad \text{or} \quad I \subseteq D^c,$$

then the required estimate is evident. In the sequel, we always assume

$$I \cap D \neq \emptyset \quad \& \quad I \subseteq D^c \neq \emptyset.$$

Up to taking a new cube centered in $I \cap D$ with edge-length $2\ell(I)$, so that it contains I, we may assume that the center of I belongs to D. We may also assume $\ell(I) \leq 1$. Indeed,

$$\begin{aligned}
\ell(I) \geq 1 &\Rightarrow D \subseteq I(0,1) \subseteq 3I \\
&\Rightarrow \Psi_\alpha(f, I \cap D) \leq \int_D\int_D \frac{|f(x)-f(y)|^2}{|x-y|^{n+2\alpha}}\,dy\,dx = \Psi_\alpha(f, I(0,1) \cap D) \\
&\Rightarrow \Psi_\alpha(f, I \cap D) \lesssim \ell(I)^{n-2\alpha}\|f\|^2_{\mathcal{Q}_\alpha(D)},
\end{aligned}$$

as desired.

Given a cube I with its center being in D plus

$$\ell(I) \leq 1 \quad \& \quad I \cap D^c \neq \emptyset.$$

Let

$$\mathscr{S}_k(I) = \{S_{k,i}\}_i$$

be as above. Set

$$\Psi_\alpha(f, I\cap D) = 2\sum_{k\ge -\log_2 \ell(I)+7}\ \sum_{i\in\mathscr{S}_k(I)}\ \sum_{m\ge k}\ \sum_{j\in\mathscr{S}_m(I)} \int_{S_{ki}}\int_{S_{mj}} \frac{|f(x)-f(y)|^2}{|x-y|^{n+2\alpha}}\,dy\,dx = \mathrm{I}_1+\mathrm{I}_2,$$

where

$$\begin{cases} \mathrm{I}_1 = 2\sum\limits_{k\ge -\log_2 \ell(I)+7}\ \sum\limits_{i\in\mathscr{S}_k(I)}\ \sum\limits_{m\ge k}\ \sum\limits_{\substack{j\in\mathscr{S}_m(I)\\ S_{mj}\cap S_{ki}\neq\emptyset}} \int_{S_{ki}}\int_{S_{mj}} \frac{|f(x)-f(y)|^2}{|x-y|^{n+2\alpha}}\,dy\,dx\\ \mathrm{I}_2 = 2\sum\limits_{k\ge -\log_2 \ell(I)+7}\ \sum\limits_{i\in\mathscr{S}_k(I)}\ \sum\limits_{m\ge k}\ \sum\limits_{\substack{j\in\mathscr{S}_m(I)\\ S_{mj}\cap S_{ki}=\emptyset}} \int_{S_{ki}}\int_{S_{mj}} \frac{|f(x)-f(y)|^2}{|x-y|^{n+2\alpha}}\,dy\,dx. \end{cases}$$

▷ Observe that

$$\begin{aligned} & m\ge k,\quad j\in\mathscr{S}_m(I)\quad \&\quad S_{mj}\cap S_{ki}\neq\emptyset\\ &\quad \Rightarrow |k-m|\le 2\quad \&\quad S_{mj}\subseteq 4S_{ki}\subseteq D\\ &\quad \Rightarrow \sum_{m\ge k}\ \sum_{\substack{j\in\mathscr{S}_m(I)\\ S_{mj}\cap S_{ki}\neq\emptyset}} \int_{S_{ki}}\int_{S_{mj}} \frac{|f(x)-f(y)|^2}{|x-y|^{n+2\alpha}}\,dy\,dx \lesssim \int_{4S_{ki}}\int_{4S_{ki}} \frac{|f(x)-f(y)|^2}{|x-y|^{n+2\alpha}}\,dy\,dx. \end{aligned}$$

So, by Lemma 8.4 we get

$$\begin{aligned} \mathrm{I}_1 &\lesssim \sum_{k\ge -\log_2 \ell(I)+7}\ \sum_{i\in\mathscr{S}_k(I)} \mathrm{m}(S_{ki})^{\frac{n-2\alpha}{n}} \|f\|^2_{\mathcal{Q}_\alpha(D)}\\ &\lesssim \sum_{k\ge -\log_2 \ell(I)+7} \big(2^k\ell(I)\big)^{\overline{\dim}_{\mathrm{lsM}}\partial D+\epsilon} 2^{-k(n-2\alpha)} \|f\|^2_{\mathcal{Q}_\alpha(D)}. \end{aligned}$$

Owing to

$$n-2\alpha-\overline{\dim}_{\mathrm{lsM}}\partial D-\epsilon>0,$$

we reach

$$\mathrm{I}_1\lesssim \ell(I)^{n-2\alpha}\|f\|^2_{\mathcal{Q}_\alpha(D)}.$$

▷ Now we turn to control I_2. For

$$\begin{cases} m\ge k\\ i\in\mathscr{S}_k(I)\\ j\in\mathscr{S}_m(I)\\ S_{mj}\cap S_{ki}=\emptyset, \end{cases}$$

we compute

$$\int_{S_{ki}}\int_{S_{mj}} \frac{|f(x)-f(y)|^2}{|x-y|^{n+2\alpha}}\,dy\,dx \lesssim \frac{m(S_{mj})m(S_{ki})}{(\operatorname{dist}(S_{ki},S_{mj}))^{2\alpha+n}} \fint_{S_{ki}}\fint_{S_{mj}} |f(x)-f(y)|^2\,dy\,dx.$$

According to Lemma 8.4(i), there exists a Whitney chain

$$F(S_{ki},S_{mj}) = \{S_{k_1,i_1},\dots,S_{k_tj_t}\} \subseteq W_D$$

joining S_{ki} and S_{mj} with its length

$$\begin{aligned} t &\lesssim \log_2 \frac{\ell(S_{mj})}{\ell(S_{ki})} + \log_2\left(\frac{\operatorname{dist}(S_{ki},S_{mj})}{\ell(S_{ki})+\ell(S_{mj})}+2\right) \\ &\lesssim \log_2 \frac{\ell(S_{mj})}{\ell(S_{ki})} + \log_2\left(\frac{\operatorname{dist}(S_{ki},S_{mj})}{\ell(S_{mj})}+2\right) \\ &\lesssim \log_2\left(\frac{\operatorname{dist}(S_{ki},S_{mj})+2\ell(S_{mj})}{\ell(S_{ki})}\right). \end{aligned}$$

Thanks to

$$S_{ki}\cap S_{mj} = \emptyset \Rightarrow \operatorname{dist}(S_{ki},S_{mj}) \ge 2^{-2}\ell(S_{mj}) \ge 2^{-2}\ell(S_{ki}),$$

we have

$$t \lesssim 3 + \log_2\left(\frac{\operatorname{dist}(S_{ki},S_{mj})}{\ell(S_{ki})}\right).$$

With the aid of $F(S_{ki},S_{mj})$, we obtain

$$\begin{aligned} &\fint_{S_{ki}}\fint_{S_{mj}} |f(x)-f(y)|^2\,dy\,dx \\ &\quad\lesssim \fint_{S_{ki}} |f(x)-f_{S_{ki}}|^2\,dy + |f_{S_{ki}}-f_{S_{mj}}|^2 + \fint_{S_{mj}} |f(y)-f_{S_{mj}}|^2\,dy \\ &\quad\lesssim \fint_{S_{ki}} |f(x)-f_{S_{ki}}|^2\,dy + \left(\sum_{s=1}^{t-1} |f_{S_{k_si_s}} - f_{S_{k_{s+1}i_{s+1}}}|\right)^2 + \fint_{S_{mj}} |f(y)-f_{S_{mj}}|^2\,dy. \end{aligned}$$

By virtue of the fractional Sobolev inequality (cf. Theorem 1.4 and [17]), we have

$$\fint_{S_{ki}} |f(x)-f_{S_{ki}}|^2\,dy \lesssim \ell(S_{ki})^{2\alpha-n}\int_{S_{ki}}\int_{S_{ki}} \frac{|f(x)-f(y)|^2}{|x-y|^{n+2\alpha}}\,dx\,dy \lesssim \|f\|^2_{Q_\alpha(D)}.$$

Since

$$S_{k_{s+1}i_{s+1}} \subseteq 2^4 S_{k_si_s} \subseteq D \quad \& \quad m(S_{k_{s+1}i_{s+1}}) \approx m(S_{k_si_s}),$$

we also have

$$|f_{S_{k_s i_s}} - f_{S_{k_{s+1} i_{s+1}}}| \lesssim |f_{S_{k_s i_s}} - f_{2^4 S_{k_s i_s}}| + |f_{2^4 S_{k_s i_s}} - f_{S_{k_{s+1} i_{s+1}}}| \lesssim \|f\|_{Q_\alpha(D)},$$

whence

$$\begin{cases} \displaystyle\fint_{S_{ki}} \fint_{S_{mj}} |f(x) - f(y)|^2 \, dy \, dx \lesssim \left(3 + \log_2\left(\frac{\operatorname{dist}(S_{ki}, S_{mj})}{\ell(S_{ki})}\right)\right)^2 \|f\|^2_{Q_\alpha(D)} \\ \displaystyle\int_{S_{ki}} \int_{S_{mj}} \frac{|f(x) - f(y)|^2}{|x - y|^{n+2\alpha}} \, dy \, dx \lesssim \left(3 + \log_2\left(\frac{\operatorname{dist}(S_{ki}, S_{mj})}{\ell(S_{ki})}\right)\right)^2 \frac{\mathrm{m}(S_{mj})\mathrm{m}(S_{ki})}{\operatorname{dist}(S_{ki}, S_{mj})^{n+2\alpha}} \|f\|^2_{Q_\alpha(D)}. \end{cases}$$

Notice that

$$k > -\log_2 \ell(I) + 7 \quad \& \quad S_{k,i} \in \mathscr{S}_k(I) \Rightarrow I \subseteq 2^{k+2}\ell(I) S_{ki}.$$

So, for

$$m \geq k \quad \text{and} \quad j \in \mathscr{S}_m(I) \quad \text{with } S_{mj} \cap S_{ki} \neq \emptyset$$

there exists exactly one

$$\delta \in \{1, \ldots, k + 2 + \log_2 \ell(I)\}$$

such that

$$\begin{cases} S_{mj} \cap (2^\delta S_{ki} \setminus 2^{\delta-1} S_{ki}) \neq \emptyset \\ \operatorname{dist}(S_{ki}, S_{mj}) \lesssim 2^{\delta-k}. \end{cases}$$

Thus

$$\begin{aligned} &\left(\mathrm{m}(S_{ki}) \|f\|^2_{Q_\alpha(D)}\right)^{-1} \sum_{m \geq k} \sum_{\substack{S_{mj} \in \mathscr{S}_m(I) \\ S_{mj} \cap S_{ki} = \emptyset}} \int_{S_{ki}} \int_{S_{mj}} \frac{|f(x) - f(y)|^2}{|x - y|^{n+2\alpha}} \, dy \, dx \\ &\lesssim \sum_{\delta=1}^{k+2+\log_2 \ell(I)} \sum_{m \geq k} \sum_{\substack{S_{mj} \in \mathscr{S}_m(I) \\ S_{mj} \cap (2^\delta S_{ki} \setminus 2^{\delta-1} S_{ki}) \neq \emptyset}} \left(3 + \log_2 \frac{\operatorname{dist}(S_{ki}, S_{mj})}{\ell(S_{mj})}\right)^2 \frac{\mathrm{m}(S_{mj})}{\operatorname{dist}(S_{ki}, S_{mj})^{2\alpha+n}} \\ &\lesssim \sum_{\delta=1}^{k+2+\log_2 \ell(I)} \sum_{m \geq k} \sum_{\substack{S_{mj} \in \mathscr{S}_m(I) \\ S_{mj} \cap (2^\delta S_{ki} \setminus 2^{\delta-1} S_{ki}) \neq \emptyset}} (3 + \delta + m - k)^2 2^{(-\delta+k)(2\alpha+n)} 2^{-mn} \\ &\lesssim \sum_{m \geq k} \sum_{\delta=1}^{k+2+\log_2 \ell(I)} \frac{(1 + m - k + \delta)^2 \sharp\{S_{mj} \in \mathscr{S}_m(I) : S_{mj} \cap (2^\delta S_{ki} \setminus 2^{\delta-1} S_{ki}) \neq \emptyset\}}{2^{(2\alpha+n)(m-k+\delta)-2m\alpha}}. \end{aligned}$$

Applying Lemma 8.4(ii), we obtain that if

$$m \geq k \quad \& \quad \delta \in \{1, \ldots, k + 2 + \log_2 \ell(I)\}$$

then

$$\sharp\{S_{mj} \in \mathscr{S}_m(I) : S_{mj} \cap (2^\delta S_{ki} \setminus 2^{\delta-1} S_{ki}) \neq \emptyset\} = \sharp\mathscr{S}_m(2^\delta S_{ki}) \lesssim 2^{(m-k+\delta)(\overline{\dim}_{\mathrm{lsM}}\partial D+\epsilon)}.$$

So by

$$\overline{\dim}_{\mathrm{lsM}}\partial D + \epsilon < n - 2\alpha,$$

we gain

$$\begin{aligned}
&\sum_{m\geq k} \sum_{\substack{S_{mj}\in\mathscr{S}_m(I)\\ S_{mj}\cap 2S_{ki}=\emptyset}} \int_{S_{ki}} \int_{S_{mj}} \frac{|f(x) - f(y)|^2}{|x - y|^{n+2\alpha}} \, dy \, dx \\
&\quad \lesssim \mathrm{m}(S_{ki}) \|f\|^2_{\mathrm{Q}_\alpha(D)} \sum_{m\geq k} \sum_{\delta=1}^{k+2+\log_2 \ell(I)} \frac{2^{(m-k+\delta)(\overline{\dim}_{\mathrm{lsM}}\partial D+\epsilon)}}{2^{(2\alpha+n)(m-k+\delta)-2\alpha m}} ((1 + m - k)^2 + \delta^2),
\end{aligned}$$

whence

$$\begin{aligned}
&\sum_{m\geq k} \sum_{\substack{S_{mj}\in\mathscr{S}_m(I)\\ S_{mj}\cap 2S_{ki}=\emptyset}} \int_{S_{ki}} \int_{S_{mj}} \frac{|f(x) - f(y)|^2}{|x - y|^{n+2\alpha}} \, dy \, dx \\
&\quad \lesssim \mathrm{m}(S_{ki}) \|f\|^2_{\mathrm{Q}_\alpha(D)} \sum_{m\geq k} 2^{(m-k)(\overline{\dim}_{\mathrm{lsM}}\partial D+\epsilon)} (1 + m - k)^2 2^{(-2\alpha-n)(m-k)} 2^{2\alpha m} \\
&\quad \lesssim \mathrm{m}(S_{ki}) \|f\|^2_{\mathrm{Q}_\alpha(D)} 2^{2\alpha k}.
\end{aligned}$$

Accordingly, an application of

$$\overline{\dim}_{\mathrm{lsM}}\partial D + \epsilon < n - 2\alpha \quad \& \quad \ell(I) \leq 1$$

derives the desired estimation

$$\begin{aligned}
\mathrm{I}_2 &\lesssim \|f\|^2_{\mathrm{Q}_\alpha(D)} \sum_{k\geq -\log_2 \ell(I)+7} \sum_{i\in\mathscr{S}_k(I)} |S_{ki}| 2^{2\alpha k} \\
&\lesssim \|f\|^2_{\mathrm{Q}_\alpha(D)} \sum_{k\geq -\log_2 \ell(I)+7} (2^k \ell(I))^{\overline{\dim}_{\mathrm{lsM}}\partial D+\epsilon} 2^{-(n-2\alpha)k} \\
&\lesssim 2^{(n-2\alpha-\overline{\dim}_{\mathrm{lsM}}\partial D-\epsilon)\log_2 \ell(I)} \ell(I)^{\overline{\dim}_{\mathrm{lsM}}\partial D+\epsilon} \|f\|^2_{\mathrm{Q}_\alpha(D)} \\
&\approx \ell(I)^{n-2\alpha} \|f\|^2_{\mathrm{Q}_\alpha(D)}.
\end{aligned}$$

The second case: $\operatorname{diam} \partial D < \infty$ & $\operatorname{diam} D = \infty$.

With no loss of generality, we may assume $\partial D \subseteq B(o,1) = B(0,1)$.

We are required to control $\Psi_\alpha(f, I \cap D)$ for any $I \in \mathscr{C}(\mathbb{R}^n)$ and so for such an $I \in \mathscr{C}(\mathbb{R}^n)$ that its center is in D and it enjoys $I \cap D^c \neq \emptyset$.

▷ If $\ell(I) \leq 2^3$, by an argument similar as above, we have the desired estimation for $\Psi_\alpha(f, I \cap D)$.

▷ If $\ell(I) > 2^3$, then it may be assumed that

$$2^{2^{k_0}-1} \leq \ell(I) \leq 2^{2^{k_0}} \quad \text{for some } k_0 \geq 4.$$

This assumption ensures

$$I \subseteq I(0, 2^{k_0+1}),$$

and so we may further assume

$$I = I(0, 2^{k_0+1}) \quad \text{for some } k_0 \geq 4.$$

Set

$$\Psi_\alpha(f, I \cap D) = \int\limits_{I\cap D}\int\limits_{I\cap D} \frac{|f(x)-f(y)|^2}{|x-y|^{n+2\alpha}}\,dy\,dx = \Psi_\alpha(f, I(0,8)\cap D) + \mathsf{I}_3 + \mathsf{I}_4 + \mathsf{I}_5,$$

where

$$\begin{cases} \mathsf{I}_3 = \displaystyle\int\limits_{I(0,4)\cap D}\int\limits_{I\setminus I(0,8)} \frac{|f(x)-f(y)|^2}{|x-y|^{n+2\alpha}}\,dy\,dx \\ \mathsf{I}_4 = \displaystyle\int\limits_{I\setminus I(0,4)}\int\limits_{I(0,2)\cap D} \frac{|f(x)-f(y)|^2}{|x-y|^{n+2\alpha}}\,dy\,dx \\ \mathsf{I}_5 = \displaystyle\int\limits_{I\setminus I(0,4)}\int\limits_{I\setminus I(0,2)} \frac{|f(x)-f(y)|^2}{|x-y|^{n+2\alpha}}\,dy\,dx. \end{cases}$$

Due to

$$\Psi_\alpha(f, I(0,8)\cap D) \lesssim \|f\|^2_{\mathcal{Q}_\alpha(D)} \lesssim \ell(I)^{n-2\alpha}\|f\|^2_{\mathcal{Q}_\alpha(D)},$$

it remains to deal with $\mathsf{I}_3 - \mathsf{I}_4 - \mathsf{I}_5$.

– Regarding I_3, we find a Whitney cube

$$J \subseteq I(0,7) \setminus I(0,5).$$

An application of

$$2^{-9}(\sqrt{n})^{-1}\ell(J) < 1$$

deduces

$$
\begin{aligned}
\mathrm{I}_3 &\le \int_{I(0,4)\cap D}\int_{I\setminus I(0,8)} \frac{|f(x)-f_J|^2}{|x-y|^{n+2\alpha}}\,dy\,dx + \int_{I(0,4)\cap D}\int_{I\setminus I(0,8)} \frac{|f_J-f(y)|^2}{|x-y|^{n+2\alpha}}\,dy\,dx \\
&\lesssim \ell(I)^{-2\alpha}\int_{I(0,4)\cap D} |f(x)-f_J|^2\,dx + \int_{I\setminus I(0,8)} \frac{|f_J-f(y)|^2}{|y|^{n+2\alpha}}\,dy \\
&= \mathrm{I}_{3,1} + \mathrm{I}_{3,2}.
\end{aligned}
$$

Obviously, we have

$$
\mathrm{I}_{3,1} \lesssim \ell(I)^{-2\alpha}\int_{I(0,4)\cap D}\int_J |f(x)-f(y)|^2\,dy\,dx \lesssim \ell(I)^{n-2\alpha}\|f\|^2_{\mathcal{Q}_\alpha(D)}.
$$

Also, if

$$
2^{k_0-1} \le \ell(I) \le 2^{k_0}
$$

is valid for some $k_0 \ge 4$, then

$$
I \subseteq I(0, 2^{k_0+1}),
$$

and hence

$$
\begin{aligned}
\mathrm{I}_{3,2} &\le \sum_{k=3}^{k_0+2}\int_{I(0,2^{k+1})\setminus I(0,2^k)} \frac{|f_J-f(y)|^2}{|y|^{n+2\alpha}}\,dy \\
&\lesssim \sum_{k=3}^{k_0+2} 2^{-k(n+2\alpha)}\int_{I(0,2^{k+1})\setminus I(0,2^k)} |f_J-f(y)|^2\,dy.
\end{aligned}
$$

Denote by $\widetilde{\mathscr{F}_k}$ the collection of S_i satisfying

$$
S_i \cap \left(I(0,2^{k+1}) \setminus I(0,2^k)\right) \ne \emptyset.
$$

Then

$$
\begin{cases}
2^{-k+7} \le \ell(S_i) \le 2^{-k+2} \quad \forall\, S_i \in \widetilde{\mathscr{F}_k} \\
\sharp\widetilde{\mathscr{F}_k} \lesssim 2^{k(n-1)} \\
\mathrm{I}_{3,2} \le \displaystyle\sum_{k=3}^{k_0+2} 2^{-2k\alpha}\sum_{S_i\in\widetilde{\mathscr{F}_k}} ⨍_{S_i} |f_J - f(y)|^2\,dy.
\end{cases}
$$

For each $S_i \in \widetilde{\mathscr{F}_k}$, by Lemma 8.4(i) there exists a Whitney chain joining J and S_i with its length $\lesssim k$. So by the fractional Poincaré inequality, we have

$$
⨍_{S_i} |f_J - f(y)|^2\,dy \lesssim k\|f\|^2_{\mathcal{Q}_\alpha(D)},
$$

whence

$$\mathsf{I}_{3,2} \lesssim \sum_{k=3}^{k_0+2} 2^{-2k\alpha} k \|f\|^2_{\mathcal{Q}_\alpha(D)} \lesssim \|f\|^2_{\mathcal{Q}_\alpha(D)} \lesssim \ell(I)^{n-2\alpha} \|f\|^2_{\mathcal{Q}_\alpha(D)}.$$

- For I_4, we can obtain the desired estimate via using the argument for I_3.
- For I_5, we consider any cube

$$J \subseteq I(0, 2^{k_0+4})$$

with

$$\begin{cases} \ell(J) = 2^{-k_0+2} \\ J \cap I(0, 2^{-k_0+3}) = \emptyset, \end{cases}$$

thereby finding

$$\mathsf{I}_5 \lesssim \int\limits_{I \setminus I(0,2)} \int\limits_{I \setminus I(0,2)} \frac{|f(x) - f_J|^2}{|x-y|^{n+2\alpha}} \, dy \, dx \lesssim \ell(I)^{-2\alpha} \int\limits_{I \setminus I(0,2)} |f(x) - f_J|^2 \, dx.$$

Upon covering

$$I \setminus I(0,2) \quad \text{by } 2^n \text{ cubes} \quad \{R_j\}_{j=1}^{2^n} \text{ with } \begin{cases} \ell(R_j) = 2^{k_0+2} \\ R_j \cap I(0,2) = \emptyset, \end{cases}$$

we get

$$\mathsf{I}_5 \lesssim \ell(I)^{n-2\alpha} \sum_{j=1}^{2^n} \fint_{R_j} |f(x) - f_J|^2 \, dx.$$

Evidently, for each R_j we can get finitely many cubes $\{R_{js}\}_{s=1}^{N}$ such that

$$\begin{cases} \ell(R_{js}) = 2^{k_0+2} \\ R_{j1} = J \\ R_{jN} = R_j \\ R_{js} \cap R_{j(s+1)} \neq \emptyset & \text{as} \quad 1 < s < N-1 \\ 2R_{j(s+1)} \cap I(0,2) = \emptyset & \text{as} \quad 1 < s < N-1. \end{cases}$$

So, under this situation we have

$$\fint_{R_j} |f(x) - f_J|^2 \, dx \lesssim N^2 \sum_{s=1}^{N} \fint_{2R_{js}} |f(x) - f_{2R_{js}}|^2 \, dx \lesssim \|f\|^2_{\mathcal{Q}_\alpha(D)},$$

whence

$$\mathsf{I}_5 \lesssim \ell(I)^{n-2\alpha} \|f\|^2_{\mathcal{Q}_\alpha(D)}.$$

$$\boxed{\text{The third case:}\quad \operatorname{diam}\partial D = \infty.}$$

Since there is no restriction on the edge-length of I or δ in Lemma 8.4(ii), upon employing the argument for

$$\boxed{\text{The first case:}\quad \operatorname{diam}\partial D < \infty \quad \& \quad \operatorname{diam} D < \infty}$$

we can establish the desired estimation. □

8.4 Reverse restriction principle

In order to establish an opposite assertion of Theorem 8.5, we are suggested to consider the concept of a linearly locally connected domain.

Definition 8.6. A domain $D \subseteq \mathbb{R}^n$ is said to be linearly locally connected (l. l. c.) provided that there is a constant $b \in (0, 1]$ such that for any point $(z, r) \in \mathbb{R}^n \times (0, \infty)$ one has:

- ▷ l. l. c.-a – points in $D \setminus B(z, r)$ can be connected by a rectifiable curve that lies in $D \setminus B(z, br)$;
- ▷ l. l. c.-b – points in $D \cap B(z, r)$ can be connected by a rectifiable curve that lies in $D \cap B(z, b^{-1}r)$.

According to Definition 8.6 and [28], we have that if $D \subseteq \mathbb{R}^n$ is either simply connected under $n = 2$ or quasiconformally equivalent to a uniform domain under $n \geq 3$ then D is l. l. c. if and only if D is a uniform domain. So, in order to get a partial converse to Theorem 8.5, we need two lemmas below.

Lemma 8.7. *Given $(\alpha, n-1) \in [0, 1) \times \mathbb{N}$ and a pair of disjoint sets $E, F \subseteq D$ let*

$$C_{Q_\alpha}(E, F, D) = \inf\{\|f\|^2_{Q_\alpha(\mathbb{R}^n)} : f \in \mho(E, F, D)\},$$

where $\mho(E, F, D)$ is the class of all continuous functions $f \in Q_\alpha(\mathbb{R}^n)$ such that

$$\begin{cases} 0 \leq f \leq 1 & \text{on } D \\ f = 0 & \text{on } E \\ f = 1 & \text{on } F. \end{cases}$$

(i) *If*

$$\begin{cases} \tilde{E} \cap \tilde{F} = \emptyset \\ E \subseteq \tilde{E} \\ F \subseteq \tilde{F}, \end{cases}$$

then

$$C_{Q_\alpha}(E, F, D) \leq C_{Q_\alpha}(\tilde{E}, \tilde{F}, \mathbb{R}^n).$$

(ii) *If $D \subseteq \mathbb{R}^n$ is a Q_α-extension domain, then:*
(ii)-a *For any pair of disjoint subsets $\{E, F\}$ of D one has*

$$\mathrm{C}_{\mathrm{Q}_\alpha}(E, F, D) \le \mathrm{C}_{\mathrm{Q}_\alpha}(E, F, \mathbb{R}^n) \lesssim \mathrm{C}_{\mathrm{Q}_\alpha}(E, F, D);$$

(ii)-b *For any $\delta > 0$, there is a positive constant c depending only on n and δ such that for every pair of disjoint sets*

$$E, F \subseteq D \cap B(x_0, r) \quad \textit{for some } r > 0$$

one has

$$\min\{\mathrm{m}(E), \mathrm{m}(F)\} \ge \delta r^n \Rightarrow \mathrm{C}_{\mathrm{Q}_\alpha}(E, F, D) \ge c.$$

Proof. (i) This follows from the definition of $\mathrm{C}_{\mathrm{Q}_\alpha}(\cdot, \cdot, \cdot)$.

(ii) Clearly, (ii)-a follows from the definition of a Q_α-extension domain. In order to validate (ii)-b, we may assume that $D = \mathbb{R}^n$. Let

$$f \in \mho(E, F, \mathbb{R}^n).$$

▷ If $f_{B(x_0,r)} \ge 2^{-1}$, then an application of the following

$$\frac{\mathrm{m}(E)}{2\mathrm{m}(I(x_0, r))} \le \fint_{I(x_0,r)} |f(x) - f_{I(x_0,r)}|\, \mathrm{d}x$$
$$\lesssim \left(\int_{I(x_0,r)} \int_{I(x_0,r)} \frac{|f(x) - f(y)|^2}{r^{n-2\alpha}|x - y|^{n-2\alpha}}\, \mathrm{d}x\mathrm{d}y \right)^{\frac{1}{2}},$$

yields the desired implication.

▷ If $f_{B(x_0,r)} < 2^{-1}$, then the desired implication follows from replacing E by F in the above chain of inequalities. □

Lemma 8.8. *For $\alpha \in [0, 1)$ let K be a closed subset of $\mathbb{R}^n$ and D be a Q_α-extension domain in $\mathbb{R}^n$. Then for every component D_0 of $D \setminus K$ and for all*

$$x \in D_0 \quad \& \quad 0 < r \le \min\{\operatorname{dist}(x, \partial D_0 \setminus \partial D), \operatorname{diam} D\}$$

one has

$$\mathrm{m}(D_0 \cap I(x, r)) \gtrsim r^n.$$

Proof. Without loss of generality, we may assume $r < \infty$. Let

$$b_0 = 1 \quad \& \quad b_j \in (0, 1) \quad \forall j \in \mathbb{N}$$

obey

$$\mathsf{m}\big(I(x,b_jr)\cap D_0\big)=2^{-1}\mathsf{m}\big(I(x,b_{j-1}r)\cap D_0\big)=2^{-j}\mathsf{m}\big(I(x,r)\cap D_0\big).$$

Set

$$f(y)=\begin{cases}1 & \forall y\in I(x,b_2r)\cap D_0\\ \frac{b_1r-|y-x|}{(b_1-b_2)r} & \forall y\in (I(x,b_1r)\setminus I(x,b_2r))\cap D_0\\ 0 & \text{otherwise.}\end{cases}$$

Since

$$I(x,r)\cap D_0\cap K=\emptyset$$

and $I(x,r)$ has empty intersection with other components of $D\setminus K$, the above chosen function f is meaningful. Accordingly,

$$\|f\|_{\dot{W}^{1,n}(D)}\lesssim\frac{\mathsf{m}((I(x,b_1r)\setminus I(x,b_2r))\cap D_0)^{\frac{1}{n}}}{(b_1-b_2)r}\lesssim\frac{\mathsf{m}(I(x,r)\cap D_0)^{\frac{1}{n}}}{(b_1-b_2)r}.$$

Since D is a Q_α-extension domain, we have

$$\|f\|_{\mathcal{Q}_\alpha(D)}\lesssim\|f\|_{\mathsf{Q}_\alpha(\mathbb{R}^n)}\lesssim\|f\|_{\dot{W}^{1,n}(D)}\lesssim\frac{\mathsf{m}(I(x,r)\cap D_0)^{\frac{1}{n}}}{(b_1-b_2)r},$$

thereby getting a constant $\kappa>0$ such that

$$\inf_{c\in\mathbb{R}}\int\limits_{I(x,r)\cap D_0}\exp\left(\frac{\kappa|f(y)-c|(b_1-b_2)r}{\mathsf{m}(I(x,b_1r)\cap D_0)^{\frac{1}{n}}}\right)dy\lesssim r^n.$$

Observe that $c\in\mathbb{R}$ ensures

$$|f(y)-c|\geq 2^{-1}\quad\forall y\in\big((I(x,r)\setminus I(x,b_1r))\cap D_0\big)\cup\big(I(x,b_2r)\cap D_0\big).$$

So

$$\begin{cases}\mathsf{m}(I(x,b_1r)\cap D_0)\exp\left(\frac{\kappa(b_1-b_2)r}{\mathsf{m}(I(x,b_1r)\cap D_0)^{\frac{1}{n}}}\right)\lesssim r^n\\ (b_1-b_2)r\lesssim\mathsf{m}(I(x,b_1r)\cap D_0)^{\frac{1}{n}}\log_2\left(\frac{\kappa r^n}{\mathsf{m}(I(x,b_1r)\cap D_0)}\right).\end{cases}$$

The similar inequalities also hold for

$$(b_j-b_{j+1})r\quad\text{with } j\geq 2.$$

Accordingly,

$$b_1r=\sum_{j\in\mathbb{N}}(b_j-b_{j+1})r$$

$$\lesssim \sum_{j\in\mathbb{N}} \mathrm{m}(I(x,b_jr)\cap D_0)^{\frac{1}{n}} \log_2\Big(\frac{\kappa(b_{j-1}r)^n}{\mathrm{m}(B(z,b_jr)\cap D_0)}\Big)$$

$$\lesssim \sum_{j\in\mathbb{R}} 2^{-\frac{j}{n}} \mathrm{m}(I(x,r)\cap D_0)^{\frac{1}{n}} \log_2\Big(\frac{\kappa 2^j r^n}{\mathrm{m}(I(x,r)\cap D_0)}\Big)$$

$$\lesssim \mathrm{m}(I(x,r)\cap D_0)^{\frac{1}{n}} \log_2\Big(\frac{\kappa r^n}{\mathrm{m}(I(x,r)\cap D_0)}\Big).$$

If $b_1 \geq (10)^{-1}$, then

$$t\log_2 t^{-1} \geq 1 \Rightarrow t \gtrsim 1 \Rightarrow \mathrm{m}(I(x,r)\cap D_0)^{\frac{1}{n}} \gtrsim r.$$

But, if $b_1 \leq (10)^{-1}$, then choosing $\rho = 2r5^{-1}$ and a point

$$y \in I(x,r)\cap D_0 \quad \text{with} \quad |y-x| = b_1r + 2^{-1}\rho$$

we obtain

$$I(x,b_1r) \subseteq B(y,\rho) \subseteq I(x,r) \quad \text{but} \quad B(y,2^{-1}\rho)\cap I(x,b_1r) = \emptyset.$$

Thus, if

$$\mathrm{m}(I(y,\tilde{b}_1\rho)\cap D_0) = 2^{-1}\mathrm{m}(I(y,\rho)\cap D_0),$$

then

$$\mathrm{m}(I(x,b_1r)\cap D_0) \geq 2^{-1}\mathrm{m}(I(y,\rho)\cap D_0)$$

derives $\tilde{b}_1 \geq 2^{-1}$. Applying the result for $b_1 \geq (10)^{-1}$, we conclude

$$\mathrm{m}(I(y,\rho)\cap D_0) \gtrsim \rho^n,$$

thereby reaching

$$\mathrm{m}(I(x,r)\cap D_0) \gtrsim r^n. \qquad \square$$

Theorem 8.9. *For $(\alpha, n-1) \in [0,1)\times\mathbb{N}$ let D be either simply connected domain in $\mathbb{R}^2$ or quasiconformally equivalent to a uniform domain in $\mathbb{R}^{n\geq 3}$. If*

$$\mathcal{Q}_\alpha(\mathbb{R}^n)|_D = \mathcal{Q}_\alpha(D),$$

then D is uniform.

Proof. The conclusion follows from using the description of a uniform domain given immediately after Definition 8.6 and validating that D is l. l. c. in the sequel.

▷ 1. l. c.-a – Let $x_1, x_2 \in D \setminus B(x_0, r)$ for some $x_0 \in D$ and $r > 0$. Without loss of generality, we may assume

$$x_1, x_2 \in \partial B(x_0, r) \cap D.$$

Indeed, we can find a rectifiable curve

$$[0,1] \ni t \mapsto \zeta(t) \subseteq D$$

that joins x_1 and x_2. Let

$$\tilde{x}_1 = \zeta(t_1) \quad \& \quad \tilde{x}_2 = \zeta(t_2)$$

be the points in $\partial B(x_0, r)$ such that

$$\zeta_1 = \zeta([0, t_1]) \quad \& \quad \zeta_2 = \zeta([t_2, 1])$$

intersect with $\partial B(x_0, r)$ at $\tilde{x}_1$ and $\tilde{x}_2$, respectively. If we can find a rectifiable curve

$$\zeta_3 \subseteq D \setminus B(x_0, cr)$$

for some $c \in (0,1)$ linking $\tilde{x}_1$ and $\tilde{x}_2$, then the rectifiable curve

$$\zeta_1 \cup \zeta_3 \cup \zeta_2 \subseteq D \setminus B(x_0, cr)$$

joins x_1 and x_2.
Suppose that x_1 and x_2 are not in the same component of $D \setminus B(x_0, br)$ but in the same component of

$$D \setminus B(x_0, b2^{-1}r) \quad \text{for some} \quad b \in (0,1).$$

If $b \geq 2^4$, then we can find a curve ζ such that

$$\zeta \subseteq D \setminus B(x_0, 2^{-5}r)$$

joining x_1 and x_2. So we may assume $b < 2^{-4}$. It then suffices to prove that b is bounded away from 0 uniformly for the set $\{x_1, x_2, x_0, r\}$. For $i = 1, 2$ denote by D_i the component of

$$D \setminus \overline{B(x_0, br)} \ni x_i.$$

Then $D_1 \cap D_2 = \emptyset$.
Let

$$F_i = D_i \cap B(x_i, 2^{-1}r) \quad \forall i = 1, 2.$$

Obviously, we have

$$\begin{cases} F_1, F_2 \subseteq B(x_0, 2r) \\ F_1 \cap F_2 = \emptyset \\ \operatorname{dist}(x_i, \partial D_i \setminus \partial D) > 2^{-1}r \\ F_i \cap B(x_0, 2^{-1}r) = \emptyset. \end{cases}$$

By Lemma 8.8, we get

$$\mathrm{m}(F_i) \gtrsim r^n \quad \forall i = 1, 2.$$

Applying Lemma 8.7, we obtain

$$1 \lesssim C_{Q_\alpha}(F_1, F_2, D) \lesssim \big(\log_2(2b)^{-1}\big)^{\frac{1-n}{n}} \quad \& \quad b \gtrsim 1.$$

Moreover, if

$$f(x) = \begin{cases} 1 & \forall\, x \in D_2 \setminus B(x_0, 2^{-1}r) \\ (\log_2(2b)^{-1})^{-1} \log_2 \frac{|x-x_0|}{b_0 r} & \forall\, x \in D_2 \cap (B(x_0, 2^{-1}r) \setminus B(x_0, b_0 r)) \\ 0 & \text{otherwise,} \end{cases}$$

then this function is well-defined and satisfies

$$|\nabla f(x)| = \big(\log_2(2b)^{-1}\big)^{-1} |x - x_0|^{-1} \mathbb{1}_{D_2 \cap (B(x_0, 2^{-1}r) \setminus B(x_0, b_0 r))},$$

and hence

$$\begin{aligned} \|f\|_{Q_\alpha(\mathbb{R}^n)} &\lesssim \|f\|_{\dot{W}^{1,n}(D)} \\ &\lesssim \big(\log_2(2b)^{-1}\big)^{-1} \Big(\int\limits_{B(x_0, 2^{-1}r) \setminus B(x_0, b_0 r)} |z - x_0|^{-n}\, dz \Big)^{\frac{1}{n}} \\ &\lesssim \big(\log_2(2b)^{-1}\big)^{\frac{1-n}{n}}. \end{aligned}$$

Accordingly, $f \in \mho(F_1, F_2, D)$.

▷ l. l. c.-b. Let x_1, x_2 be an pair of points in D and

$$x_1, x_2 \in B(z, r) \quad \text{for some} \quad (z, r) \in \mathbb{R}^n \times (0, \infty).$$

Then

$$2r > |x_1 - x_2|.$$

We are about to show the existence of a rectifiable curve

$$\zeta \subseteq B(z, rb^{-1})$$

that connects x_1 and x_2 for some constant $b < 1$.
If

$$|x_1 - x_2| \le \max\{\operatorname{dist}(x_1, D^c), \operatorname{dist}(x_2, D^c)\},$$

then the line segment joining x_1 and x_2 gives the desired curve.
However, if

$$|x_1 - x_2| > \max\{\operatorname{dist}(x_1, D^c), \operatorname{dist}(x_2, D^c)\},$$

then upon letting D_i be the connected component of

$$D \cap B(x_i, 2^{-3}|x_1 - x_2|) = D \setminus B(x_i, 2^{-3}|x_1 - x_2|)^c$$

containing x_i for $i = 1, 2$, and using Lemma 8.8, we have

$$\mathsf{m}(D_i) \gtrsim |x_1 - x_2|^n \quad \forall i = 1, 2.$$

Obviously,

$$D_1 \cap D_2 = \emptyset \quad \& \quad D_1, D_2 \subseteq B(x_1, 2|x_1 - x_2|).$$

Upon employing Lemma 8.7, we obtain

$$\mathsf{C}_{Q_\alpha}(D_1, D_2, D) \gtrsim 1.$$

Next, we claim that there exists a positive constant $N_0 \ge 2$ independent of $\{x_1, x_2, D_1, D_2\}$ such that D_1 and D_2 are in the same component of

$$D \cap B(x_1, N_0|x_1 - x_2|).$$

To see this, suppose that D_1, D_2 are not in the same component of

$$D \cap B(x_1, N|x_1 - x_2|)$$

for some $N > 2$, say, in two different components $\widetilde{D}_1$ and $\widetilde{D}_2$, respectively. Define

$$f(y) = \begin{cases} 1 & \forall y \in B(x_1, 2^{-3}|x_1 - x_2|) \cap \widetilde{D}_1 \\ \frac{\log_2(N|x_1-x_2|)-\log_2(|y-x_1|)}{\log_2(2^3 N)} & \forall y \in (B(x_1, N|x_1 - x_2|) \setminus B(x_1, 2^{-3}|x_1 - x_2|)) \cap \widetilde{D}_1 \\ 0 & \text{otherwise.} \end{cases}$$

Then

$$\|f\|_{Q_\alpha(\mathbb{R}^n)} \lesssim \|f\|_{\dot{W}^{1,n}(D)} \lesssim (\log_2 N)^{\frac{1-n}{n}}.$$

Upon observing

$$f = \begin{cases} 1 & \text{on } D_1 \\ 0 & \text{on } D_2, \end{cases}$$

we have

$$C_{Q_\alpha}(E, F, D) \lesssim (\log_2 N)^{\frac{1-n}{n}}.$$

This inequality ensures $N \lesssim 1$ and shows the existence of N_0 as claimed. As a consequence, there exists a rectifiable curve

$$\zeta_0 \subseteq D \cap B(x_1, N_0|x_1 - x_2|)$$

joining D_1 and D_2. Let

$$\tilde{x}_1, \tilde{x}_2 \in D_i \cap \zeta_0.$$

Since

$$x_i, \tilde{x}_i \in D_i \quad \forall i = 1, 2$$

and D_i is connected, we can find a rectifiable curve

$$\zeta_i \subseteq D_i \subseteq D \cap B(x_1, N_0|x_1 - x_2|)$$

connecting x_i and $\tilde{x}_i$ for $i = 1, 2$. Now

$$D \cap B(x_1, N_1|x_1 - x_2|) \supseteq \zeta = \zeta_0 \cup \zeta_1 \cup \zeta_2.$$

links x_1 and x_2. So, upon noticing

$$\begin{aligned} B(x_1, N_0|x_1 - x_2|) &\subseteq B(z, N_0|x_1 - x_2| + |z - x_1|) \\ &\subseteq B(z, (N_0 + 1)|x_1 - x_2|) \\ &\subseteq B(z, 2(N_0 + 1)r), \end{aligned}$$

we obtain

$$\zeta \subseteq D \cap B(z, rb^{-1}) \quad \& \quad b = 2^{-1}(N_0 + 1),$$

as required. □

8.5 Notes

Below are some historical remarks on Chapter 8.

Section 8.1. Theorem 8.1 (partially modified by investigating the weighted Hardy operators in [64, 90, 83]) appears first in this chapter, and its the argument, along with *Step 1* of the argument for Theorem 4.5, actually shows that $\mathcal{M}$ is bounded on

$$\mathrm{C}^{2,2\lambda}(\mathbb{R}^n) \supseteq \begin{cases} \mathrm{C}^{-\lambda}(\mathbb{R}^n) & \text{as } 2\lambda \in [-1,0) \\ \mathrm{BMO}(\mathbb{R}^n) & \text{as } 2\lambda = 0 \\ \mathrm{M}^{2,2\lambda}(\mathbb{R}^n) & \text{as } 2\lambda \in (0,1], \end{cases}$$

whose "$\supseteq$" actually becomes "$=$" under $2\lambda \in [-1,0]$ (cf. Lemma 7.1). In particular, we have the following results.

▷
$$\|\mathcal{M}f\|_{\mathrm{BMO}}(\mathbb{R}^n) \lesssim \|f\|_{\mathrm{BMO}}(\mathbb{R}^n).$$

This boundedness was established initially in [5] and recently proved again in [73] via the following inequality:

$$\begin{aligned}&\int_I |\mathcal{M}f(x) - (\mathcal{M}f)_I|\,\mathrm{d}x \\ &\lesssim \int_I |\mathcal{M}((f - f_{3I})\mathbb{1}_{3I})|\,\mathrm{d}x + \int_{I\setminus I\cap\{\mathcal{M}f=\mathcal{M}(f\mathbb{1}_{3I})\}} \Big(\sup_{x\in J\ \&\ \mathrm{m}(J)>\mathrm{m}(I)} |f_J - f_{2J}|\Big)\,\mathrm{d}x\end{aligned}$$

that holds for any

$$(f, I) \in \mathrm{BMO}(\mathbb{R}^n) \times \mathscr{C}(\mathbb{R}^n).$$

▷
$$\|\mathcal{M}f\|_{\mathrm{Lip}_{0<-\lambda\le1}(\mathbb{R}^n)} \lesssim \|f\|_{\mathrm{Lip}_{0<-\lambda\le1}(\mathbb{R}^n)}.$$

This estimation was found in [10].

▷
$$\|\mathcal{M}f\|_{\mathrm{CIS}(\mathbb{R}^n)} \lesssim \|f\|_{\mathrm{CIS}(\mathbb{R}^n)}.$$

This inequality can be obtained by the boundedness of $\mathcal{M}$ on $\mathrm{M}^{2,2}(\mathbb{R}^n)$ (cf. [13]) and the following estimation for $I \in \mathscr{C}(\mathbb{R}^n)$

$$\ell(I)^{2-n} \int_I |\nabla\mathcal{M}f(x)|^2\,\mathrm{d}x \lesssim \ell(I)^{2-n} \int_I |\mathcal{M}|\nabla f|(x)|^2\,\mathrm{d}x \lesssim \|\nabla f\|^2_{\mathrm{M}^{2,2}(\mathbb{R}^n)}.$$

The case $n = 2$ of this last boundedness is a special situation of the boundedness (cf. [44]) of $\mathcal{M}$ on the Sobolev space $\dot{\mathrm{W}}^{1,p}(\mathbb{R}^n)$ of all weakly differentiable functions f with

$$\int_{\mathbb{R}^n} |\nabla f(x)|^p\,\mathrm{d}x < \infty.$$

Section 8.2. Theorem 8.3 comes from a suitable modification of [89, Theorem 1.1], and has been appropriately pushed in [51, Theorem 1.5] to $\mathtt{CIS}(\mathbb{R}^n)$ and its local space $\mathtt{CIS}(D)$ of all weakly-differentiable functions f on a given domain $D \subseteq \mathbb{R}^n$ with

$$\sup_{I\in\mathscr{C}(\mathbb{R}^n)} \ell(I)^{2-n} \int_{I\cap D} |\nabla f(x)|^2 \, dx < \infty.$$

More precisely, every uniform domain $D \subseteq \mathbb{R}^n$ is $\mathtt{CIS}$-extension domain, namely, there is a bounded linear operator

$$\mathcal{E} : \mathtt{CIS}(D) \to \mathtt{CIS}(\mathbb{R}^n)$$

such that

$$f = \mathcal{E}f|_D \quad \forall f \in \mathtt{CIS}(D)$$

conversely, every $\mathtt{CIS}$-extension domain is linearly locally connected—especially—D (being quasiconformally equivalent to a uniform domain) is a $\mathtt{CIS}$-extension domain if and only if it is uniform.

Section 8.3. Theorem 8.5 is a variant of the first part of [89, Theorem 1.3(i)]. The second part of [89, Theorem 1.3(i)] essentially says that for $\alpha \in [0, 1)$ there is a uniform domain D_α such that

$$n = \begin{cases} 2\alpha + \overline{\dim}_{\mathrm{lsM}} \partial D_\alpha & \text{as } \partial D_\alpha \text{ is bounded} \\ 2\alpha + \overline{\dim}_{\mathrm{gsM}} \partial D_\alpha & \text{as } \partial D_\alpha \text{ is unbounded,} \end{cases}$$

however,

$$\mathcal{Q}_\beta(\mathbb{R}^n)|_{D_\alpha} \neq \mathcal{Q}_\beta(D_\alpha) \quad \forall \beta \in [\alpha, 1).$$

Moreover, the case $\alpha \in (-\infty, 0)$ of [89, Theorem 1.3] lies in $\mathsf{BMO}(\mathbb{R}^n)$-setting: in accordance with [41, Theorem 1], a function enjoying

$$\|f\|_{\mathsf{BMO}(D)} = \sup_{B(x_0,r_0)\subseteq D} \fint_{B(x_0,r_0)} |f(x) - f_{B(x_0,r_0)}| \, dx < \infty,$$

is the restriction to D of some function $\tilde{f} \in \mathsf{BMO}(\mathbb{R}^n)$; that is,

$$\mathsf{BMO}(D) = \mathsf{BMO}(\mathbb{R}^n)|_D,$$

when and only when there are two distance functions d_1, d_2 on the family of Whitney cubes $\mathscr{W} = \{I_j\}$ such that

$$d_1(I_j, I_k) \lesssim d_2(I_j, I_k) \quad \forall I_j, I_k \in \mathscr{W}.$$

Especially, under D being the interior or exterior domain complementary to a Jordan curve $\zeta \subseteq \mathbb{R}^2$,

$$\mathrm{BMO}(D) = \mathrm{BMO}(\mathbb{R}^2)|_D$$

if and only if $D \subseteq \mathbb{R}^2$ is a uniform domain – i. e. – ζ is a quasicircle but this is not valid for $\mathbb{R}^{n\geq 3}$. More precisely, it is basically proved in Riemann's work [72] that if ∂D is a quasisphere (i. e., the image of the unit sphere $\mathbb{S}^{n-1}$ under a globally quasiconformal self-homeomorphism of $\mathbb{R}^{n\geq 3}$) then

$$\mathrm{BMO}(D) = \mathrm{BMO}(\mathbb{R}^n)|_D,$$

but not conversely as shown in [41, Theorem 3].

Section 8.4. Theorem 8.9 is taken from [89, Theorem 1.3(ii)] and its demonstration presented in the final section of [89]. Additionally, according to [95, Theorem 1.2], under

$$\begin{cases} (\alpha, n-1) \in (0,1) \times \mathbb{N} \\ d_D(x) = \operatorname{dist}(x, \partial D) \quad \& \quad d_{\mathbb{R}^n}(x) = \infty, \end{cases}$$

if $D \subseteq \mathbb{R}^n$ is a uniform domain and $\mathcal{Q}_\alpha(D)$ comprises all $f \in \mathrm{L}^2_{\mathrm{loc}}(D)$ with

$$\sup_{(x_0,r_0)\in D\times(0,\infty)} \left(\int_{B_D(x_0,r_0)} \int_{|y-x|<\min\{r_0, d_D(x)\}} \left(\frac{|f(x)-f(y)|^2}{|x-y|^{n+2\alpha}} \right) \frac{dydx}{r_0^{n-2\alpha}} \right)^{\frac{1}{2}} < \infty,$$

then

$$\mathrm{Q}_\alpha(\mathbb{R}^n)|_D = \mathcal{Q}_\alpha(D).$$

Conversely, if this last set-theoretic equation holds under the hypothesis that D is a simply connected bounded domain of $\mathbb{R}^2$ or quasiconformally equivalent to a uniform domain in $\mathbb{R}^{n\geq 3}$, then D is a uniform domain.

Bibliography

[1] D. R. Adams, A note on Choquet integral with respect to Hausdorff capacity. In *Function Spaces and Applications*, Lund 1986. Lecture Notes in Math., vol. 1302, Springer-Verlag, 1988, pp. 115–124. 41, 59

[2] L. Ambrosio, Transport equation and Cauchy problem for BV vector fields. *Invent. Math.* 158(2004)227–260. 167, 171

[3] G. Bao and H. Wulan, Q_K spaces of several real variables. *Abstr. Appl. Anal.* 2014(2014)931937, 14 pages. 82

[4] A. El. Baraka, Littlewood–Paley characterization for Campanato spaces. *J. Funct. Spaces Appl.* 4 (2006)193–220. 78, 83, 116

[5] C. Bennett, R. A. DeVore and R. Sharpley, Weak L^{∞} and BMO. *Ann. Math.* 113(1981)601–611. 212

[6] G. Bourdaud, M. Lanza de Cristoforis and W. Sickel. Functional calculus on *BMO* and related spaces. *J. Funct. Anal.* 189(2002)515–538. 169

[7] J. Bourgain and H. Brézis, On the equation div $Y = f$ and application to control of phases. *J. Am. Math. Soc.* 16(2002)393–426. 103

[8] J. Bourgain, H. Brézis and P. Mironescu, Another look at Sobolev spaces. In J. L. Menaldi, E. Rofman and A. Sulem (Eds.), *Optimal Control and Partial Differential Equations, a Volume in Honor of A. Bensoussan's 60th Birthday*, IOS Press, 2001, pp. 439–455. 7

[9] J. Bourgain and N. Paviović, Ill-posedness of the Navier–Stokes equation in a critical space in 3D. *J. Funct. Anal.* 255(2008)22–33. 129

[10] S. M. Buckley, Is the maximal function of a Lipschitz function continuous? *Ann. Acad. Sci. Fenn., Math.* 24(1999)519–528. 212

[11] S. M. Buckley, Estimates for operator norms on weighted spaces and reverse Jensen inequalities. *Trans. Am. Math. Soc.* 340(1993)253–272. 134

[12] M. Cannone, Y. Meyer and F. Planchon, *Solutions Auto-similaires des Équations de Navier–Stokes in* $\mathbb{R}^3$. Exposé n. VIII, Séminaire X-EDP, Ecole Polytechnique, 1994. 107

[13] F. Chiarenza and M. Frasca, Morrey spaces and Hardy–Littlewood maximal function. *Rend. Mat. Appl. (7)* 7(3–4)(1987)273–279(1988). 125, 212

[14] A. Clop, R. Jiang, J. Mateu and J. Orobitg, Flows for non-smooth vector fields with sub-exponentially integrable divergence. *J. Differ. Equ.* 261(2016)1237–1263. 169

[15] A. Clop, R. Jiang, J. Mateu and J. Orobitg, A note on transport equations in quasiconformally invariant spaces. *Adv. Calc. Var.* 11(2018)193–202. 170

[16] R. R. Coifman, Y. Meyer and E. M. Stein, Some new function spaces and their applications to harmonic analysis. *J. Funct. Anal.* 62(1985)304–335. 77

[17] G. Dafni and J. Xiao, Some new tent spaces and duality theorems for fractional Carleson measures and $Q_{\alpha}(\mathbb{R}^n)$. *J. Funct. Anal.* 208(2004) 377–422. 50, 59, 95, 96, 102, 198

[18] G. Dafni and J. Xiao, The dyadic structure and atomic decomposition of *Q* spaces in several variables. *Tohoku Math. J.* 57(2005)119–145. 22

[19] R. J. DiPerna and P. L. Lions, Ordinary differential equations, transport theory and Sobolev spaces. *Invent. Math.* 98(1989)511–547. 168, 171

[20] J.-G. Dong and T.-J. Xiao, Notes on interpolation inequalities. *Adv. Differ. Equ.* 2011(2011)913403, 6 pages. 38

[21] S. S. Dragomir, *Some Gronwall Type Inequalities and Applications*. Nova Science Publishers, Inc., Hauppauge, NY, 2003. 166

[22] M. Essén, S. Janson, L. Peng and J. Xiao, *Q* spaces of several real variables. *Indiana Univ. Math. J.* 49(2000)575–615. V, VI, 6, 22, 25, 31, 39, 59, 63, 64, 80, 82, 87, 131, 173, 195

https://doi.org/10.1515/9783110600285-009

[23] F. B. Fabes, R. L. Janson and U. Neri, Spaces of harmonic functions representable by Poisson integrals of functions in *BMO* and $\mathcal{L}_{p,\lambda}$. *Indiana Univ. Math. J.* 25(1976)159–170. 71
[24] A. Farhat, Z. Grujić and K. Leitmeyer, The space $B^{-1}_{\infty,\infty}$, volumetric sparseness, and 3D NSE. *J. Math. Fluid Mech.* 19(2017)515–523, doi: 10.1007/s00021-016-288-z; Erratum: *J. Math. Fluid Mech.* 19(2017)525–527. 129
[25] C. Fefferman and E. Stein, H^p spaces of several variables. *Acta Math.* 129(1972)137–193. 103
[26] M. Frazier, B. Jawerth and G. Weiss, *Littlewood–Paley Theory and the Study of Function Spaces*. CBMS Regional Conference Series in Mathematics, vol. 79, 1991. 84
[27] J. Garcia-Cuerva and J. L. Rubio de Francia, *Weighted Norm Inequalities and Related Topics*. Math. Studies vol. 116, North Holland, Amsterdam, 1985. 42
[28] F. W. Gehring and O. Martio, Quasiextremal distance domains and extension of quasiconformal mappings. *J. Anal. Math.* 45(1985)181–206. 204
[29] M. Giaquinta, *Multiple Integrals in the Calculus of Variations and Nonlinear Elliptic Systems*. Annals of Math. Studies, vol. 105, Princeton University Press, Princeton, N. J., 1983. 70, 169
[30] A. Gogatishvili and R. Ch. Mustafayev, A note on boundedness of the Hardy–Littlewood maximal operator on Morrey spaces. *Mediterr. J. Math.* 13(2016)1885–1891. 126
[31] Y. Gotoh, On composition operators which preserve BMO. *Pac. J. Math.* 201(2001)289–307. 155
[32] L. Grafakos, *Classical and Modern Fourier Analysis*, Pearson Education Inc. 2008. 100, 104, 126
[33] P. Hajłasz, P. Koskela and H. Tuominen, Measure density and extendability of Sobolev functions, *Rev. Mat. Iberoam.* 24(2008)645–669. 178
[34] J. Heinonen, T. Kilpeläinen and O. Martio, *Nonlinear Potential Theory of Degenerate Elliptic Equations*. Dover, 2006. 161
[35] T. Hmidi and D. Li, Small $\dot{B}^{-1}_{\infty,\infty}$ implies regularity, *Dyn. Partial Differ. Equ.* 14(2017)1–4. 129
[36] S. Janson, On functions with conditions on the mean oscillation. *Ark. Mat.* 14(1976)189–196. 103
[37] R. Jiang, J. Xiao and D. Yang, Towards spaces of harmonic functions with traces in square Campanato spaces and their scaling invariants. *Anal. Appl.* 14(2010)679–703. 83
[38] F. John and L. Nirenberg, On functions of bounded mean oscillation. *Commun. Pure Appl. Math.* 18(1965)415–426. V, 3, 25
[39] R. L. Johnson and C. J. Neugebauer, Homeomorphisms preserving A_p. *Rev. Mat. Iberoam.* 3(1987)249–273. 169
[40] R. L. Johnson and C. J. Neugebauer, Properties of BMO functions whose reciprocals are also BMO. *Z. Anal. Anwend.* 12(1993)3–11. 135, 169
[41] P. W. Jones, Extension theorems for BMO. *Indiana Univ. Math. J.* 29(1980)41–66. 173, 213, 214
[42] T. Kato, Strong L^p-solutions of the Navier–Stokes equation in $\mathbb{R}^m$, with applications to weak solutions. *Math. Z.* 187(1984)471–480. 106
[43] T. Kato, Strong solutions of the Navier–Stokes equations in Morrey spaces. *Bol. Soc. Bras. Math.* 22(1992)127–155. 129
[44] J. Kinnunen, The Hardy–Littlewood maximal function of a Sobolev function. *Isr. J. Math.* 100(1997)117–124. 212
[45] H. Koch, P. Koskela, E. Saksman and T. Soto, Bounded compositions on scaling invariant Besov spaces. *J. Funct. Anal.* 266(2014)2765–2788. 166
[46] H. Koch and D. Tataru, Well-posedness for the Navier–Stokes equations. *Adv. Math.* 157(2001)22–35. 83, 110, 111
[47] Y. Komori and S. Shirai, Weighted Morrey spaces and a singular integral operator. *Math. Nachr.* 282(2009)219–231. 174

[48] P. Koskela, *Lectures on Quasiconformal and Quasisymmetrical Mappings*, University of Jyväskylä, 2011. 150

[49] P. Koskela, J. Xiao, Y. Zhang and Y. Zhou, A quasiconformal composition problem for the Q-spaces. *J. Eur. Math. Soc.* 19(2017)1159–1187. 169

[50] P. Koskela, D. Yang and Y. Zhou, Pointwise characterizations of Besov and Triebel–Lizorkin spaces and quasiconformal mappings. *Adv. Math.* 226 (2011)3579–3621. 145

[51] P. Koskela, Y. Zhang and Y. Zhou, Morrey–Sobolev extension domains. *J. Geom. Anal.* 27(2017)1413–1434. 213

[52] A. Kufner and I.-E. Persson, *Weighted Inequalities of Hardy Type*. World Sci. Pub., River Edge, NJ, USA, 2003. 82

[53] W. Kwon, Easy proof of John–Nirenberg inequality. http://willkwon.dothome.co.kr/index.php/archives/618. 38

[54] J. M. Lee, *Riemannian Manifolds*. GTM, vol. 176, Springer, 1997. 110

[55] P. G. Lemarié-Rieusset, *Recent Developments in the Navier–Stokes Problem*. Chapman and Hall/CRC, Boca Raton, Fla, 2002. 111, 117, 118

[56] P. G. Lemarié-Rieusset, The Navier–Stokes equations in the critical Morrey–Campanato space. *Rev. Mat. Iberoam.* 23(2007)897–930. 119, 128, 129

[57] P. G. Lemarié-Rieusset, Small data in an optimal Banach space for the parabolic-parabolic and parabolic-elliptic Keller–Segel equations in the whole space. *Adv. Differ. Equ.* 18(2013)1189–1208. 129

[58] P. Li and Z. Zhai, Well-posedness and regularity of generalized Navier–Stokes equations in some critical Q-spaces. *J. Funct. Anal.* 259(2010)2457–2519. 126

[59] Q. Li, Littlewood–Paley characterization for $Q_\alpha(\mathbb{R}^n)$ spaces. arXiv:0908.4380v2[math.CA]31Aug2009. 83, 116

[60] Q. Lin, J. Zhao and S. Cui, Existence and regularizing rate estimates of mild solutions to a generalized magneto-hydrodynamic system in Q-spaces. *Asian-Eur. J. Math.* 08(2015)1550049, 20 pages. 126

[61] Q. Liu, Well-posedness for the nematic liquid crystal flow with rough initial data. *Chin. Ann. Math., Ser. A* 35(2014)591–612. 127

[62] V. Maz'ya and T. Shaposhnikova, On the Bourgain, Brezis, and Mironescu theorem concerning limiting embeddings of fractional Sobolev spaces. *J. Funct. Anal.* 195(2002)230–238; Erratum: *J. Funct. Anal.* 201(2003)298–300. 7

[63] Y. Meyer, *Wavelets and Operators*. Cambridge Univ. Press, 1992. 16, 21

[64] M. Milman, A note on reversed Hardy inequalities and Gehring's lemma. *Commun. Pure Appl. Math.* L(1997)0311–0315. 212

[65] C. B. Morrey, On the solutions of quasi-linear elliptic partial differential equations. *Trans. Am. Math. Soc.* 43(1938)126–166. 70

[66] P. B. Mucha, Transport equation: extension of classical results for $\operatorname{div} b \in BMO$. *J. Differ. Equ.* 249(2010)1871–1883. 171

[67] J. Orobitg and J. Verdera, Choquet integrals, Hausdorff content and the Hardy–Littlewood maximal operator. *Bull. Lond. Math. Soc.* 30(1998)145–150. 59

[68] J. Peetre, On the theory of $\mathcal{L}_{p,\lambda}$ spaces. *J. Funct. Anal.* 4(1969)71–87. 70

[69] J. Peetre, *New Thoughts on Besov Spaces*. Duke Univ. Math. Series I, 1975. 8, 9

[70] H. M. Reimann, Functions of bounded mean oscillation and quasiconformal mappings. *Comment. Math. Helv.* 49(1974)260–276. 147, 154, 155, 161

[71] H. M. Reimann, Ordinary differential equations and quasiconformal mappings. *Invent. Math.* 33(1976)247–270. 166

[72] V. S. Rychkov, On restrictions and extensions of the Besov and Triebel–Lizorkin spaces with respect to Lipschitz domains. *J. Lond. Math. Soc.* 60(1999)237–257. 214

[73] O. Saari, Poincaré inequalities for the maximal function. arXiv:1605.05176v2[math.CA]16Feb2017. 212

[74] A. Schikorra, D. Spector and J. V. Schaftingen, An L^1-type estimate for Riesz potentials. *Rev. Mat. Iberoam.* 33(2017)291–303. 96

[75] G. Seregin, Liouville type theorem for stationary Navier–Stokes equations. *Nonlinearity* 29(2016)2191–2195. 122, 129

[76] J. Shatah and M. Struwe, *Geometric Wave Equations*. Courant Lecture Notes in Math., vol. 2, Courant Institute of Mathematical Sciences, New York Univ., Amer. Math. Soc., Providence, RI, 1998. 22

[77] H. Smith, Lecture 14: The John-Nirenberg inequality. https://sites.math.washington.edu/hart/m582/Lecture14.pdf. 38

[78] G. Stampacchia, The spaces $\mathcal{L}^{(p,\lambda)}$, $N^{(p,\lambda)}$ and interpolation. *Ann. Sc. Norm. Super. Pisa* 19(1965)443–462. 169

[79] E. M. Stein and G. Weiss, On the theory of harmonic functions of several variables. I. The theory of H^p-spaces. *Acta Math.* 103(1960)25–62. 97

[80] R. S. Strichartz, Bounded mean oscillation and Sobolev spaces. *Indiana Univ. Math. J.* 29(1980)539–558. 22, 84

[81] Y. Sun, On the existence and boundedness of square function operators on Campanato spaces. *Nagoya Math. J.* 173(2004)130–151. 83

[82] C. Sweezy, The maximal function on spaces that lie between L^∞ and $\mathrm{BMO}(\mathbb{R}^n)$. In *Proceedings of the 10th WSEAS International Conference on Appl. Math*, Dallas, Texas, USA, November 1–3, 2006, pp. 433–438. 38

[83] C. Tang and Z. Zhai, Generalized Poincaré embeddings and weighted Hardy operator on $Q_p^{\alpha,q}$ spaces. *J. Math. Anal. Appl.* 371(2010)665–675. 212

[84] M. E. Taylor, Analysis on Morrey spaces and applications to Navier–Stokes and other evolution equations. *Commun. Partial Differ. Equ.*, 17(1992)1407–1456. 106

[85] A. Torchinsky, *Real Variable Methods in Harmonic Analysis*. Pure and Appl. Math. Series, vol. 123, Academic Press, 1986. 134

[86] A. Uchiyama, A constructive proof of the Fefferman–Stein decomposition of $BMO(\mathbb{R}^n)$. *Acta Math.* 148(1982)215–241. 23, 103

[87] C. Wang, Well-posedness for the heat flow of harmonic maps and the liquid crystal flow with rough initial data. *Arch. Ration. Mech. Anal.* 200(2011)1–19. 127

[88] Y. Wang and J. Xiao, Homogeneous Campanato–Sobolev classes. *Appl. Comput. Harmon. Anal.* 39(2015)214–247. 85

[89] Z. Wang, J. Xiao and Y. Zhou, The Q_α-restriction problem. *Asian J. Math.* (2019, in press), 34 pages. 213, 214

[90] J. Xiao, L^p and *BMO* bounds of weighted Hardy–Littlewood averages. *J. Math. Anal. Appl.* 262(2001)660–666. 212

[91] J. Xiao, A sharp Sobolev trace inequality for the fractional-order derivatives. *Bull. Sci. Math.* 130(2006)86–96. 22, 38, 96

[92] J. Xiao, Homothetic variant of fractional Sobolev space with application to Navier–Stokes system. *Dyn. Partial Differ. Equ.* 4(2007)227–245. 83, 126

[93] J. Xiao, Homothetic variant of fractional Sobolev space with application to Navier–Stokes system revisited. *Dyn. Partial Differ. Equ.* 11(2014)167–181. 83, 128, 129

[94] J. Xiao, The transport equation in the scaling invariant Besov or Essén–Janson–Peng–Xiao space. *J. Differential Equations* (2018), doi:10.1016/j.jde.2018.11.027. 169, 170

[95] J. Xiao, D. Yang, J. Zhang and Y. Zhou, Stokes system and uniform trace for BMO-Q. *Preprint* (2017). 214

[96] J. Xiao and J. Zhang, N-S systems via $\mathcal{Q}$-$\mathcal{Q}^{-1}$ spaces. *J. Geom. Anal.* (2018) doi: 10.1007/s12220-018-0048-9. 82, 83, 126, 129

[97] J. Xiao and J. Zhang, Predual forms, harmonic maps & liquid crystals of $(BMO - Q)$ & $(BMO - Q)^{-1}$. *Preprint* (2017). 102, 103, 127

[98] J. Xiao and Y. Zhou, A reverse quasiconformal composition problem for $Q_\alpha(\mathbb{R}^n)$. *Preprint*(2018). 170

[99] M. Yamazaki, The Navier–Stokes equation in various function spaces. *Am. Math. Soc. Transl.* 204(2001)111–132. 106

[100] D. Yang and W. Yuan, A note on dyadic Hausdorff capacities. *Bull. Sci. Math.* 132(2008)500–509. 59

[101] Q. Yang, T. Qian and P. Li, Fefferman–Stein decomposition for Q-spaces and micro-local quantities. *Nonlinear Anal.* 145(2016)24–48. 103

[102] X. F. Ye and T. F. Wang, Two-weighted norm inequality on weighted Morrey spaces. *Turk. J. Math.* 38(2014)426–435. 174

[103] H. Yu and G. Dafni, A John–Nirenberg type inequality for $Q_\alpha(\mathbb{R}^n)$. *J. Math. Anal. Appl.* 351(2009)428–439. 38

[104] Y. Zhou, Fractional Sobolev extension and imbedding. *Trans. Am. Math. Soc.* 367(2015)959–979. 177, 178

[105] C. Zhuo, W. Sickel, D. Yang and W. Yuan, Characterizations of Besov-type and Triebel–Lizorkin-type spaces via averages on balls. *Can. Math. Bull.* 60(2017)655–672. 22

Index

Bei Fragen zur Produktsicherheit wenden Sie sich bitte an:
If you have any questions regarding product safety,
please contact:

Walter de Gruyter GmbH
Genthiner Straße 13
10785 Berlin
productsafety@degruyterbrill.com